POLICE PATROL
OPERATIONS AND MANAGEMENT

Second Edition

Charles D. Hale

President, Resource Management Associates
South Holland, Illinois

Prentice Hall Career & Technology
Englewood Cliffs, NJ 07632

Library of Congress Cataloging-in-Publication Data

Hale, Charles D.

Police patrol : operations and management / by Charles D. Hale. — 2nd ed.

p.cm.

Includes bibliographical references and index.

ISBN 0-13-814484-2

1. Police patrol—United States.2. Police administration—United States

.I. Title.

HV8138.H351994

363.2'32'0973—dc20 93-29722 CIP

Editorial/production supervision, interior design,
and electronic production: ***Eileen O'Sullivan***
Managing Editor: ***Mary Carnis***
Cover design: ***Yes Graphics***
Manufacturing buyer: ***Ed O'Dougherty***
Acquisitions editor: ***Robin Baliszewski***
Director of Manufacturing and Production: ***David Riccardi***
Editorial assistant: ***Rose Mary Florio***
Production Assistant: ***Bunnie Neuman***
Page Layout Assistant: ***Stephen Hartner***

Prentice-Hall, Inc.
A Paramount Communications Company
Englewood Cliffs, New Jersey 07632

Printed in the United States of America

10 9 8 7 6 5 4 3 2 1

ISBN 0-13-814484-2

Prentice-Hall International (UK) Limited, *London*
Prentice-Hall of Australia Pty. Limited, *Sydney*
Prentice-Hall Canada Inc., *Toronto*
Prentice-Hall Hispanoamericana, S.A., *Mexico*
Prentice-Hall of India Private Limited, *New Delhi*
Prentice-Hall of Japan, Inc., *Tokyo*
Simon & Schuster Asia Pte. Ltd., *Singapore*
Editora Prentice-Hall do Brasil, Ltda., *Rio de Janeiro*

To my children

Mark, Kathleen, Michael, Steven, and Kimberly

CONTENTS

[1] Jeffrey T. Mitchell, "Development and Functions of a Critical Incident Stress Debriefing Team," JEMS, December 1988, pp. 43–46.

PREFACE

The purpose of a second edition to any text is to update and enhance the original. In the present case, however, this second edition of *Police Patrol: Operations and Management* goes beyond what is usually considered to be an update of the original text. This is simply because so much has happened in the field of law enforcement, and police patrol operations in particular, since the first edition of this book was published in 1981. Community-oriented policing programs and problem-oriented policing programs and all their various forms by other names have become the trend in police agencies large and small throughout the United States and in several foreign countries. Increasing attention is being devoted to the victims of crime, and to victims of domestic violence in particular. Major technological developments have changed the ways police agencis gather, store, and analyze data; investigate crimes; and maintain control of field units.

These changes merely indicate the dynamic and fast-paced environment in which the police operate. Nothing is static. What worked yesterday may be outmoded today. Nothing can be taken for granted. Our attention must constantly be focused on identifying new ways to solve problems. To the extent possible, we must commit ourselves to a proactive approach to serving the needs of our constituents. We must continue to train and educate our police officers and enable them to be more sensitive to the needs of the community. Police supervisors and command officers can no longer afford to rely upon traditional and commonly accepted means of managing the resources available to them. Creativity and imagination must be the watchwords of our police managers.

This book is intended to bring the reader up to date on the most contemporary thinking concerning police patrol operations and methods. it examines the new and emerging issues facing the police administrator today as well as some of the more routine, but nonetheless important issues of managing patrol resources. Although it does not provide many ultimate solutions, it is hoped that it does provide the reader with a better insight into how the myriad problems facing the police administrator and patrol manager may be identified, analyzed, and solved.

This book is intended to serve the interests of both the practitioner and the student of police operations. It is written from the viewpoint of the practitioner, but ample attention has been given to the theoretical foundation upon which police patrol operations are conducted. Thus, this book should find a useful place in the police training library as well as in the personal library of the patrol supervisor and command officer. Although I have attempted to avoid the heavily academic orientation that is so common in many college textbooks, I have nevertheless attempted to ensure that the organization of the text, the learning objectives, and the end-of-chapter review and study material lend themselves to the needs of the college instructor.

It is, of course, impossible to predict what changes may take place in the field of law enforcement in the years ahead, but we may be assured that they will be no less rapid than has been the case in the last decade. This text will enable the student and the practitioner to prepare for and adapt to those changes in a manner that will ensure the highest level and quality of law enforcement services.

Acknowledgments

The author of a book is not the source of all knowledge or ideas that the book contains, but is rather the instrument by which the information contained in the book is assembled and takes form. Accordingly, whatever credit is accrued by the book must be shared by all those who, in one way or another, made it possible. As is customary, whatever blame must be directed toward the book is borne only by the author. I must thank, first of all, Robin Baliszewski, who, as my editor, gave me the confidence to proceed with what turned out to be a bigger chore than I had anticipated. Nevertheless, Robin saw the need for doing the book and inspired me to tackle the problem.

Sam Chapman, formerly of the University of Oklahoma and now Adjunct Professor at the University of Nevada at Reno, was a technical advisor to me on the first edition of this book. As such, many of the ideas in that first version were attributable to Sam. I cannot charge Sam with the same responsibility for this edition of the book, but I can credit him with being a constant source of encouragement and I am indebted to him for graciously agreeing to write the Foreword. I would also like to thank Dan McCollum, Director of Training of the Elgin, Illinois, Police Department, and good personal friend, for providing a portion of Chapter 5 dealing with firearms training and police liability. One day Dan will write his own book and I may be fortunate enough to help him when he does. In addition, I wish to acknowledge the kind assistance provided by Dr. Chuck Milofsky concerning the role of the police psychologist in Chapter 10 as well as the information on critical incident intervention in Chapter 4.

I am also indebted to my son, Steven, who worked with me through much of the revision process, and who helped me type and rework many of the chapters. Steve willingly took on many of the routine, plodding, albeit extremely important tasks required in completing a revision as extensive as this. To Steven fell the task of typing and retyping portions of the first version, checking out the bibliography, putting together many of the illustrations, helping me with correspondence, and a variety of other necessary and time-consuming tasks. Much of the credit for completing the manuscript is owed to Steve, whose assistance was invaluable. I am grateful to Crystal Zajeski who assisted in the final editing and revision of the manuscript.

I would like to thank the following reviewers: Lois A. Wims, Ph.D. - Salve Regina University, Fred Hayes, Chief of Police - Gainesville, Georgia, and Stephen Caruso, Law Enforcement Consultant.

Finally, I would like to thank those thousands of unnamed but important police administrators, supervisors, managers, patrol officers, dispatchers, and clerical employees who have welcomed me into their offices, meeting rooms, and squad cars over the years. As a consultant to many police agencies throughout the United States since 1974, I have learned much more from these people than they have learned from me. It is the knowledge that I have gained from them that has helped to shape the ideas and theories in this text. These people, then, deserve much of the credit for whatever value this book represents.

Charles D. Hale

FOREWORD

While hundreds of publications have addressed police patrol, few have gone beyond spelling out the nuts and bolts, or the how-to-do-it of this important enterprise. In fact, most are dull and repetitive in their coverage of patrol issues, like "old wine in new bottles."

Laudably, in 1981, Charles D. Hale took patrol well beyond the traditional mode with his first edition of *Police Patrol: Operations and Management.* But his second edition, written eleven years after the first, is a work of conscience as well as of professional excellence. For example, five of the thirteen chapters are new; the other eight are not just updated or revised, they are reorganized and rewritten. And while the title is the same, the content is anything but the same!

There are many high points in this book; one is Chapter 7 which discusses police patrol methods. It is an especially lucid, timely, and thorough discussion of every conceivable means of patrol and the issues inherent in each. This chapter alone would be required reading for anyone who aspires to a supervisory role in law enforcement. Chapter 9 treats the frequently overlooked matter of proportional scheduling of patrol personnel according to need for coverage.

In Chapter 10, Hale clearly and persuasively declares that a department's support-services activities are crucial to the success of the patrol force. Hale, unyielding on this matter, describes how urgent it is that all support services be programmed in a collaborative effort to enhance patrol effectiveness.

Hale's second edition emphasizes the management aspect of patrol while not overlooking applied considerations. There is abundant evidence that Hale has culled through an impressive array of materials from a wide range of sources, choosing to use the most timely ones. Moreoever, he has made a proper attribution of sources which have been quoted. Furthermore, at the end of each chapter, there are references which are timely and relevant along with many provocative review questions. The Bibliography is an impressive enumeration of important contributions made by many authorities on patrol and the Index is, as it should be, a timely roadmap into the book's extensive content.

In conclusion, to many people, police patrol is an endless succession of family squabbles, drunken persons, barking dogs, prowler calls, burglar alarms, traffic stops, and abandoned cars. Add to this boredom. But Charles Hale clearly shows there is much more. His book is for persons who seek to enrich their understanding of what goes into making patrol as effective as it can be. The book is surely an important resource for police supervisors and executives as well as for those in support-services roles who want to know more about how their work contributes to public safety. Students will find in the book a rich lode to mine as they go about studying the role of uniformed personnel in local policing. And let's not forget the officers in uniform—the ones who patrol. The book quickly points out that the critical role they play deserved the unqualified support and attention of law enforcement personnel at all levels. *Police Patrol: Operations and Management* is intended to help authorities make informed decisions about every aspect of police uniformed work.

Samuel G. Chapman, Professor
Department of Criminal Justice
University of Nevada—Reno

CHAPTER 1
INTRODUCTION

CHAPTER OUTLINE

LEARNING OBJECTIVES

After completing this chapter, the student should be able to:

1. Explain the importance of the evolution of policing in terms of contemporary policing strategies and philosophies.
2. Describe the contributions of Sir Robert Peel to modern principles of police administration.
3. Describe the role of the police patrol officer in relation to the overall mission of the police in modern society.
4. Define "police mission".
5. List the various roles played by the police patrol officer and describe the potential conflict among these roles.

Patrol is the essence of the police function. It is the core of all police operations. The basic police responsibilities of prevention of crime, enforcement of the laws, and protection of life and property fall upon the shoulders of the uniformed patrol officer. All other police functions are ancillary to the patrol function. Not surprisingly, therefore, the majority of the resources of a police department are devoted to the patrol function.

The very idea of policing, which extends back several centuries, began with patrol. All other police functions evolved later, and were designed to supplement, complement, and support the patrol function. Despite its importance, patrol does not receive the recognition it deserves.

THE EVOLUTION OF POLICING

It is often said that those who forget history are doomed to repeat it. This is certainly true in any study of the police mission and role. In order to fully understand and appreciate the impact of the changes that are shaping the police patrol function today, it is important to briefly review the evolution of policing.

When we think of "the police," we usually have an image of a uniformed officer driving around in a marked patrol car equipped with emergency lights and a siren. We remember the last time that we were stopped for a traffic violation, or when our neighbor's car was stolen, or when our child was taken to the police station for violating curfew or some other minor offense. Thus, the idea of "policing" is typically associated with a uniformed force of men and women who are called upon to exercise the powers of the state in the protection of life and property, the enforcement of laws, and the maintenance of social order.

This practice of policing is of relatively recent vintage. It was not until 1829 in London, and a few years later in the United States, that the first modern police departments were established. Prior to that time, policing was a responsibility of every citizen or, in some cases, citizen delegates.

English Origins

In ancient England, shortly after the Norman Conquest, policing was carried out under a system called *frankpledge,* or "mutual pledge," which provided that every male over the age of 12 join nine of his neighbors to form a *tything. Tythingmen,* as they were called, were sworn to apprehend and incarcerate any of their neighbors who violated the law and to deliver them to court for trial. Anyone who failed in this obligatory duty was subject himself to a very severe fine.[1] Thus, the responsibility for upholding laws and apprehending lawbreakers rested with the individual citizen.

Eventually, the frankpledge system was replaced by a system of parish constables, who were appointed to enlist the aid of local watchmen to guard the parish against lawlessness. Neither constables nor their appointed watchmen were paid, and it is unclear as to how they recovered their expenses. However, like the tythingmen before them, they were subject to fines if they refused to perform their assigned duties.

Eventually, parish constables were supervised by justices of the peace: originally wealthy and respected men who assumed the position as a matter of honor and prestige. Over the years, however, men of less dignity, respect, and financial means were appointed, and constables began to charge a fee for their services. This eventually led to a system of corruption, since persons might be convicted of a crime only to bring profit to the justice of the peace constable. The corruption eventually extended to the constables themselves.

During the eighteenth century, the parish constable system further evolved to the point that a man called upon to serve as parish constable could hire someone else to take his place. The usefulness of this system,of course, depended entirely upon the ability or desire of the person who was selected to work in place of the appointed constable. As one might expect, those who volunteered to serve in this capacity were largely underpaid, unskilled, and of dubious qualification.

Modern Antecedents

Centuries of trial, error, and experimentation seem to have demonstrated the ineffectiveness of "citizen policing" as a legitimate or successful way of protecting the community from lawbreakers. Clearly, a more professional and well-organized approach was needed. This need became even greater with the advent of the Industrial Revolution, as greater numbers of people moved from the farms and rural countryside into the cities. Crime became more prevalent, and a more organized manner of dealing with it was needed.

The problem of crime and disorder soon gained the attention of Sir Robert Peel, the British home secretary, who in 1822 observed that, while better policing might not eliminate crime, poor policing contributed to social disorder.[2] In 1829, Peel intro-

[1]Carl B. Klockars, *The Idea of Police* (Newbury Park, Ca.: Sage Publications, 1985), p. 23.

[2]President's Commission on Law Enforcement and Administration of Justice, *Task Force Report: The Police* (Washington, D.C.: U.S. Government Printing Office, 1967), p. 4.

duced in Parliament an "Act for Improving the Police In and Near the Metropolis." This act led to the creation of the first organized police force, which became a forerunner of the modern American police system.

Peel's concept of policing included a number of principles which, although radical for that time, have become imbedded in the fabric and framework of modern police organizations:

1. The police should be organized along military lines.
2. Emphasis should be placed on screening and training police personnel.
3. Police officers should be hired on a probationary basis.
4. The police should be under governmental control.
5. Police officers should be deployed by time and by area.
6. Police headquarters should be centrally located.
7. Police record-keeping is essential.[3]

Evolution of American Policing

Policing in the American colonies followed the model developed in England. In most towns and villages, constables and sheriffs were responsible for enforcing laws and maintaining peace and order in the community. Many of these early law enforcement officers were elected, thus ensuring their popularity, if not their efficiency. In larger cities, such as Boston, New York, and Philadelphia, a night-watch system was established. Members of the night-watch patrolled the streets and alleys of the cities, alert for trouble or criminal activity, and prepared to chase away would-be lawbreakers. This system worked reasonably well, and the night watches were later supplemented with a day-watch system.

In 1844, New York City created what is commonly regarded as being the first regular police force by consolidating the night watch and the day watch into a single organization. This example was followed in Boston a few years later. These became the models upon which other police forces were created in other cities in the early years of the new nation.

The earliest organized police forces in this country did not inspire a great deal of public confidence. They lacked much in the way of a formal organizational structure, and training of their members was almost nonexistent. Appointments to the forces were made on the basis of friendship and political patronage, with no regard to merit or qualification. It was common practice for local politicians to control appointments to the police force, for which they charged very handsome sums of money. Promotions were bought and paid for in very much the same way. As a result,

[3]Robert Sheehan and Gary W. Cordner, *Introduction to Police Administration*, 2nd. ed. (Cincinnati: Anderson Publishing Co., 1989), p. 5.

many of these early police systems became mired in graft and corruption and became objects of public ridicule and criticism.

There were numerous attempts to modernize and professionalize American police forces during the latter part of the nineteenth century, but few were successful. In some cities, independent boards and commissions were set up to oversee the organization and management of police departments. These efforts were largely unsuccessful since persons appointed to these boards and commissions were not schooled in police administration. In some states, the police forces of large cities were actually placed under state control as a means of insulating the police from local political influences. State appointed boards still exercise control over the police forces in Kansas City and St. Louis.

Also during the waning years of the nineteenth century, civil service systems were introduced to the police service. These systems, organized either at the state or local level, had little to do with the direct operations of the police departments, but rather controlled the appointment and promotion of persons within those systems. The theory of the civil service concept was to ensure efficient and professional police administration through merit and qualification rather than friendship and patronage. Most of these civil service systems still exist today and for essentially the same purposes. One of the unfortunate byproducts of civil service then and now is that it tends to protect the guilty from discipline and, in many cases, promotes and protects substandard performance.

From time to time during the twentieth century, various attempts were made to focus attention on the problems of American police systems. In 1931, the Wickersham Commission examined the problems of American police forces and found that one of the most significant problems was ineffective leadership. The root of this problem, according to the commission, was that police chiefs were too dependent upon local political forces to be effective and that their limited tenure precluded continuity of operations.

If there is a beginning to what may be called the reform era of American policing, it was during the 1960s, when the police again came under scrutiny for their role in urban riots, campus demonstrations, and civil rights protests. A crisis in American policing was recognized at the national level. In 1967, the President's Commission on Law Enforcement and Administration of Justice issued its report which contained one of the most comprehensive and detailed analyses of American policing ever conducted. Through its report, entitled *A Challenge of Crime in a Free Society*, the commission examined the root cause of crime and identified ways that it could be dealt with at the local, state, and national level. The commission also issued a series of reports, each dealing with a different branch of the criminal justice system, including the police, the courts, and the corrections institutions. The commission's task force report on the police contained a very probing analysis of the problems and needs of American police systems and contained a series of recommendations for improvement.

The commission's work was followed a few years later by the National Advisory Commission on Criminal Justice Standards and Goals. The Goals Commission also issued a series of reports, each one dealing with a different part of the criminal justice system. These reports were issued in 1973 and differed from the earlier commission reports in that they contained a series of very specific recommendations for improvement in the various agencies of criminal justice. For the first time, a national agenda had been proposed for improving the police service. Moreover, the recommendations were stated in terms of easily understood and achievable goals and objectives.

Not too long thereafter, the Commission for Accreditation of Law Enforcement Agencies (COLEA) was created through the joint efforts of the International Association of Chiefs of Police, the National Sheriffs' Association and the National Organization of Black Law Enforcement Executives. The goal of COLEA was to establish reasonable standards by which police agencies could be evaluated and to develop a program of national accreditation that would place police agencies on the same level as hospitals and educational institutions.

Policing in Recent Years

In recent years, the police mission and role has evolved significantly, as new demands are being placed upon the police and as traditional police practices are coming under greater scrutiny and criticism. During the 1970's and 1980's, police operations and management practices came under more scientific study and evaluation than ever before in the history of policing. Aided with substantial amounts of federal funding, scientists and researchers began to examine all facets of police operations. Traditional beliefs about preventive patrol, criminal investigation and community relations began to be evaluated. New strategies for making the police more effective were developed and attempted.

During the 1970's, team policing came into vogue and was widely adopted by police agencies throughout the United States. During that same era, studies were conducted to evaluate the effectiveness of random patrol, field interviews, one-person versus two-person patrol, foot patrol versus automobile patrol, the criminal investigation process and alternatives ways of handling calls for service. The role of the police in dealing with domestic violence, the elderly, and victims in general was studied. As a result of this considerable research, a number of new proposals were offered to change the manner in which police operations were conducted.

For the most part, very little definitive evidence has been obtained to change the basic way in which the police perform their duties. With the obvious exception of the technology being employed (that is, computer-aided dispatch, enhanced communications systems, mobile data terminals, etc.), the police today are performing their mission very much as they did two decades ago.

This does not mean that the police have not progressed during the last two decades, for they surely have. Nonetheless, the basic role of the police is not that much different than it was in years past, and the manner in which it is performed has not changed much either. This is not to say that the police have not benefited by the millions of dollars that have been funneled into research, but rather that the way the police approach their mission has changed only in degrees. To the casual observer, this change is quite subtle.

Law enforcement is a dynamic and evolving occupation and one in which there must be continual research, experimentation, trial, and error in order to discover what works and what doesn't in the provision of police services. Unfortunately, we seem to know more about what does not work than what does work, but this knowledge itself is valuable in designing new programs and evaluating existing ones. We must continually strive to find new and better ways of fulfilling our responsibilities to the public and to improving the level and quality of the services they expect.[4]

Notwithstanding the rather slow development of the police over the years, we may be on the threshold of a major evolution in the way police do their job and in the police mission itself. Research conducted during the 1970s has yielded some rather startling information about the traditional approach to policing. We have learned, for example, that[5]

- Simply spending more money on the police and adding more police officers to the payroll will not solve the crime problem, nor will it make people be less fearful of crime.
- Routine, random patrol does not have a significant deterrent effect on crime.
- Rapid response by the police to crime scenes does not ensure apprehension or conviction of offenders, nor does it necessarily ensure greater citizen satisfaction with the police.

In short, many of our fundamental views about how police operations ought to be conducted have been challenged by the research. Perhaps even more far-reaching has been the realization that the police themselves are not, or should not be, the sole arbiters of what the police do or how they should do it. Since the police have been found to be relatively ineffective in solving the crime problem, they have come to the conclusion that they need to develop a closer partnership with the people in the community. Such interaction could help identify problems needing police attention, and devise ways of dealing with those problems.

As a result of these developments, we may see in the next decade a significant shift in the way the police operate. This subject is addressed in more detail in Chapter 6.

IMPORTANCE OF THE PATROL FUNCTION

The police are among the most visible and active providers of public service in contemporary society. They operate on a 24-hour-a-day basis and are called upon when

[4]George L. Kelling, "What Works—Research and the Police," National Institute of Justice, *Crime File Study Guide*, no date.

[5]Jerome H. Skolnick and David H. Bayley, *The New Blue Line: Police Innovation in Six American Cities* (New York: The Free Press, 1985), pp. 4–5.

no other municipal agencies are available or qualified to assist. The duties they perform often represent the difference between life and death, peace and disorder, and tranquility or tragedy. They are viewed, at one time or another, as brutal, heroic, lazy, indifferent, resourceful, crafty, dull, and brave. They are seen as public servants, law enforcers, information providers, protectors of public safety, crime preventers, skilled investigators, mediators, counselors and crime fighters.

Over the years, a great deal of lip service has been paid to the importance of the patrol function. Patrol officers, for example, are often described as being the "backbone" of the police department. At the same time, they occupy the bottom rung of the organizational hierarchy. Even though they have the least formal authority in the police organization, they are expected to exercise the greatest amount of discretion in the manner in which they perform their duties. Patrol officers are expected to make split-second decisions that will later be reviewed, studied, debated, and examined by their superior officers, lawyers, judges, juries, and the general public. Even though the patrol officer is the lowest-paid member of the department, has the least amount of formal authority, and is given little recognition by police executives, it is the patrol officer who makes the decisions which have the most immediate and direct impact upon the lives and welfare of the residents of the community.

The lack of recognition given to the patrol function is due to several things. First, patrol is seen as the bottom rung of the ladder to success in the police profession. Upon graduating from the police academy, police officers are routinely assigned to the patrol force. In many cases, it is several years before they may aspire to be promoted to more seemingly "important" or challenging positions, such as investigations or drug enforcement.

Because of this, few police departments make any effort to recognize patrol as a satisfying career. Indeed, officers who remain on patrol for the entire length of their career are often seen as unambitious or lazy. They are not encouraged to spend their entire careers as patrol officers, and are often looked down upon if they choose to do so. Moreover, few police departments have formalized mechanisms for rewarding conscientious performance by patrol officers. Only those actions which are heroic or extraordinary in nature receive attention. In other words, there are few incentives to motivate patrol officers to perform their duties in an exceptional manner. "Few of the rewards they have to offer have anything to do with skills of doing patrol work well."[6]

Second, there are many other police functions which are much more glamorous and seemingly more exciting than patrol. With the exception of the *Adam 12* television series of the 1960s and the *Police Story* television series of the 1970s, few television series, motion pictures, books, or magazines have focused on the police patrol function as being exciting or even interesting. Instead, such vehicles of entertainment have focused on the more specialized aspects of policing, such as criminal investigation. By doing so, they have failed to recognize that detective work is often dull and plodding and that the real success of any criminal investigation often relies upon the

[6]Klockars, *The Idea of Police*, p. 57.

initial work done by the patrol officer at the scene. However, this kind of reality rarely attracts the interest of readers and viewers.

Third, patrol officers are often confronted with the worst part of policing. They risk their lives each time they make a traffic stop or check an open door late at night; they arrest drunken drivers and issue speeding tickets to a thankless public; they work rotating shifts which often affect their physical health; they are constantly confronted with adversity, hostility, and tragedy which they carry with them long after a particular incident has ended; and they are often criticized by an uncaring, unknowing, and unappreciative public because they do not always smile and act friendly.

The actions of the patrol officer have a direct impact on the quality of life in the community. The police officer on the beat plays a pivotal role in protecting the lives and property of the local residents and in ensuring that justice is administered fairly. It is the patrol officer on the beat to whom citizens turn for advice or assistance and they expect to be protected from criminal attack.

Organizational policies and management philosophies often contribute to the lack of recognition and respect enjoyed by the patrol officer. The emphasis on specialization, for example, has taken away from the patrol officer many duties which he is capable of performing and given them over to detectives, crime-scene technicians, traffic officers and other specialists. As a result, in many police agencies, the patrol officer is seen as little more than a report taker and first responder. At a major crime scene, the patrol officer is expected to do little but protect the scene until the arrival of "trained" investigators. Even when a patrol officer has a lead about a possible burglary suspect, department policy often requires this information to be given over to a detective rather than to allow the officer to follow up.

Despite all this, the patrol force is the backbone of any police department. It is responsible for the initial response in almost all requests for police service or action. It is expected to maintain a visibility and a sense of omnipresence to deter would-be lawbreakers and to instill a feeling of comfort and security among the members of the community.

Patrol operations in many police departments are often taken for granted. Even though we should know better by now, we still cling to the belief that the mere presence of a uniformed police officer in a marked car is sufficient to deter crime. Despite good evidence to the contrary, we still act as if there is a direct correlation between the size of the patrol force and numbers of crimes committed. If this were true, an obvious solution to an increase in crime would be to increase the size of the patrol force.

Traditionally, it has been assumed that the number of patrol officers was directly related to the volume of crime. High visibility of the patrol force was assumed to be a deterrent to crime. Times have changed, however, and it is now recognized that traditional patrol methods (that is, routine, random patrol) are no longer as effective as they were once thought to be. Clearly, there is a great need for innovative and creative ways in which to deploy and utilize the patrol force.

Over time, these traditional beliefs about the value and purpose of police patrol are being torn down. Slowly but surely, police administrators are being forced to recognize that it is not just the number of patrol officers, but also the way that these officers are managed, that can have a direct impact on the deterrence of crime.

What is needed is not more and better-equipped patrol officers, but more reactive and effective ways to utilize them. We are just beginning to devise new and creative strategies for dealing with crime and community problems which the police have traditionally been called upon to handle. We are just beginning to recognize that highly visible police patrol and rapid response to crimes in progress, while important, are not the real answers to preventing or solving crime in the community.

We are witnessing today a historic and fascinating period in the evolution of modern policing. It may be fair to refer to this period of policing as "back to the basics." Some could even argue that we are returning to the methods of policing that characterized the late nineteenth and early twentieth centuries.

Police patrol officers are being taken out of the police cars—the metal cocoons that have isolated them for so long—and returned to the neighborhoods of the communities they serve. We are finding more and more police agencies today experimenting with the "novel" idea of putting officers on foot-patrol beats—a concept that had been all but abandoned in many communities since the advent of the police vehicle as a standard mode of police transportation.

Not only are patrol officers returning to the neighborhoods of the community, they are beginning to establish dialogue with community groups. Storefront centers and neighborhood precincts are beginning to reappear, decades after they were eliminated in an effort to centralize police services as a means of achieving greater efficiency of operations.

Patrol officers, who for decades were relegated to the bottom rung of the police hierarchy, and who were subject to close supervision and scrutiny because they were thought not capable of making intelligent decisions or exercising good judgment, are now being allowed to identify and solve problems within their assigned areas of responsibility. Community relations, once thought to be the province of a highly trained and specialized unit of the department, is now being accepted as the responsibility of each officer in the department, and especially the patrol officer.

Truly, we are witnessing a rebirth of the patrol function in America. Change, innovation, experimentation, and adaptation are replacing tradition and conventional wisdom. Today, more than at any time during the last fifty years, the patrol function is beginning to be recognized for the contribution it makes to the police operation. Now, more than ever before, patrol officers are beginning to attain their just status in the police organization. It is truly a most interesting and challenging period in the evolution of American policing.

Of course, there will be more changes and innovations in the years ahead, and no one can accurately predict what the next evolution in policing will entail. It is not the purpose of this book to predict the future, but rather to explore the various alternatives, strategies, and concepts that are now possible in the managing of the patrol force.

DEFINING THE POLICE MISSION

All organizations, regardless of their type, function according to a defined mission or purpose. At first glance it would seem that the mission of the police is self-evident. It is commonly accepted that the police exist for the purpose of enforcing criminal laws and ordinances; investigating criminal offenses; identifying and apprehending lawbreakers and assisting in their prosecution; protecting life and property; and providing a number of service-related activities, such as aiding in the search for missing persons, intermediating domestic disputes, and assisting stranded motorists.

When examined more closely, however, it can be seen that the true mission of the police in contemporary society is not clear at all. There are several reasons for this. First, people define the mission of the police in terms of their own needs and expectations, and these often depend upon the particular situation in which they find themselves.

Misconceptions

The police function in contemporary society is highly romanticized. Newspapers, television, and motion pictures have contributed to the sensationalizing of police operations. As a result, the general public often has a very distorted view of what the police do and how they do it. Ultimately this can be damaging to good police-community relations. For example, the general public might be shocked to learn that, on the average, only about one out of five **serious** crimes are ever solved by the police. The rate of clearances for crimes against persons is higher, but the rate for crimes against property is lower than 20 percent.

Unfortunately, the police do not do much to dispel this myth. It would not be good public relations for a police officer or detective to admit to a homeowner that the person who broke into his house and stole his stereo and television set would probably not be apprehended and that the property would probably not be recovered, even though this is more than likely the case. In addition, few police officials are ready to admit publicly that, despite their best efforts, crime does pay and that there is very little that the police can actually do to prevent crime.

Sometimes we expect too much of the police. We wonder, for example, why so many crimes can occur despite the obvious presence of the police. While we do not blame the police for the incidence of crime, we may question why they cannot do more to stop crime. This view, however, does not take into consideration the realities of crime and the limitations of the police.

> The police did not create and cannot resolve the social conditions that stimulate crime. They did not start and cannot stop the convulsive social change that is taking place in America.[7]

[7]President's Commission on Law Enforcement and Administration of Justice, *The Challenge of Crime in a Free Society* (Washington, D.C.: U.S. Government Printing Office, 1967), p. 92.

A false public perception about the ability of the police creates conflict when these expectations are not met. Would it not make sense for police administrations to admit the realities of life so that public expectations would not be so unrealistic? Would it not be better for police executives to publicly proclaim that only certain types of crimes can usually be solved? Would it not be refreshing for police administrators to admit that the limited resources of the police department dictate that only those cases that appear promising will receive much attention from investigators?

Lack of Public Consensus

The police owe their authority to enforce the laws to the public they serve. The laws that the police are expected to enforce are enacted by legislatures which are, at least in theory, representative of society. Thus, it can be argued that the law enforcement function of the police is consistent with society's demands and expectations. However, this is not always the case. As Banton has indicated, public consensus regarding the police role is high in small communities with integrated values and social norms.[8] However, society today is much less integrated than it was in the first half of the twentieth century. Society is much more complex and dynamic, much less homogeneous, and there is much less consensus of social values. Thus, even though the laws which the police enforce are enacted by the representatives of the people, there is not nearly the high level of public support for those laws today as there was years ago. Accordingly, the police themselves do not enjoy the same level of public support.

Ironically, the police enjoy a much greater level of public support than they think. Police officers, because they interact with the worst parts of society, and deal with society's worst aspects, tend to develop a very pessimistic view of the world and tend to see society at its worst. As a result, they develop rather negative views and assume that their own experiences are a reflection of the entire society. In fact, however, this is not the case and the general public is often much more supportive of the police than the police themselves might expect. "National opinion polls have shown that the vast majority of citizens have favorable attitudes toward the police."[9]

To some, the police are expected to be primarily crime prevention specialists. Storekeepers and business owners, for example, expect the police to patrol the downtown area and check their stores regularly in order to prevent them from being robbed or broken into. The same people, on the other hand, are not often anxious to have the police enforcing parking or traffic regulations near their business because such activity may adversely affect their customers.

Parents often expect the police to aggressively patrol their neighborhood streets and the areas around neighborhood schools in order to keep their children safe from speeding vehicles, child molesters, and drug pushers. These same people, however,

[8]Michael Banton, *The Policeman in the Community* (New York: Basic Books, Inc., 1964), p.3.

[9]James Q. Wilson, *Varieties of Police Behavior: The Management of Law and Order in Eight Communities* (Cambridge, Mass.: Harvard University Press, 1968), p. 28.

often become agitated when they themselves are stopped for traffic violations or when they are called by the police to come down to police headquarters to pick up a child who has been arrested for shoplifting or some other minor offense.

Local politicians, who themselves enacted the laws and set forth the policies under which the police are expected to operate, are also ambivalent about the role of the police. They expect certain laws to be enforced more rigorously than others, or certain classes of people to be protected more vigilantly than others, simply as a result of their perceived obligations to their constituents.

The views and perceptions of each of these classes of people is certainly understandable and they are also legitimate. Nevertheless, they do compound the problem of defining the proper role and mission of the police.

Conflicting Responsibilities

The responsibilities assigned to the police are many and varied, and they are often conflicting. For example, are police officers to be law officers or peace officers? The duties of these two roles are quite different and sometimes in conflict with each other. A peace officer, for example, is one whose contacts with the public are generally nonpunitive and nonadversarial in nature. The peace officer is the one who assists stranded motorists, looks for missing persons, and does what is necessary to maintain public order. The law officer, on the other hand, uses tactics reminiscent of the days of the frontier marshal, when force and the rule of the law were the chief tools of the trade.[10]

In fact, the police officer is both a peace officer and a law officer. The emphasis placed on each of these roles depends upon a number of things including community values, local administrative procedures, and political considerations. Moreover, these conflicting duties and responsibilities often require opposing attitudes and philosophical orientations on the part of the police officer. An orientation toward law enforcement and criminal investigation, for example, may be in direct contradiction to an officer's obligation to provide care and assistance to the destitute and homeless.

A police officer may see a homeless person as a burglary or theft suspect rather than a person in need of assistance, for example. An officer who has a strong orientation toward traffic enforcement may have little interest in routine preventive patrol aimed at crime reduction.

THE POLICE MISSION AND COMMUNITY VALUES

It is also true that an individual's (or an organization's) role in society is shaped and influenced by social customs, traditions, values, and norms of behavior. The role of the police is often impacted by social values and expectations, which change with

[10]See Banton, *The Policeman in the Community,* p. 7 for a discussion of the distinction between the "law officer" and the "peace officer."

increasing regularity. We live in a society characterized by changing fads and trends. Attitudes toward living, the family, religion, and government have changed radically in the last two decades, yet the police endeavor to perform their duties in much the same way that they have always done. As a result, their actions often come into conflict with societal values. This is not, however, a fault of the police, since they are simply reacting to conditions under the policies and rules that have been set for them. Nevertheless, this conflict between police actions and societal values makes the definition of the police mission even more difficult.

To a large extent, the police role is nothing more and nothing less than a reflection of local community values. The police are not, or should not be, an autonomous arm of local government, but rather an integral component of the local government's "service delivery system." Local police departments must be sensitive to and responsive to local community values and expectations. Police goals and objectives, likewise, must be developed with an appreciation for what the community wants and expects from the police.

Just as no two communities are exactly alike, there is no reason to believe that any single style of policing is better than another. Similarly, the operating philosophy of a police department in a large industrialized city will obviously differ from that of a small, rural community, since the problems in the two municipalities are dramatically different and the needs and expectations of the residents are not the same. The orientation of the police in any jurisdiction must be shaped by public opinion and local values.

> A police department is intimately related to the community in which it is located and in which it serves.... The men on the job are responsible to the public definitions of behavior. The nature of the community determines many of the problems that the police department must meet. The political structure of the community may have an important influence on the actions of the department and the areas of law enforcement that it emphasizes.[11]

As Banton points out, a police officer uses his experience in society in exercising his discretionary powers (Banton, p. 145). A police officer who lives in the community where he works has a better understanding of how people in the community think and what they expect of the police. A successful officer is one who understands the mood of the community and who behaves according to the norms and expectations of those who reside in the community. Conversely, an officer who adopts a style of official conduct that is contrary to community values provokes conflict and mistrust among the people he or she is sworn to protect.

This does not mean that police officers can only enforce laws that the public wants enforced, or that they must seek public approval for every action they take. Indeed, the police must be objective and independent in the exercise of their discre-

[11]William A. Westley, *Violence and the Police: A Sociological Study of Law, Custom, and Morality* (Boston: The Massachusetts Institute of Technology, 1970), p. 19.

tion and consider the common good when making difficult decisions. It does mean, though, that police officers must be conscious of community values and must be sensitive to public opinion in their actions. They must keep in mind that they are agents of the people, not guardians over them. They must see themselves as service providers rather than just law enforcers.

It is for this reason that it is not practical to develop universal standards by which the performance of one police agency can be measured against another unless the two communities they represent are quite similar.

PURIFYING THE POLICE MISSION

Over the years, the police role has evolved from a narrowly defined one (that is, law enforcer and peace keeper) to a proliferation of duties that no other public agency has the time, resources, or inclination to perform. Klockars has suggested that the police have inherited these responsibilities because the streets are their domain and that "whatever goes on in the streets is their business."[12] For the most part, the police have performed these duties ably, thus rewarding the public's confidence in their ability.

Some authorities have argued that one solution to the problem of more accurately defining the police mission lies in narrowing the range of police responsibilities. For example, if the police were assigned only to enforce traffic laws, or only to prevent crime and search out and prosecute criminals, there would be little confusion as to their proper role in society. A number of proposals have been advanced to "purify" the police role by relieving them of many duties which are not directly related to the traditional tasks of law enforcement and crime prevention. The major thrust of these proposals has been to eliminate from the police role many of the service functions that occupy much of the patrol officer's time. In this way, it is argued, the police will be free to pursue more productive and presumably important tasks.

It cannot be denied that many of the duties typically performed by the police have little in common with the traditional and historic role of the police in society. What must be recognized, however, is that these service functions are actually very important in contemporary society. Indeed, in many communities, they outweigh the importance of the more traditional police responsibilities of law enforcement and criminal apprehension. Although these service functions may seem like an unnecessary bother to the individual police officer, they assume much greater importance to the local community resident.

The notion that police effectiveness may be improved by transferring these "nonpolice" duties to other governmental agencies is based upon several untested assumptions which were discussed at some length in the American Bar Association's analysis of the urban police function. Among other things, the Association's report

[12] Klockars, *The Idea of Police*, p. 57.

questions the validity of several basic assumptions upon which the move to purify the police role is based.[13]

1. **An assumption as to what should be the primary role of the police**. For example, are the police to be primarily law enforcement officers, crime fighters, traffic regulators, preventers of disorder, or the providers of social services? This question is not answered easily, and depends upon the values, expectations, social conditions, and political expectations of the individual community.
2. **An assumption regarding the potential effectiveness of the police.** It has not been clearly demonstrated, for example, that the police can be more effective in their crime prevention and law enforcement roles if they are relieved of other duties. There is no clear evidence to support this contention, nor have there been any attempts to demonstrate the validity of this proposition.
3. **An assumption that police activities as they now exist are separable.** Is it really feasible and practical, for example,to relieve the police of the responsibility of enforcing traffic regulations, controlling the flow of traffic, and investigating traffic accidents, when a great number of crimes involve the use of automobiles? Similarly, is it practical to relieve the police of the responsibility of responding to domestic quarrels in view of the significant potential for violence posed by such events?
4. **An assumption that it is both desirable and feasible to reduce the conflict that arises by virtue of the police acting in both a helping and punitive role.** While it is certainly true that the police role often creates conflict and stress, so long as the police must deal with human problems, this conflict is inevitable. No practical means of reducing this conflict has yet been devised.
5. **An assumption that private or other governmental agencies can perform many existing police functions more effectively than the police.** It is certainly possible to transfer certain nonessential duties, such as rounding up stray animals, to other governmental agencies. However, the fact remains that the police are often the single government agency with trained and experienced personnel on duty when human emergencies arise.

The fact of the matter is that, for the most part, the police perform these duties quite well, even though they may privately complain that these are not "real" police duties. Moreover, these "nonpolice" duties actually occupy the greatest portion of the average patrol officer's time. Even though some officers may complain that they would rather be doing "real police work," the reality is that these same duties make the police officer's job interesting and rewarding. The opportunity to peacefully settle a domestic dispute, to rescue a drowning child, or to aid a stranded motorist who has

[13] *Project on Standards for Criminal Justice, Standards Relating to the Urban Police Function* (Chicago: American Bar Association, 1972), pp. 39-42

no one else to call for assistance can be the highlight of a police officer's tour of duty. It is difficult to imagine how a police officer would react to being faced with the responsibility of only enforcing laws and apprehending criminals.

> ...the police role in performing these tasks may, from the public's standpoint, be more important than dealing with some aspects of crime...[14]

In truth, the police have always been expected to deal with a variety of problems, many of them totally unrelated to the commonly accepted duties of law enforcement and crime prevention. Indeed, the peace-keeping and service-oriented functions of the police are deeply imbedded in the history of the police function. Since the earliest days of recorded history, the police have been charged with performing tasks not directly associated with what is commonly believed to be their primary function. Early English constables, for example, were expected to protect citizens from all manner of public dangers, including, among others, plague, fire, and commercial fraud.[15]

It was not until the creation of the Metropolitan Police in London in 1829—the system upon which the American police were modeled—that responsibility for prevention of crime and law enforcement was formally assigned to the police. Prior to that time, citizens were largely left to their own devices to protect themselves from criminal attack.

In this country, the police continue to spend a great deal of their time performing service-oriented activities. This is particularly true in small towns and rural areas where the incidence of crime is relatively low and where there are few other agencies to provide for the needs of the citizens. While those duties may be seen by some as an unnecessary and unwarranted drain on police resources, they are nevertheless important public services. If not for the police, these needs might go unmet.

This does not mean, however, that the police cannot and should not be relieved of some functions that have been given to them over the years. These are duties in which the police have no legitimate interest and that can and should be performed by other public agencies which are better equipped to perform them. Various kinds of inspection and regulatory functions, for example, could be better performed by a city's inspections division rather than by the police department. Animal control, as another example, could be performed much better by a private agency rather than by the police. Parking enforcement, which actually is more oriented toward the production of revenue rather than to parking regulation, is, in some cities, under the control of the finance department. These functions, and others, are not critical to public safety, nor are the police the most suitable agency to perform them.

Nevertheless, the attempt to "purify" the police role by narrowing the focus of police activities should proceed with caution for several reasons. First, it must

[14] Herman Goldstein, *Policing a Free Society* (Cambridge, Mass.: Ballinger Publishing Co., 1977), p. 34.

[15] Jonathan Rubenstein, *City Police* (New York: Farrar, Straus, and Giroux, 1973),.pp. 3-4.

be remembered that many of the service duties performed by the police are inextricably interwoven with the more essential aspects of the police mission. Thus, it may not be entirely possible, or desirable, to completely remove them from police responsibility.

Second, it is through the performance of many of these nonessential duties that the police have an important opportunity to interact with citizens on a positive rather than negative basis. These service functions provide an important opportunity to create and maintain favorable public support and appreciation for the police. This support and understanding is essential if the police are to carry out their other duties and responsibilities effectively. For example, we now recognize that the prevention of crime is not just a police responsibility, but a function that can only be performed effectively through widespread community support. The police need to use every means at their disposal to cultivate community support for their programs.

Finally, the self-image of the individual police officer must be considered. Too often, police officers see themselves as primarily "crime fighters" whose only responsibility is to enforce the law and put criminals in jail. In actuality, however, police officers spend relatively little time performing law enforcement functions compared to public service activities. It is likely that a totally different kind of police officer, in terms of attitude and personality, would be required if law enforcement and criminal apprehension became the primary mission of the police. This, in turn, could produce a negative impact on police-community relations.[16]

All the reliable evidence, combined with the experience of the past, points to the fact that the police role in contemporary society will continue to be multifaceted. Moreover, there can be no simple definition of the police mission, since the social, political, and cultural factors that influence the definition of the police mission are themselves dynamic. This fact must be recognized by the police themselves, as well as by the communities they serve.

Defining the proper mission of the police must then take into consideration local political, economic, social, and cultural factors present in the local community. Thus, the police mission may be different in each community. This is already the case, even though it is not formally recognized. Policing styles vary considerably from one community to the next and are based, in large part, upon community expectations and needs. However, these factors are usually not translated into a formal mission statement of the police. This is the weakness that needs to be addressed.

Police officials, political leaders, and community representatives must recognize that the police mission is a reflection of community values and should work together to develop mission statements that accurately reflect these values. This mission statement should be flexible enough to meet new and changing conditions, but should be drafted in clear and unambiguous language so as to leave no doubt in anyone's mind as to the purpose and function of the police in the individual community. This mission statement will serve to provide police officers with a clear

[16]Gordon E. Misner, "Enforcement: Illusion of Security," *The Nation,* (April 21, 1969), p. 488.

sense of purpose and direction so that they may clearly understand their role in the community.

SUMMARY

It is the intent of this book to put the role of the police patrol and the importance of police patrol into different perspective than has been the case in the past. The remaining chapters of this book are intended to challenge the reader to begin thinking differently about the patrol force and to recognize that it is, in fact, the nucleus of the police department and that the success of the police mission depends directly upon the success of the patrol effort. In the past, public attitudes, as well as the attitudes of the police themselves, have frequently failed to recognize the true importance of the patrol force.

Patrol work is too often seen as routine, dull, boring and suitable only for people who have no aspirations for more challenging work. As a result, the patrol officer is seen as occupying the bottom rung of the occupational ladder in the police organization, even though the decisions the patrol officer makes have greater potential impact on the community than those of the leadership of the department.

The focus now being placed on community-oriented policing and problem-oriented policing has turned attention once again to the "basics" of policing, in which the patrol officer plays a central role. Like team policing, directed deterrent patrol, and other innovations in policing, community-oriented policing and problem-oriented policing may only be temporary programs and may eventually give way to new ideas and policing strategies. This is not only expected, but it is healthy, since change is necessary to respond to new and different problems which the police face.

The ability to recognize the importance of police patrol and to reject conventional wisdom and tradition in favor of innovation and imaginative concepts is the underlying theme of this book. The following chapters are intended to reinforce this message and to motivate the reader to discover still other methods by which the patrol effort can be redesigned and energized. If a single new idea about police patrol—its purpose, its use, and its methods—results from reading this book, then it will have accomplished its goal.

REVIEW QUESTIONS

1. Explain briefly why the police mission is difficult to define.
2. What things should be taken into consideration in defining the mission of the police?
3. What is meant by the "reform era" of American policing and when did it begin?

4. Why is it just as important to know what does not work as it is to know what does work in policing?
5. What were some of the more significant findings that emerged from research on policing during the 1970s?
6. What are some of the reasons for the lack of recognition given to the patrol officer?
7. In what ways are we witnessing a rebirth of the patrol function in America?
8. In what ways might the various roles played by the police patrol officer sometimes conflict with each other?
9. Explain how the police mission is shaped by community values.
10. What is meant by the attempt to "purify" the police role?

REVIEW EXERCISES

The following exercises are designed to reinforce or amplify comprehension of the material contained in this chapter.

1. Conduct library research on Sir Robert Peel and prepare a short paper (no more than 1500 words) describing the significance of Peel's reforms on modern policing.
2. In small group sessions, or as a general class discussion, explore the various roles played by the police in modern society and potential points of conflict.
3. In small group sessions, assign teams to debate the merits of, as well as the practical obstacles to "purifying" the role of the police in society.
4. Discuss how and why various aspects of the police role may differ from one community to the next.

REFERENCES

Banton, Michael, *The Policeman in the Community*. New York: Basic Books, Inc.

Kelling, George L., "What Works—Research and the Police," National Institute of Justice, *Crime File Study Guide*, no date.

Klockars, Carl B., *The Idea of Police*. Newbury Park, Ca.: Sage Publications, 1985.

Misner, Gordon E., "Enforcement: Illusion of Security," *The Nation,* (April 21, 1969).

President's Commission on Law Enforcement and Administration of Justice, *The Challenge of Crime in a Free Society*. Washington, D.C.: U.S. Government Printing Office, 1967.

President's Commission on Law Enforcement and Administration of Justice, *Task Force Report: The Police*. Washington, D.C.: U.S. Government Printing Office, 1967.

Project on Standards for Criminal Justice, Standards Relating to the Urban Police Function. Chicago: American Bar Association, 1972.

Rubenstein, Jonathan, *City Police*. New York: Farrar, Straus, and Giroux, 1973.

Sheehan, Robert, and Gary W. Cordner, *Introduction to Police Administration* (2nd. ed.). Cincinnati: Anderson Publishing Co., 1989.

Skolnick, Jerome H., and David H. Bayley, *The New Blue Line:Police Innovation in Six American Cities* . pp. 4-5. New York: The Free Press, 1985.

Westley, William A., *Violence and the Police: A Sociological Study of Law, Custom, and Morality*. Boston: The Massachusetts Institute of Technology, 1970.

Wilson, James Q., *Varieties of Police Behavior: The Management of Law and Order in Eight Communities*. Cambridge, Mass.: Harvard University Press, 1968.

CHAPTER 2
PATROL GOALS AND OBJECTIVES

CHAPTER OUTLINE

LEARNING OBJECTIVES

After completing this chapter, the student should be able to

1. Explain the importance of goals and objectives as they apply to the municipal police function.
2. Describe how police goals and objectives are formulated and used in the police service.
3. List the principal functions of the police and describe how police performance can be measured.
4. Give examples of the ways community values and characteristics affect the development of police goals and objectives.

If it is difficult to precisely define the role of the police in contemporary society, it is equally difficult to reach consensus concerning police goals and objectives. Just as all organizations should have a mission or sense of purpose, they also need goals and objectives by which their performance can be measured. An organization without a goal or objective is like a ship without a rudder; it weaves aimlessly upon the seas, without destination or means to guide it in its travels.

Defined goals and objectives are essential for the police, if they are to provide effective public service and maintain good public relations. Political leaders, city administrators, and the general public are usually not very well informed about what the police do or how they do it. As indicated in Chapter 1, some people have very realistic expectations about police capability. They may, depending upon the circumstances, feel instinctively that the police are doing a "pretty good job," or that the police are highly efficient, or that the police in their community are not as good as those in other communities. However, they usually do not have very good measures upon which to make these judgments. This is usually because the police have failed to provide accurate and realistic means by which to measure their performance. In other words, the police have not taken the initiative to develop for themselves a set of realistic goals and objectives.

Without established goals, the police operate in a vacuum and lack specific guidance or direction. Articulated goals and objectives provide a framework with which operating policies, procedures, and programs can be developed. When properly developed and defined, goals and objectives help the police to monitor their own performance and assess their own effectiveness.

The words *goals* and *objectives* are sometimes used interchangeably; however, they are not synonymous. A goal, for example, is a statement of a desired state of affairs, described in general, rather than specific, terms. An objective, on the other hand, can be stated in rather specific and measurable terms. Objectives can be used to help us measure our progress toward the accomplishment of a goal. In this respect, then, objectives are subordinate to and supportive of goals.[1]

To illustrate, a goal for next Sunday afternoon might be to enjoy a pleasant drive in the country. Use your own imagination to describe what a pleasant drive in the country might be like. In order to have a more specific means of describing the goal, or measuring attainment of it, set forth a series of objectives. To make the goal more precise, and to measure the extent to which our goal is achieved, set forth certain supportive objectives such as the following:

1. To experience weather with temperature between 75 and 80 degrees.
2. To have blue skies with no more than 25 percent clouds.
3. To experience no traffic congestions, road construction, or mechanical breakdowns.

Each of these objectives can be measured in rather precise fashion and can help us to determine whether our goal has been achieved.

Police goals and objectives can be stated in somewhat the same way. A legitimate goal for the police, for example, would be to provide an environment free from crime and the fear of crime. While this is a very lofty goal, we can set several objectives which will help us determine whether or not we are making progress toward our goal. For example, the following objectives would tend to support the goal of an environment free from crime and the fear of crime:

1. To achieve a 25 percent reduction in the number of crimes of violence during the next 12 months.
2. To increase the rate of apprehension for robbery, aggravated assault, and sexual assault by 20 percent during the next 18 months.
3. To decrease citizens' fear of crime and to increase their perception of safety by 15 percent over the next year.

Each of these objectives is intimately linked to the goal of creating an environment free of crime and the fear of crime. Local community standards need to be considered when fashioning police goals and associated objectives.

Police goals and objectives, the importance attached to them, and the means by which they are achieved, vary considerably among police agencies. This is as it should be, for the philosophical foundation of the American police system is based upon the concept of local control and autonomy. For this reason, police goals and objectives, and operating procedures will be influenced to a great extent by local community characteristics and expectations. Some communities, for example, may place a high value on the maintenance of social order, while others may not. In some

[1]See National Commission on Criminal Justice Standards and Goals, *Police* (Washington, D.C.: U.S. Government Printing Office, 1973), p. 50.

towns and villages, the police may have a high service orientation, while this may not be the case in others. In still other communities, aggressive enforcement of traffic laws may be a top priority for the police, while this may be of lesser importance in other communities.[2]

As a result, any discussion of police goals and objectives must take into consideration the influence placed upon police conduct by local community values and expectations. The police exist solely to serve the community and their methods of operation must take this fact into consideration. Nevertheless, there are a number of functions that are common to all local government police agencies, even though the emphasis placed on these functions will vary considerably. Within each of these functional areas, it is possible to define police responsibilities and to set goals and objectives that are consistent with local community values. Generally, these functions consist of the following: (a) crime prevention; (b) criminal apprehension; (c) law enforcement; (d) order maintenance; (e) traffic enforcement; and (f) public service. It is largely through the efforts of the patrol force that these functions are performed.

CRIME PREVENTION

The detection, suppression, and prevention of crime has traditionally been accepted as one of the primary goals of the local police force. Some authorities have even defined crime control as the "core mission" of the police.[3]

This responsibility falls directly on the patrol force. Unfortunately, police performance is often judged on the basis of the incidence of crime in the community. When sharp increases in the rate of crime occur, the assumption is sometimes made that the police have failed to accomplish one of their primary objectives. What many people fail to realize, however, is that crime is not just a police problem, but is rather a community problem and that its causes and cures are beyond the control of the local police.

All too often, the public holds a rather narrow view of the police mission and perceives the police as being primarily concerned with the control of crime. In some instances, the police share the belief that crime control is their most important responsibility. This misconception of the police function is often supported by the romantic portrayal of police work in the news media and in television and motion pictures. In reality, this misconception has little basis in fact.

During the "reform" era of policing, the police claimed total responsibility for crime control, perhaps not anticipating the futility of this effort, or their own limitations in controlling crime. As the police have moved toward a more open relationship

[2]For a detailed discussion of police goals, see V.A. Leonard and Harry W. More, *Police Organization and Management,* 7th ed. Rev, (Mineola, New York: The Foundation Press, Inc., 1987), pp. 45-48.

[3]See Mark H. Moore, Robert C. Trojanowicz, and George L. Kelling, "Crime and Policing," in National Institute of Justice, *Perspectives on Policing* (June, 1988).

with the community, they have begun to recognize the importance of community involvement and support in controlling crime. Not only are the police not the "first line of defense" in the war against crime, they are not wholly responsible for controlling crime in their community. Other social institutions, such as the family, churches, and schools share in this responsibility.[4]

Part of the problem is that police executives have failed to speak out on this issue; when they have spoken out, their words have not been heeded by political leaders and community members. Few chiefs want to openly admit that they are largely ineffective in controlling crime and that some type of crime will occur despite the best efforts of the police. What they should be saying is that crime is a community problem and that it requires the concerted and coordinated efforts of all segments of the community to deal with it.

In reality, crime prevention is not the single most important function of the police in most communities. Police spend relatively little time in activities directly related to the prevention and investigation of crime and the enforcement of laws.[5] In addition, crime is the result of many factors and conditions over which the police have little direct control. Unemployment, poor educational opportunities, alcoholism, drug abuse, lack of adequate housing, and community attitudes contribute more to the incidence of crime than perhaps any other factors. Thus, the police are but one element in any comprehensive crime control strategy.

This does not mean, of course, that crime prevention should not be a major goal of the police, for clearly it is and should be. It simply means that we must recognize the limitations under which the police work and not assess their total performance on the incidence of crime in the community. Clearly, there are many things the police can and should do to control crime. Some of these methods are discussed at length in other chapters of this book.

It is also important to distinguish different types of crime and to realistically assess the ability of the police to prevent them. Some types of crime, such as robbery, burglary, and auto theft, because of the conditions and circumstances in which they occur, are more likely to be prevented or deterred through aggressive and vigilant police patrol efforts. Other types of crime, such as bank fraud, embezzlement, and other kinds of white-collar crime are much less susceptible to preventive efforts by the police. As Goldstein has observed, such distinctions are necessary if we are to be able to realistically judge police effectiveness.

> It is, of course, absurd to argue that the police have the same responsibilities regarding a robbery, for example, as they would have regarding disorderly conduct, simply because both are defined by the legislature as criminal offenses.[6]

[4]Robert Sheehan and Gary W. Cordner, *Introduction to Police Administration,* 2nd ed. (Cincinnati: Anderson Publishing Co., 1989), p. 10.

[5]Herman Goldstein, *Policing a Free Society* (Cambridge, Mass.: Ballinger Publishing Co., 1977), p. 24.

[6]Goldstein, *Policing a Free Society*, p. 24.

In evaluating the effectiveness of the police in their crime prevention role, it is important to understand the nature of crime and its causes as well as the limitations of the police. Constitutional guarantees which benefit all of society place significant restrictions on the police. These restrictions are most burdensome in detecting and preventing those types of crimes which are less apparent to the casual observer, such as the white-collar crimes referred to earlier. Street crimes, on the other hand, are more visible and can be detected and interdicted much more easily by the police.

In recent years, the police have begun to recognize that traditional methods of crime prevention such as routine, random patrol are of limited value. They have begun to search for more effective methods of deterring crime. It is difficult to measure the effectiveness of these methods, however, since it is impossible to determine how many crimes are not committed due to the efforts of the police.[7] Moreover, the usual means of comparing results of various crime control methods under similar conditions are not reliable, since the causes of crime are so many and varied. It is thus difficult to establish precise cause-and-effect relationships.[8]

Clearly, more effective ways of controlling crime are necessary. The police cannot continue to rely on traditional, reactive methods of attacking crime in the community. Beefed-up patrols, quicker response times, and a wider variety of tactical strategies are only part of the solution. The police must not necessarily work harder but rather think harder to become more effective in the means they employ. For example:

> ...improved crime control can be achieved by (1) diagnosing crimes, (2) fostering closer relationships with the community to facilitate crime solving, and (3) building self-defense capabilities within the community itself.[9]

Although the police cannot control the root causes of crime, they can improve their ability to detect and deter certain types of crime. They can also develop and implement various kinds of target-hardening programs, designed to acquaint members of the community with the means to better safeguard themselves and their property. Security surveys, for example, can be used by police to instruct homeowners and business owners in the use of reliable antitheft and anti-intrusion devices. The police can also cooperate with private vendors who are marketing various proven and reliable means of safeguarding property, such as the Lojack system, and other systems designed to prevent automobile theft. Programs such as Operation Identification, Block Watchers, Citizen Patrol, and other activities designed to promote greater citi-

[7]Harry P. Hatry, "Wrestling with Police Crime Control Productivity Measurement," *Readings on Productivity in Policing* (Washington, D.C.: Police Foundation, 1975), pp. 86-128.

[8]Jerome H. Skolnick and David H. Bayley, *Community Policing: Issues and Practices Around the World* (Washington, D.C.: National Institute of Justice, 1988), p. 68.

[9]Mark H. Moore, Robert C. Trojanowicz, and George L. Kelling, "Crime and Policing," *Police Practice in the '90s: Key Management Issues* (Washington, D.C.: International City Management Association, 1989), p. 32.

zen awareness of and participation in crime prevention efforts are equally useful. Crime prevention must be a cooperative effort between the police and the public.

The police can also learn to better understand the **precipitating causes** of certain kinds of crime and develop more practical means of dealing with those causes. For example, there is a strong correlation between certain types of street crime, such as robbery and aggravated assault, and the use of drugs. Some sections of large cities have been havens for drug dealers and users, and serious crimes occur frequently in such areas. Some cities have taken innovative steps to eliminate these hard-core drug havens by combining forces with other city agencies to declare the buildings where drug sales and use take place as public nuisances and having them torn down. In Hallandale, Florida, the city has formed a private corporation to provide low-interest loans to building owners to erect new buildings on the sites where drug houses once predominated, thereby providing much-needed housing for the poor.

One of the most powerful tools available to the police to gain public participation in crime prevention is the local news media. Radio, television, and newspapers can play an important part in informing the public of crime trends and crime prevention techniques. The use of television by the police to alert the general public to criminal activity and to solicit their assistance in identifying and locating criminal suspects has become very fashionable during the last decade. A number of syndicated television programs, watched by millions of viewers, have aided in the capture of some of the most wanted criminals in America. Local news media, including cable television, have been used very effectively by local police agencies to warn the public about particular kinds of crime.

If the police are to be successful in their crime prevention responsibilities, they must learn more about the nature, scope and causes of crime and the means by which crime is committed. They must better understand the criminal mind and be more adept at predicting crime trends from available information. They must utilize available technology as well as their own instincts and experience in order to develop ways to "harden the target." They must expand their knowledge and understanding of when, where, how, and under what conditions certain kinds of crime are committed. They must also have a much better understanding of which kinds of crime are most likely to be deterred through their efforts and which ones are not. This will help them to develop much more effective strategies to be employed by the patrol force.

Police administrators must recognize that patrol officers must not be simply turned out to "go and fight crime," but must be given much more definitive direction. Patrol supervisors and command officers must be encouraged to use discretion and imagination, rather than tradition and custom when deploying their forces. They must reject the notion that simply assigning more officers to patrol will reduce crime. They must continue to seek out new ways of dealing with both traditional and emerging kinds of criminal activity.

The police must also become more receptive to the notion of enlisting and accepting the aid of the private sector in providing services traditionally reserved for the police. In recent years, citizens have begun to contract with private security firms to provide supplemental police protection to their neighborhoods, arguing that the police cannot provide services sufficient to their needs.

The police must also redouble their efforts to enlist the aid of the public in the prevention of crime. To this end, they must admit publicly that, without widespread community support, they will not have a significant impact on crime. As the former Chief of Police in Los Angeles stated:

> The police themselves cannot control crime... However, the police play a major role as the catalytic agency in society to assist the process of "feeding back" to the rest of society information on what is happening in terms of crime and disorder. No one else can perform this function but the police. No one else is in contact with crime and disorder in its totality. No one else has the machinery or perception or access to the basic facts as do the police.[10]

In recent years, the use of volunteer citizen patrols has become popular in many American cities. These patrols serve as the eyes and ears of the police and are cost-effective ways of supplementing police manpower. These patrols have been used effectively in a number of large cities including New York, Boston, Los Angeles, St. Louis, and Detroit. In one neighborhood in New York City, crime dropped sharply after the institution of the volunteer citizen patrols.[11]

Crime Prevention Goals and Objectives

Notwithstanding the fact that it is difficult, if not impossible, to accurately assess the effectiveness of a police agency's crime prevention efforts, it should nevertheless be possible to establish reasonable goals and objectives for crime prevention. These goals would be defined as **process** goals rather than as **outcome** goals, since they refer to activities contributing to an outcome rather than to the actual outcome.

There are a number of activities for which the patrol force can be responsible which are directly related to the prevention of crime. One reasonable objective would be to set aside a specific portion of time during each tour of duty to *specific* patrol activities. Such activities might include, for example, patrolling parking lots in retail shopping areas where a high incidence of auto theft has been reported; conducting foot patrol in neighborhood parks and around school playgrounds where a high number of incidents involving crimes against children have been reported; and conducting bicycle patrol in parks and recreation areas where strong-armed robberies and assaults have been reported.

The amount of time devoted to various nontraditional patrol activities by a particular shift or group of officers over a period of time can be measured and compared with increases or decreases in target crimes. While it may not be possible to establish a direct cause-and-effect relationship between these activities and the incidence of specific types of crimes, some correlation between them may be possible. Moreover,

[10]Edward M. Davis, "Professional Police Principles," *Federal Probation,* 35 (March, 1971), p. 29.

[11]Robert W. Poole, Jr., *Cutting Back at City Hall* (New York: Universe Books, 1980), pp. 38-39.

as the time devoted to these activities is recorded and measured, it should be possible to alter, or more precisely define, future objectives. For example, if it is discovered that an average of three hours of bicycle patrol between the hours of 4:00 PM and 8:00 PM during summer months has had a noticeable impact on the incidence of targeted crimes in the affected area, it may be desirable to determine whether a slight increase or decrease in bicycle patrol in the same area will affect further change in criminal activity.

The identification of specific types of crime prevention activities by the patrol force must be made by the individual patrol supervisor or patrol commander, based upon information provided by the department's crime analysis section or crime prevention officer. Not all types of crimes are susceptible to preventive efforts by the patrol force, but many are. These need to be identified and specific means for attacking them should be developed and implemented on a trial basis. If the desired effect is not achieved, alternative measures can be developed and tried. Developing innovative ways in which to deploy and direct the patrol force should aid the prevention of selected types of criminal activity.

CRIMINAL APPREHENSION

When crimes occur despite the best efforts of the police to prevent them, it then becomes a police responsibility to identify, locate, and apprehend the person(s) responsible. In some instances, a patrol officer may be present when the crime occurs and may be able to apprehend the criminal at the scene of the crime, or soon after the crime has been committed. In other cases, the perpetrator may be identified only after a long and exhaustive investigation. In the majority of cases, however, the criminal will never be apprehended. Only about one in five serious crimes will be cleared by the arrest of the offender or through some other means. The rate of apprehension is considerably greater for crimes against persons (homicide, rape, robbery, and aggravated assault), and substantially lower for crimes against property where there are few reliable leads to aid in identifying the persons responsible.

The primary means by which criminals not apprehended in the commission of an offense are eventually brought to justice by the police is through the careful and meticulous investigation of the circumstances surrounding the crime. Often, this investigation is conducted by plainclothes detectives, who act upon information supplied by the patrol officer at the conclusion of a brief preliminary investigation. Unfortunately, many police agencies continue to ignore and misunderstand the vital role of the patrol officer in the investigation of crimes and the subsequent apprehension of offenders.

All too often the patrol officer's responsibility is limited to preparing a brief report containing the basic information about the crime. This report is then forwarded to detectives, where it will probably be screened to determine whether there is sufficient information to warrant further investigation. When it is determined that a follow-up investigation is warranted, the case will be assigned to a detective, who will often retrace some of the same ground covered in the preliminary investigation by the patrol officer. In some instances, the patrol officer may be required to stand by and protect the crime scene to await the arrival of the detective. Often, other specialists, such as crime-scene technicians, are also called in to gather up and preserve important items of evidence.

This strategy often ignores the training, experience, and interest of the patrol officer. It is assumed that the patrol officer is too busy to conduct a thorough crime-scene investigation, or lacks the basic skills to do so. It is also assumed that victims will be placated by seeing the case turned over to plainclothes detectives, since our society has been led to believe that detectives are more expert at conducting investigations than are uniformed officers. All too often, the patrol officer who conducted the preliminary investigation and who has a vested interest in the outcome of the case, is never advised regarding the progress of the investigation. Moreover, the contributions of the patrol officer toward the successful conclusion of the case and the apprehension of the offender are often ignored by the agency. This is perhaps one of the most important factors contributing to poor relationships between patrol officers and plainclothes detectives.

In more progressive police departments, however, attempts have been made to broaden the role of the patrol officer in the investigation of crime and the apprehension of offenders. This trend has been prompted by several factors, among them economic pressures forcing police administrators to make critical decisions about the utilization of limited resources. As a result, they have begun to realize that it is simply not possible to assign detectives to investigate all crimes that warrant follow-up work. They have begun to explore alternative means of conducting criminal investigations. In some departments, patrol officers are being given greater latitude in conducting follow-up investigations, thus allowing detectives to concern themselves with only the more serious cases, or those which, due to their complexity, cannot be adequately investigated by the beat officer.

In addition, police administrators are beginning to realize that something must be done to elevate the professional image of the patrol officer. As indicated in the previous chapter, the important contribution of the patrol officer toward the overall mission of the police has often been overlooked. To counteract the trend toward police specialization, there has been a concerted effort in some police agencies to broaden the role of the patrol officer through a more generalized approach to policing. This broadening effect also helps to ensure that patrol officers are able to take advantage of the training and education which has prepared them for their responsibilities.

Two major research efforts helped to demonstrate the important contribution made by patrol officers in the investigative process. These studies have shown that, notwithstanding the efforts of detective specialists, it is often the information developed by the patrol officer during the preliminary investigation of the crime that determines whether a case will eventually be solved. In contrast, the painstaking efforts of detectives—checking modus operandi files; interviewing victims, witnesses and suspects; tracing vehicle license numbers—often contribute little to the final outcome of a case.

> The single most important determination of whether or not a case will be solved is the information the victim supplies to the immediately responding patrol officer. If information that uniquely identifies the perpetrator is not presented at the time the crime is reported, the perpetrator, by and large, will not be subsequently identified.[12]

Studies conducted by the Stanford Research Institute produced similar findings. The SRI studies revealed that, for specific types of crimes (robbery, rape, assault with a deadly weapon, and auto theft), the chances of a case being solved by detectives were minimal unless relevant information concerning the offense and the offender was obtained by the patrol officer during the preliminary investigation.

> A conclusion to be drawn from this observation is that the roles of the patrol officer and the investigator cannot be viewed as separate and distinct functions. We view patrol as fulfilling not only a crime suppression role but also as performing an investigative function. How effectively a patrol officer documents the event of a crime to which he responds will have a definite impact on the case outcome when investigators attempt to pursue the case.[13]

This does not mean that detectives are not instrumental in solving crimes, or that all crimes can be successfully investigated by patrol officers, for this is clearly not the case. Nor does it mean that specialist skills beyond those possessed by patrol officers are not important, for clearly they are. These findings do suggest, however, that greater recognition needs to be made of the skills and abilities of patrol officers and that the traditional dichotomy between the patrol officer and the plainclothes investigator needs to be reassessed. Police administrators have an opportunity to maximize the resources available to them by allowing patrol officers greater latitude in the criminal investigation process.

Several changes must be made in police operating procedures if the investigative skills of patrol personnel are to be employed to their full potential. First, patrol officers should be allowed and encouraged to devote more time to the investigation of crimes to which they respond. Patrol supervisors should be encouraged to allow their subordinates to pursue investigative leads on their own initiative, rather than be content to simply turn a case over to detectives. In most cases, patrol officers have both the time and the ability to do this, despite traditional beliefs to the contrary. In some cases, this may require that another officer be assigned to handle the investigating officer's calls while the investigation is in progress. The patrol supervisor should be able to provide the patrol officer with the assistance necessary to determine whether a case merits follow-up investigation by the officer, and whether that investigation can be successfully pursued by the officer.

[12]Peter W. Greenwood and Joan Petersilia, *The Criminal Investigation Process, Vol. 1 - Summary and Policy Implications* (Santa Monica, Ca.: The Rand Corporation, 1975), p. viii.

[13]Bernard Greenberg, and others, *Felony Investigation Decision Model - An Analysis of Investigative Elements of Information* (Menlo Park, Ca.: Stanford Research Institute, 1975), p. xx.

Second, patrol officers should be better trained and equipped to conduct routine investigations. Even in situations where specialists are required, there are many basic investigative functions that patrol officers can perform. If a burglary suspect has been identified, it may be possible for the patrol officer to locate and question that person concerning the case. When witnesses are available and can be contacted, their statements should be taken as soon as possible, not three days later when the case is assigned to a detective. When physical evidence is available, it should be identified, marked, and taken into custody.

The investigative skills of patrol officers should also be reflected in their regular performance evaluation reports. The ability to gather physical evidence at a crime scene, to locate and skillfully interview victims and witnesses, to prepare a complete and accurate report on a criminal investigation, to correctly analyze and interpret the pieces of an investigative puzzle, and to pursue investigative leads to their logical conclusion should all be taken into consideration when evaluating the performance of patrol officers.

Finally, patrol officers should be given credit and encouragement for their accomplishments in the investigation of crime and apprehension of criminals. Too often, credit for the solution of a major case or the apprehension of a notorious criminal goes to the detectives who make the arrest, even though the arrest might not have been possible without the contribution of the uniformed patrol officer. Monthly activity reports, which report on the number of cases cleared and persons arrested, rarely distinguish the contributions of patrol officers from those of detectives. Too often, case clearances are viewed as a measure of investigative performance, even when patrol officers contributed significantly to the solution of those cases.

Criminal Apprehension Goals and Objectives

Criminal apprehension goals must be based upon reasonable expectations of the mission of the patrol force. Moreover, any specific goals should focus not merely upon the number of arrests effected, but the quality of those arrests, both in terms of the ultimate disposition of the case and its overall impact upon the police mission. The type of arrests made and the circumstances under which they were made are also important. Arrest efforts should be geared toward the reduction of specific types of criminal activity. In addition, arrests made other than through the initiative of the individual officer should be given considerably less weight than those resulting directly from the officer's own initiative.

For example, an officer assigned to a patrol beat in which there are a number of retail establishments may make a number of arrests of shoplifters. Many, if not all, of these arrests may be actually effected by store security personnel or clerks, in which case the officer is simply responsible for transporting and booking the persons arrested. In addition, these arrests may or may not result in conviction and may or may not have a deterrent impact on criminal activity.

Police planners and patrol supervisors need to identify certain kinds of criminal activity which can realistically be impacted by increased arrest rates and should set forth reasonable goals for the patrol force in terms of arrests for those offenses. Once again, these goals should be developed very carefully, giving due consideration to other patrol responsibilities. In addition, the quality of the arrests, as indicated by their final disposition, should also be considered.

Reporting systems should be designed to account for the contributions of patrol officers toward arrests made by detectives. As is the case in the sport of basketball, an officer should be given credit for an "assist" even though he may not have actually made the arrest. Similarly, those arrests made by an officer which are not the direct result of the officer's own initiative should be given considerably less consideration.

Establishing reasonable and realistic criminal apprehension goals for the patrol force will aid in the monitoring and evaluating of patrol performance and provide officers with greater incentive by acknowledging their contributions to the overall police mission.

LAW ENFORCEMENT

It is a basic responsibility of the police to enforce the law. This responsibility is especially important due to the potential impact upon the community if laws are not enforced vigorously enough, or if they are enforced too rigidly. If the police fail to enforce the laws in a fair and reasonable manner, public confidence in the police will be seriously weakened. Conversely, police authority may be seriously undermined if laws are enforced in an arbitrary and prejudicial manner.

That the patrol officer has a statutory obligation to enforce the law is obvious. However, the manner in which this obligation is fulfilled raises a number of questions and poses many potential problems. There are many more laws on the books than can ever be reasonably enforced. Indeed, in enacting some of these laws, it is questionable whether the legislature ever meant for them to be wholly enforced. Instead, laws are often enacted by lawmakers as a means of establishing public policy. How those laws will actually be enforced, or whether they should be at all, may not be a real consideration to those who enact the laws.

In addition, there are many laws on the books today which are hopelessly outdated and which have no relevance to modern society. Rarely, however, are such laws repealed, except for those which are so diametrically opposed to the public interest or popular appeal (i.e., prohibition). Nevertheless, these outmoded and unnecessary laws remain in effect and the police are, at least in theory, responsible for their enforcement. In practice, however, the police themselves usually decide which laws to enforce, based upon their own sense of necessity and practicality. These decisions are most often made by the patrol officer on the street, based upon his or her understanding of departmental policy, public need, and practical necessity. These decisions are subsequently reviewed, critiqued, and sometimes reversed by the general public, superior officers, and the courts.

Compounding the problem of the police in deciding which laws to enforce is the fact that the public is often ambivalent in its attitude toward the law. Many people are interested in seeing certain types of laws enforced vigorously when it serves their own interests. For example, people who do not own pets (as well as some who do) are often anxious to see leash laws and other laws directed toward the control of animals enforced because they view such animals as a public nuisance. Animal lovers often do not share this view. Similarly, elderly people, or those who live alone, or those without children, may be anxious to have the police enforce local ordinances regarding loud parties, curfew, and speeding vehicles on neighborhood streets; younger members of society are usually much more tolerant of such occurrences. Depending upon the person's point of view, the police may be seen as being either too lenient or overly strict. They can rarely satisfy everyone.

The police are also called upon to enforce laws which are very unpopular with some parts of society. These laws often have more to do with moral conduct than they have to do with antisocial behavior. Such laws are often enacted in response to various interest groups, without regard to their full impact or enforcement. This creates problems for the police as well as for the other elements of the criminal justice system and could be considered, in some cases, an abuse of the legislative process. As Morris and Hawkins have pointed out, "The criminal law is an inefficient instrument for imposing the good life on others."[14]

Police patrol officers, unfortunately, often have little to guide them in their law enforcement responsibilities. In the police academy, they are introduced to the state criminal code. Later on, they are given a set of local ordinances which they are expected to know, understand, and—when necessary or practical—enforce. However, they are usually not told which of these thousands of laws they are expected to enforce, nor are they told how much discretion they are expected or allowed to use in enforcing them. In addition, local departmental policies rarely offer any useful guidance to the individual officer.

As a result, officers must rely upon their own judgment, what they have learned from their peers, or what they have learned from their own experience on the street. They learn, for example, which laws can be enforced without incurring the wrath of their superiors, the public, the news media, and the courts. In this way, they formulate their own informal standards which guide them in their exercise of discretion. Thus, the enforcement of law in this country varies not only from department to department, and from one community to the next, but also from one officer to another.

The exercise of discretion creates problems for police administrators, since it means that all laws are not enforced equally and that officers must use their own good sense in determining which laws to enforce and in what way. This can make a police administrator very uncomfortable, since it exposes him to criticism when an officer's good sense leads him to make a decision which later proves to be a problem. A police chief's "comfort zone" would be considerably wider if the range of discretion afforded to officers could be considerably limited.

[14]Norval Morris and Gordon Hawkins, *The Honest Politician's Guide to Crime Control* (Chicago: University of Chicago Press, 1970), p. 2.

> Many police administrators believe that an open acknowledgment that all laws are not inflexibly enforced in all situations would be tantamount to admitting that police agencies are shirking their duty—that they are ignoring the rule of law.[15]

The exercise of discretion, both in the enforcement of laws and in the performance of their many other duties, is an awesome police responsibility. Ironically, it is the patrol officer, who holds the least formal authority in the police organization, who exercises the greatest discretion on matters of critical importance. Surely, the discretion afforded the police is a great responsibility that must not be taken lightly. Patrol officers must be made to understand the importance of their role, not just in the enforcement of laws, but in defining, through the application of their own discretion, what is legal and illegal behavior. As Niederhoffer has indicated:

> ... because the application of the law depends to a large degree on the definition of the situation and the decision reached by the patrolman, he, in effect, makes the law; it is his decision that establishes the boundary between legal and illegal.[16]

Due to the kinds of conditions under which the police operate—uncertainty, danger, crisis, tragedy, hostility, anger, and frustration—the individual police officer is confronted with a wide range of alternative courses of action in any situation. In some instances, and under some conditions, a violation of the law may be ignored altogether. In other situations, and under other conditions, a person may be arrested for committing the same offense. In some circumstances, a verbal warning may be an appropriate response to a violation of the law, while in others, and for different reasons, the officer may elect to make an arrest for the same infraction. In making these decisions, the officer must rely almost entirely upon his or her own judgment.

Discretion on the part of the individual officer is not absolute, however. Even when the limits of power and authority are only loosely controlled by the law or by department policy, other limitations directly or indirectly influence the options available to the officer.

First, the officer's immediate supervisor may set down informal rules or guidelines governing the officer's performance. These rules or guidelines are usually learned by the officer over time. Officers learn through experience what their supervisors expect of them. They learn what kinds of arrests will be approved or condoned by a particular supervisor and which ones will not. Some supervisors, like some officers, encourage their officers to be aggressive in their law enforcement responsibilities. These supervisors will support their officers when they make good and reasonable arrests.

Other supervisors tend to be more passive and do not encourage their officers to be aggressive in enforcing the law. Through their own actions and attitudes, they con-

[15]National Commission on Criminal Justice Standards and Goals, *Police* (Washington, D.C.: U.S. Government Printing Office, 1973), p. 22.

[16]Arthur Niederhoffer, *Behind the Shield: The Police in Urban Society* (Garden City, N.Y.: Anchor Books, Doubleday Co., Inc., 1967), p. 11.

vey an impression to their subordinates that inaction is to be preferred over action. They like to keep things quiet on their shift and prefer to have officers working for them who don't "rock the boat." These are usually supervisors who are in the twilight years of their career and who are more concerned with their plans for retirement than they are with their supervisory responsibilities. Unfortunately for those they supervise, retirement never seems to come soon enough.

Another factor affecting police discretion in the enforcement of laws is what is known as the "unwritten code," which prescribes acceptable police conduct in particular situations. Although the code is unwritten, it nevertheless exists in most police agencies and officers soon learn that they must live by it if they are going to be able to "get along" in the department. One article of the code, for example, dictates that officers will extend professional courtesy to fellow law enforcement officers who are stopped for violations of the law. Professional courtesy usually means that the officer is given a "pass" and that no formal enforcement action is taken. In more serious cases, such as drunken driving, the offending officer may be turned over to a supervisor or fellow officer rather than be arrested. Police officers are expected to comply with this part of the code and may be chastised by their fellow officers if they do not.

Peer-group pressures also influence the number and type of arrests that an officer will make. Officers who share permanent shifts often develop group personalities which dictate, among other things, their attitudes toward law enforcement. Some groups of officers tend to be very traffic oriented, while others are not. Some groups of officers tend to place a premium on searching out and arresting particular kinds of violators. Officers who work on shifts with other officers soon learn the orientation of their fellow officers and will usually learn to adapt their own performance to that style of behavior.

Finally, community values and public attitudes help to influence and limit police discretion. Most police officers, whether they recognize it or not, are guided in their actions by community values and expectations. Once they have graduated from the police academy and are introduced to the "real world" of policing, new officers soon learn what the community expects from the police. They learn what kinds of violations are tolerated by the public and which ones are not. They quickly learn that the residents of one neighborhood may have a very tolerant attitude toward parking infractions and speeding vehicles, while the residents of another neighborhood may not. They learn that the residents of one part of town may be quite concerned about juvenile problems while people living in another part of town have concerns of a very different nature. Police officers soon learn that they must adapt their style of performance to these expectations if they are to obtain and maintain community support, and they usually recognize that community support is very important to the successful performance of their duties. As a result, they learn to tailor their actions to what the community wants and expects. For the most part, this is a very legitimate response by the officer.

Police discretion in the enforcement of the law may be viewed by some as a necessary evil. Given too little latitude, police officers will not be able to effectively perform the duties expected of them. There are simply too many laws to enforce and more infractions than they can reasonably be expected to deal with uniformly. On the other hand, if they are afforded too much discretion, and are provided with too few guidelines, there is a possibility that they will abuse the authority conferred upon them. Perhaps the answer lies not in restricting the discretion of the police, but rather helping them learn to exercise their discretion in the most reasonable and appropriate manner possible under the circumstances of the moment.

It is in the enforcement of laws that the police play perhaps their most vital role in helping to influence the quality of life in the community. The police possess powers that no other agency of government, except for the military under emergency conditions, possess. They have the absolute and nearly unlimited authority to affect a person's life, liberty, and safety through their discretionary powers of arrest. Through these powers, they can convey an image to the public that will reflect upon the police department, public officials, and the community itself. The consequences of their actions, or lack of action, are far-reaching and most significant.

Law Enforcement Goals and Objectives

Due to the considerable importance the law enforcement responsibilities of the police play in influencing the quality of life in a community, it is important that reasonable goals and objectives be established for this area of police activity. The establishment of law enforcement goals must be approached carefully, however, and must take into consideration community values and expectations. Indeed, a full understanding of the community and its needs is fundamental to the successful development of law enforcement goals. This would not be necessary if all laws could be uniformly enforced, but as has already been pointed out, this is not the case. Therefore, some means must be devised to determine which laws can and should be enforced, to what extent, and under what circumstances.

Unfortunately, police administrators might be reluctant to undertake this task, or to even admit that such a task is practical. To do so is to admit that the police, through the exercise of discretion, develop their own standards of conduct. This may seem to some as an abuse of police authority, but it is rather a necessary and appropriate extension of that authority.

Some police administrators, though, have taken the necessary steps to outline reasonable limits on the use of police discretion.

> Officers may use discretion, taking into account the seriousness of the offense and other mitigating and aggravating circumstances, when deciding whether to make an arrest for violations which constitute a misdemeanor, or for municipal ordinance violations. If an officer reasonably believes that it is not in the best interests of justice and the community to effect an arrest for a misdemeanor or municipal ordinance violation, he may utilize an alternative to arrest.[17]

[17]Elgin, Illinois, Police Department, *Rules and Regulations*, undated, section 2401.

One obvious goal of the police in terms of their law enforcement authority is the development of more uniform guidelines that will prescribe how various kinds of laws are to be enforced and under what circumstances. These guidelines should be flexible enough to allow the exercise of needed discretion by patrol officers, but yet clear and precise enough to limit unreasonable and unnecessary discretion which might be construed as abuse of police authority.

While total uniformity of the law enforcement process is neither possible nor desirable, greater uniformity in the enforcement of the law than is usually provided for under department guidelines or state law is a very desirable goal. For example, some police departments have issued written policies that dictate the limits of discretion that will be allowed in the enforcement of various kinds of traffic regulations, such as speeding. These directives have the effect of limiting the discretion of the individual officer. Such directives need to be developed only after fully considering community needs.

> In determining enforcement policies and priorities, police agencies should identify and direct primary attention to those crimes which are 'serious': those that stimulate the greatest fear and cause the greatest economic losses. Beyond that, police agencies should be guided by the law, cumulative police experience, the needs and expectations of the community, and the availability of resources.[18]

ORDER MAINTENANCE

Maintaining social order is one of the most troublesome responsibilities of the police. This is due, in part, to the difficulty in defining the term order and also because of different ideas about how order should be maintained. While civilized societies share a common desire for order, there is little consensus about what this term means, or how it should be accomplished. What is an accepted level of social order in one community or neighborhood, may not be acceptable in another. As a result, the patterns of policing, the means of maintaining social order, and the level of social disorder that can be tolerated will vary from place to place. This puts an added burden on the police officer, who has few clear-cut guidelines for performing this function.

There is a thin line between law enforcement and order maintenance. The idea of public order is based in law and police use the law to ensure social order and to maintain the peace. However, the law is usually used only as a threat in order to gain voluntary compliance by persons who threaten to disturb social order or peace. While the police may invoke the law to gain compliance, they prefer that the authority of their uniform and the badge of office, if not the weapons at their disposal, will be suf-

[18]National Commission on Criminal Justice Standards and Goals, *Police* (Washington, D.C.:" U.S. Government Printing Office, 1973), p. 14.

ficient to gain peaceful compliance with the law. In this way, social order is maintained. When this fails, the role of the officer switches from that of order maintenance to law enforcement.[19]

The order-maintenance function also contains an element of danger, since it usually involves the police officer imposing (forcibly or otherwise) authority on others. Resistance by those upon whom this authority is imposed may result in conflict or bloodshed. Each year, thousands of police officers are assaulted in the process of simply trying to maintain order in the community.

The order-maintenance function of the police includes a variety of duties in which the possibility of legal sanction usually exists but is not imposed. This includes situations in which legal action clearly may be taken, but is not, as well as those in which the legal authority of the police is only presumed to exist. An example of this would be an officer's command to "move on," or to "clear out," or to "break it up," to a loiterer or group of juveniles found congregating in an area where their presence was not desired. Although the police officer may have no legal authority to order the persons out of the area, since they have committed no obvious crime, the officer's authority is presumed to exist and the orders are usually obeyed. In those few instances where someone chooses, foolishly, to challenge this authority, the officer will usually be able to find some violation of the law upon which to invoke criminal action. Even though that action may not result in formal charges or prosecution later on, the immediate objective has been achieved.

In most cases, social order is maintained by the acceptance and observance of values and norms of conduct imposed by the community upon its members. The police are only needed to maintain social order when the value system fails to operate. This is more often the case in large cities, with heterogeneous populations and widely ranging styles of behavior than it is in smaller communities where community values and norms of behavior are more widely accepted.

The order-maintenance role of the police relies heavily upon the authority that is inherent in the position occupied by the police officer. The uniformed police officer in any community is the visible symbol of "law and order." Even where legal sanctions are available, they are rarely used. Police officers learn through experience that the best solution to a problem involving social disorder is something that falls short of an arrest. Family or neighborhood disturbances—to which the police are frequently called as mediators—offer the opportunity to invoke the criminal law, but may not result in criminal sanctions being imposed by the police officer, simply because much more can sometimes be achieved through more informal means. The exception to this would be in the case of spouse abuse, in which the discretion once afforded the police officer has now been replaced by formal policies or state laws requiring the arrest of the offending person.

Order-maintenance, like law enforcement, requires the exercise of judgment on the part of the individual police officer. In most order-maintenance situations, the

[19]James Q. Wilson, *Varieties of Police Behavior: The Management of Law and Order in Eight Communities* (Cambridge, Mass.: Harvard University Press, 1968), p. 17.

officer can choose from several alternate courses of action. There are very few written guidelines to aid officers in making these kinds of decisions. Instead, they learn to rely upon their own experience and what they believe to be the most appropriate course of action under the circumstances. They make choices based upon what they hope to achieve in particular situations. If seeking only a short-term solution to a problem that is likely to reoccur in the near future, the officer will probably resort to a more informal type of action, such as the threat of arrest, even though no violation may have occurred. Other, more serious situations, or those in which the officer wishes to achieve a long-term goal (such as ensuring that there will be no reoccurrence of the problem), may require the officer to make an arrest. The arrest will send a message to other participants that the officer means business and that similar action will be taken against others unless the problem is resolved.

While the order-maintenance function of the police occupies a great deal of the patrol officer's time, the importance of this function is often ignored or treated lightly by police administrators. Few police departments have written policies which define the limits of police authority or discretion in order-maintenance activities. Little effort has been made to systematically analyze the order-maintenance duties of the police and the impact they play on the overall role of police in the community. Little research has been conducted to determine the proper bounds of police conduct in order-maintenance situations. The exception to this statement is in the area of domestic violence, where a considerable amount of study has resulted in some very specific recommendations, guidelines, and even changes in state laws. This subject will be discussed in greater detail in Chapter 12.

By failing to recognize the importance of order-maintenance responsibilities, training programs for police officers also fail to provide officers with proper insight or the interpersonal skills necessary to handle those situations in a sensitive and enlightened manner. Police officers need to be trained to recognize the consequences of their actions and the range of alternatives available to them to effectively resolve these problems without formally invoking their arrest authority.

The order-maintenance function of the police must be recognized for the important part it plays in the police role in the community. While social order is important in any community, it is the community, not the police, that determines how social order is defined. The police must fashion their order-maintenance activities around their own understanding of what level of public order the community expects and demands.

> The ultimate test of police in a democratic society is their capacity to safeguard life and to maintain order and to do so in ways that preserve confidence in democratic processes. While there is no easy measure of police performance, there is abundant evidence that some police fail to comprehend the delicate nature of a role that commits them not just to safeguarding life and maintaining order, but to doing it in ways that advance and support democratic values.[20]

[20]*Project on Standards for Criminal Justice, Standards Relating to the Urban Police Function.* (Chicago: American Bar Association, 1972), p. 76.

Police administrators must ensure that personnel under their control understand that the means to achieve social order are just as important as the goal itself. Patrol officers need to appreciate the delicate balance that must be maintained between individual freedoms and community expectations concerning social order. Police policies need to be established that will provide clear-cut guidelines addressing the various social-order duties of the police, while at the same time allowing for the application of reasonable limits of discretion by the officer.

Police training programs and supervisory practices need to be altered to address the long-ignored order-maintenance responsibilities of the police. Training programs must emphasize the importance that democratic values play in determining what constitutes social order and how it should be maintained in the community. Police administrators need to recognize that community values and social norms are the determining factors in defining what constitutes social order in a community. It is these values, rather than the rule of law, that must guide the police in their efforts to maintain social order.

Order-Maintenance Goals and Objectives

Of all police responsibilities, order maintenance is perhaps the most difficult of all to translate into measurable goals and objectives. Since there are no reliable definitions of social order, how is it possible for the police to know what level of order is desired by the community? Without having direct community input, it is impossible for the police to know what is expected of them in terms of their order-maintenance responsibilities and how well they have fulfilled those responsibilities.

Even if social order is defined as the absence of disorder (for example, relatively few incidents involving domestic disputes, drunken or rowdy persons, etc.), can this be taken as a measure of police performance? Probably not, since police control over these occurrences comes only after the fact, not before. There is little that the police can do, except through their sheer presence, to prevent violations of social order.

Perhaps the true measure of police effectiveness in terms of order-maintenance duties is the absence of police invocation of formal controls (for example, arrest) to maintain social order. Assuming that a workable definition of *social order* can be achieved (an assumption not yet tested), it should then be possible to identify those incidents which represent an actual or potential risk to social order (for example, domestic disputes, public loitering, drunkenness). A realistic measure of police effectiveness, then, might be the number or magnitude of these incidents and the ability of the police to handle them without invoking formal sanctions. Another measure might be the level of police resources devoted to order-maintenance activities relative to the number or magnitude of such incidents over time.

There are no clear-cut measures of police effectiveness in this potentially troublesome area of police responsibility. Since public order is defined by the community itself, only the community can realistically assess police performance in this area. Thus, another measure of police effectiveness might be through periodic samplings of community attitudes concerning police performance, not just in this area but in all spheres of police responsibility. Such samplings might provide an important and useful way of assessing police performance and could provide useful insight into how that performance might be improved.[21]

PUBLIC SERVICES

The public-service function of the police has come under growing scrutiny in recent years, in part because of the increasing financial constraints placed on local government, which have resulted in fewer resources available to the police. Police administrators are being asked to provide an increased level of service with the same or declining resources. As a result, they have been forced to find ways to reduce or eliminate some service activities in order to devote sufficient resources to more pressing duties.

Some authorities have suggested that service activities occupy too much of a police officer's time and that they interfere with the ability of the police to pursue other, more important, duties. Nevertheless, service functions continue to be an important part of the police officer's daily responsibilities.

The argument against the police performing duties that are not directly related to the more important functions of law enforcement, criminal investigation, and criminal apprehension is certainly a valid one. In many cases, there are public and private agencies equally qualified or better prepared to perform these same functions. For example, would it not be more efficient to call a locksmith to help someone retrieve their keys from their locked car? Yet, thousands of police officers are dispatched daily to perform this function.

Even though police administrators may question the value or utility of these service functions, the public has come to expect them. Through their own commitment to public service, the police have conditioned people to expect them to respond and to assist them when no one else will. Through past practice, and a desire to perform a service that few others can provide, the police have created a dependency among the public. Most people turn to the police automatically, assuming that they will be able to help, or to summon the necessary assistance in an emergency.

> When people need help, it is to a police officer that they are most likely to turn. He responds—immediately—without first ascertaining the status of the person in need. It does not matter if the person is rich or poor; he need not meet complicated criteria to qualify as a recipient of aid or as a potential client.[22]

[21]See National Commission on Criminal Justice Standards and Goals, *Police* (Washington, D.C.: U.S. Government Printing Office, 1973), p. 19, for a discussion of various methods used by police to measure community assessment of their performance.

[22]*Ibid*, p. 9.

Because the police are always available, because they possess powers not possessed by any other agency of government, and because they are the visible symbol of authority, they are called upon to assist stranded motorists, make death notifications, locate lost persons or runaway juveniles, deliver mail, provide bank and funeral escorts, and many other similar duties.

A few police agencies have begun to seriously study the full range of public-service activities performed by the police, to assess their relevance to the police mission, and to determine whether there are alternative means of providing these services. Some police departments, due to increased demands for police services and diminished resources, have begun to identify service activities which can be eliminated altogether or provided in other ways. For example, some police departments no longer will take accident reports for property-damage-only accidents occurring on private property or which are not immediately reported to the police. Instead, citizens are given information packets which contain the necessary forms and instructions to enable them to make the necessary reports and notifications themselves. Other police departments have established policies providing for routine reports to be taken directly over the telephone, thereby eliminating the need to send a uniformed officer to the location. Still other departments have either eliminated altogether or modified their traditional response to such service requests as lockouts and vacant-house checks.

It is likely that providing a variety of public services will continue to occupy a great deal of the patrol officer's time in the future. It is also possible that the scope of those services, the priorities attached to them, and the manner in which they are provided may be radically altered as police administrators are forced to become more budget conscious. It may eventually come to pass that the police will not be able to be "all things to all people," and that police administrators must make hard choices concerning their service delivery priorities. Nevertheless, it is unlikely that the police will ever be relieved of their public-service responsibilities altogether, since they are an integral part of the unique mission of the police. The police are, after all, an integral element of the community's public-service delivery system.

Public Service Goals and Objectives

Although public services provided by a police department may be relatively difficult to quantify (for example, number of times the police respond to assist stranded motorists, aid in the search for a lost child, etc.), the quality with which these services are performed may also elude precise measurement.[23] In effect, any time a public service is rendered by the police, it is a positive accomplishment. People rarely complain that the police did not do a good job in helping them in their time of difficulty.

Since it seems that tight economic conditions may eventually dictate the level and scope of public services provided by the police, it may be necessary to be very

[23]For a discussion of measures of police effectiveness, see Theodore H. Schell and others, *Traditional Preventive Patrol* (Washington, D.C.: National Institute of Law Enforcement and Criminal Justice, Law Enforcement Assistance Administration, U.S. Department of Justice, 1976), pp. 15-17.

selective in setting goals and objectives for this category of police service. It may be necessary, for example, to identify those services which are considered integral to the police mission and which contribute most toward the quality of life in a community. Police executives need to critically examine their agency's ability to perform service functions and to tailor those service-delivery strategies to community needs and expectations. They should also recognize

> ... that some government services that are not essentially a police function are, under some circumstances, appropriately performed by the police. Such services include those provided in the interest of effective government or in response to established community needs.[24]

Accordingly, the police need to recognize that their public-service responsibilities are equally as important as their other duties and establish priorities accordingly. Once this is done, appropriate goals and objectives concerning public services can be developed. These goals should be focused on improving the quality of life in the community rather than simply enhancing the public image of the police.

TRAFFIC ENFORCEMENT

Police responsibility with regard to the investigation of traffic accidents and the enforcement of traffic laws creates a real dilemma for many police officers and chief executives. Some police officers and administrators are very traffic oriented and take pride in establishing and promoting a climate of aggressive traffic enforcement. Other police officers and administrators take a very passive attitude toward traffic law enforcement. There does seem to be some question as to whether a systematic and aggressive program of traffic law enforcement has any worthwhile effect on accident rates or driver behavior, according to Lundman.[25]

Those who consider an aggressive program of traffic law enforcement to be a necessary, legitimate, and proper exercise of police authority recognize the fact that millions of dollars and thousands of lives are lost each year as a result of traffic accidents. Moreover, many would argue that there are very few real traffic accidents, because most mishaps classified as "accidents" are preventable through better education, enforcement, and engineering. Moreover, since it is the duty of the police to protect life and property, and since traffic violations jeopardize lives and property, the prevention of traffic violations through the aggressive enforcement of traffic laws is an appropriate role for the police.

In most police departments, the enforcement of traffic laws and the investigation of traffic collisions are basic responsibilities of the patrol officer. In many larger

[24]National Commission on Criminal Justice Standards and Goals, *Police* (Washington, D.C.: U.S. Government Printing Office, 1973), p. 12.

[25]Richard J. Lundman, "The Traffic Violator: Organizational Norms," in Carl B. Klockars, *Thinking About Police: Contemporary Readings* (New York: McGraw-Hill Book Co., 1983), p. 287.

cities, specialized traffic units have been created to handle these duties. The creation of specialized traffic units in those departments that have sufficient resources makes sense for two reasons. First, the investigation of traffic accidents can be a very time-consuming matter and often requires special skills, particularly in accidents involving serious injury or death. Accident reconstruction is a very technical field and one requiring extensive training. It is not practical to train all officers to be specialists in accident reconstruction and investigation.

Second, many patrol officers simply do not like to work traffic. They do not enjoy ruining a person's day by giving him or her a traffic citation. It makes sense, then, to assign these duties to officers who do like to perform them.

Unfortunately, many police departments do not have the resources to form specialized traffic units, and must rely upon the patrol officer to handle both traffic enforcement and accident-investigation duties. In such cases, it may be advisable to have one officer (or more, depending upon the size of the department) on each shift designated as an accident-investigation officer. This officer could be trained in the complex skills of accident investigation and reconstruction and would be available to handle those traffic accidents requiring special skills.

Traffic enforcement, on the other hand, should be the responsibility of all members of the patrol force, even in those departments that do have designated traffic units. Traffic enforcement is an integral part of the police function and should not be ignored by officers simply because a specialized unit has been given this as a primary responsibility. As a result of their regular patrol duties, patrol officers know what parts of town, what times of the day, and what days of the week pose the most serious traffic problems. They also know what types of violations are likely to occur and which violations cause the greatest number of traffic accidents.

In addition, it has been shown that there is a strong relationship between traffic enforcement and crime prevention. First, a department that has an aggressive traffic-enforcement program will be in a much better position to detect and apprehend criminals. Most crimes involve the use of an automobile or other vehicle for the purpose of escaping the crime scene or to aid in the transportation of stolen goods or other contraband. Thousands of arrests for serious crimes are made each year simply as a result of an otherwise routine traffic stop.

In addition, a police department that has an aggressive traffic-enforcement program establishes a reputation for vigilance and a high profile. Police cars on the move, checking vehicles and pedestrians, making traffic stops at high-volume intersections, tend to make criminals nervous. They would just as soon commit their crimes somewhere where the police are not as visible and aggressive. There is a correlation, therefore, between a police department's traffic-enforcement efforts and the incidence of serious crime in the community.

At the same time, police administrators need to be wary of sending the wrong kind of message to the community. Some police departments seem to have nothing better to do than to stop drivers for minor violations such as going five miles over the speed limit or not having the registration properly displayed. Some police departments have managed to create the image that they are only in business to generate revenue for the municipality through traffic fines. Unfortunately, this image is often supported and enhanced by statements made by local elected officials. Indeed, some police officers, and some police administrators, take great pleasure on boasting about the amount of money they generate for the municipality through traffic-enforcement efforts. Unfortunately, some political leaders think along the same lines. They see the police as primarily revenue producers and make it clear that the police should "pay their own way" through the generation of traffic fines.

Of course, this is exactly the opposite of what should be the case. The very last consideration that should be made in determining whether a traffic violator should be stopped and whether a citation should be issued is the amount of revenue that might be produced for the municipality. To allow this consideration to affect the officer's decision is to demean and corrupt the integrity of the police department. Certainly, political leaders need to be concerned about balancing the budget and finding ways to provide for the services demanded by the public; however, they cannot and should not use the police department as a means of producing revenue. Police administrators should educate elected officials that this is not a proper role for the police.

Similarly, police agencies should avoid any attempt to establish a system of enforcement "quotas," either for individual officers, or for groups or shifts of officers. In some states, traffic quotas are illegal, and even where they are not illegal, they create a very bad impression with the motoring public. Where quotas do exist, people tend to believe that the only reason they were given a citation was that the officer had to make his quota for the day, even when this is not the case.

At the same time, there can be established reasonable standards of performance for the individual patrol officer. These standards should be flexible enough to account for the shift to which the officer is assigned and other types of activities that occur during the shift that might preclude the officer's ability to devote much time to traffic law enforcement. For example, an officer who, during an eight-hour tour of duty, made three shoplifting arrests, took a burglary report, assisted in handling a three-car traffic accident, and handled a domestic disturbance probably could not be criticized for not spending any time on traffic enforcement or not issuing a single traffic citation. On the other hand, the same officer on the same shift who turned in an activity log showing no arrests, reports, or assists during the eight-hour period might be questioned as to why he had devoted no time to traffic enforcement and how he had managed not to issue a single traffic warning or citation.

Perhaps more important than anything else is that a police agency encourage its officers to be consistent in their traffic-enforcement responsibilities. The public needs to know that, as a general rule, similar violations occurring in the same or similar locations will be treated in generally the same way. Even though officer discretion must be employed, there should be limits to this discretion that will ensure a reasonable degree of uniformity in the enforcement of traffic laws.

> Any traffic law enforcement program must achieve a high degree of uniformity to be acceptable to the motoring public. It is imperative that citizens committing identical offenses under similar conditions and circumstances be accorded the same treatment by

officers of the Department. To issue a summons to one person for an offense and warn another for the same offense committed to the same degree and under similar circumstances reflects inequality and partiality.[26]

Traffic Enforcement Goals and Objectives

Unlike other police responsibilities, goals and objectives for traffic enforcement can be more readily and precisely identified. The single objective of any traffic enforcement program is to reduce the frequency and seriousness of traffic accidents, thereby saving lives and reducing property loss. Thus, the specific objectives of a traffic enforcement program could be stated in the following way:

1. To reduce the number of persons injured or killed as a result of traffic accidents by ten percent a year for the next three years.
2. To reduce the amount of dollar loss as a result of traffic accidents by ten percent a year for the next three years.

Although there are other factors such as engineering, traffic volume, and construction projects that affect the incidence of traffic accidents in a community, we can assume that enforcement activity does have a considerable impact as well. Determining the exact nature of that impact and how to translate it into enforcement strategies is not as simple. It may be sufficient to simply correlate the number of traffic accidents with traffic enforcement activity over a period of time. If enforcement activity does have an impact, there should be some correlation between enforcement efforts and the number of traffic accidents over a period of time. Other measures have been developed which are much more precise. One, for example, is the **Traffic Enforcement Index,** which is the number of traffic *convictions* for hazardous moving violations, divided by the number of accidents involving personal injury and death.[27]

Using the Traffic Enforcement Index (TEI), it is possible to establish a very precise objective for enforcement efforts. This objective should be established based upon the level of enforcement activity that is required to decrease or hold stable the level of traffic accidents. For example, it may be discovered that a TEI of between 15 and 20 seems to make a difference in the number of serious injury and fatal accidents. If this can be found to be the case, then the department should establish this as a reasonable objective for the department.

Because each community is different and has its own unique problems relating to traffic safety, it is not possible or reasonable to establish uniform standards of traffic enforcement. However, it is possible, and it is highly desirable, that these standards be established by the local police agency based upon a thorough analysis and

[26]Elgin, Illinois, Police Department, *Rules and Regulations,* undated, Section 2403.

[27]*Ibid.*

understanding of the needs of the community. Once these needs are determined, realistic and relatively precise enforcement standards can be established and should be articulated within the police department.

SUMMARY

No organization can operate effectively without clearly defined and articulated goals and objectives. The same principles that contribute to managing a successful business apply equally to police organizations. For too long, police departments have operated primarily in a reactive mode, without paying proper attention to the purposes for which they were created and without devoting sufficient attention to identifying goals and objectives as guiding principles. For too long, the police and the public have taken for granted what the police are expected to do and have given too little thought to how police performance is to be evaluated.

Any definition of police goals and objectives must begin with an assessment of community values, needs, and expectations. The police are, after all, a part of the community's service-delivery system. The nature, scope, and level of the services provided by the police must be based on what the community sees as its needs, not on what the police see as best for the community. Accordingly, community input must be sought in the development of police goals and objectives.

REVIEW QUESTIONS

1. Why are goals and objectives important in planning, supervising and evaluating police patrol operations?
2. What is the difference, if any, between a *goal* and an *objective?*
3. Why do goals and objectives vary among police agencies?
4. Should the police not claim total responsibility for crime control? Why or why not?
5. In what ways are a patrol officer's responsibilities for criminal apprehension limited and why?
6. What, more than anything else, determines whether a criminal case will be solved, and why?
7. Describe some of the things that make the law enforcement role of the police officer particularly troublesome.
8. Why is the element of discretion important in the enforcement of criminal laws?
9. What are some of the public-service functions traditionally provided by the police that might be performed just as well by other agencies?
10. What should be the primary goals of any police traffic-enforcement program?

REVIEW EXERCISES

The following exercises are designed to reinforce or amplify comprehension of the material contained in this chapter.

1. As a class project, study and compare various cities of differing size, political structure, and socioeconomic composition. Using these differences or similarities, describe how police goals and objectives might vary among these cities.
2. Select one of the principle police functions described in this chapter. Prepare a detailed set of goals and objectives for this function as they might apply in your home town or in a city with which you are familiar.
3. Using either factual or fictional data, prepare a five-page case study to show how the actions of a patrol officer might be compared with specific goals for one of the principal police functions described in this chapter.

REFERENCES

Davis, Edward M., "Professional Police Principles," *Federal Probation,* 35 (March, 1971).

Elgin, Illinois, Police Department, *Rules and Regulations,* undated.

Goldstein, Herman, *Policing a Free Society.* Cambridge, Mass.: Ballinger Publishing Co., 1977.

Greenberg, Bernard,and others, *Felony Investigation Decision Model – An Analysis of Investigative Elements of Information.* Menlo Park, Ca.: Stanford Research Institute, 1975.

Greenwood, Peter W., and Joan Petersilia, *The Criminal Investigation Process, Vol. 1 – Summary and Policy Implications.* Santa Monica, Ca.: The Rand Corporation, 1975.

Hatry, Harry P., "Wrestling with Police Crime Control Productivity Measurement," *Readings on Productivity in Policing,* pp. 86-128, Washington, D.C.: Police Foundation, 1975.

Klockars, Carl B., *Thinking About Police: Contemporary Readings.* New York: McGraw-Hill Book Co., 1983.

Leonard, V. A., and Harry W. More, *Police Organization and Management* (7th ed.) Mineola, New York: The Foundation Press, Inc., 1987.

Lundman, Richard J., "The Traffic Violator: Organizational Norms," in Klockars, *Thinking About Police,* pp. 286-292.

Moore, Mark H., Robert C. Trojanowicz, and George L. Kelling, "Crime and Policing," *Police Practice in the '90s: Key Management Issues,* pp. 31–54. Washington, D. C.: International City Management Association, 1989.

Morris, Norval, and Gordon Hawkins, *The Honest Politician's Guide to Crime Control.* Chicago: University of Chicago Press, 1970.

National Commission on Criminal Justice Standards and Goals, *Police* Washington, D.C.: U.S. Government Printing Office, 1973.

Niederhoffer, Arthur, *Behind the Shield: The Police in Urban Society.* Garden City, N.Y.:

Anchor Books, Doubleday Co., Inc., 1967.

Poole, Jr., Robert W., *Cutting Back at City Hall.* New York: Universe Books, 1980.

Project on Standards for Criminal Justice, Standards Relating to the Urban Police Function. Chicago: American Bar Association, 1972.

Schell, Theodore H., and others, *Traditional Preventive Patrol.* Washington, D.C.: National Institute of Law Enforcement and Criminal Justice, Law Enforcement Assistance Administration, U. S. Department of Justice, 1976.

Sheehan, Robert and Gary W. Cordner, *Introduction to Police Administration* (2nd ed.), Cincinnati: Anderson Publishing Co., 1989.

Skolnick, Jerome,and David H. Bayley, *Community Policing: Issues and Practices Around the World.* Washington, D.C.: National Institute of Justice, 1988.

Wilson, James Q., *Varieties of Police Behavior: The Management of Law and Order in Eight Communities.* Cambridge, Mass.: Harvard University Press, 1968.

CHAPTER 3
THE PATROL ENVIRONMENT

CHAPTER OUTLINE

LEARNING OBJECTIVES

After completing this chapter, the student should be able to:

1. List the various elements in the police work environment which affect the manner in which they operate.
2. Discuss how the differences that exist from one shift to another can lead to different styles of policing.
3. Explain how community values and expectations influence the manner in which police patrol officers perform their jobs.
4. Name the basic forms of local government and explain how each one may affect the delivery of police services.
5. Describe three basic characteristics of the police organization which set it apart from most other types of organizations.

The police do not operate in a vacuum, nor are they an autonomous, independent agency of government. Instead, they are subject to a number of internal and external influences that guide their actions. Many of these influences have a direct impact upon patrol operations. The world of the patrol officer is shaped by community characteristics; public attitudes; political decisions; supervisory directives; administrative procedures; peer-group pressures; local, state, and municipal laws and ordinances; court decisions and countless other variables. In addition, the perceptions of patrol officers about the external environment influence their attitudes toward their job and the manner in which they conduct themselves on the job.[1]

Because of these numerous influences, few things are clear-cut, and decisions must often be made on the basis of incomplete and ambiguous information. Unlike other workers, who can predict with a fair degree of certainty what their day or work will be like, the patrol officer comes to work each day not knowing what to expect. The elements of excitement, fear, and apprehension that characterize the world of the patrol officer can bring with them both pleasure and anguish.

Perhaps the most important external influence on patrol operations is the community itself. As pointed out in the previous two chapters, police work in general—and patrol operations in particular—should be guided by community values and expectations. The patrol force is an arm of the community government and must be responsive to community needs. Just as no two communities are exactly alike, no two police departments are identical in the manner in which they are organized or in the way in which they operate. Patrol operations in small towns and rural villages vary considerably from those conducted in large urban centers. Patrol operations should be designed and managed in such a way that the result of their effort reflects differences in community norms, values, and expectations.

[1]Jerome H. Skolnick and David H. Bayley, *Community Policing: Issues and Practices Around the World* (Washington, D.C.: National Institute of Justice, 1988), p. 9.

Local politics—the process of determining priorities for government action—also influence police operations. Although the police are expected to be apolitical, that is, to base their actions on objective criteria rather than on partisan influences, they must nevertheless be responsive to political sentiment expressed as local policy. Politicians are, after all, policy makers, and it is policy which must guide police administrators as they design and manage police operations. It is policy (at least in theory), which reflects community sentiment and influences the nature, scope, and level of police services in the community. It is through the political process that community needs are identified. Local government is, in effect, the net result of the political process. As a result, political decisions play a very large part in shaping police operations.

The police organization possesses many unique characteristics that distinguish it from other types of organizations. As in the military, police organizations rely very heavily upon strict discipline, a system of accountability, and a rigid chain of command. For the most part, the police organization is a closed social system which shields its members from outsiders and dominates both the on- and off-duty lives of its members. Organizational characteristics and the value systems of those who dominate the organizational hierarchy tend to isolate the police from society in general, and instill in their members a "we versus they" attitude. These organizational characteristics directly impact the manner in which police operations are carried out.

It is also important to consider the police organization in its relationship to the other elements of the criminal justice system. Thus, just as the police department is part of the local government's service-delivery system, it is also part of the broader system of criminal justice in which it interacts with courts, prosecutors, and corrections officials. In this role, the police organization may have certain responsibilities that are in conflict with local community values. For example, a local prosecutor (who is an officer of the court) may put pressure on the police to aggressively enforce state laws against gambling or other so-called 'victimless' crimes, while local community values may tolerate such offenses. Moreover, each element of the criminal justice system has its own value system as well as its own set of goals and objectives which may be in conflict with those of another part of that system. Rather than harmony, the end result may be conflict.

These are but a few of the external influences which impact upon the delivery of police services and upon patrol operations. The remainder of this chapter will examine each of these influences in greater detail.

THE PHYSICAL ENVIRONMENT

One of the most important characteristics of the police patrol function is the uncertainty facing patrol officers as they carry out their assigned duties. Unlike other people who have a set routine they follow each day at work, patrol officers deal almost exclusively with human beings and their problems. As a result, each day or tour of duty brings with it a whole new set of problems and experiences. No two situations are alike, and each new problem requires a different solution. There are no textbook answers, and written policies and procedures may offer only a general guide to action. In nearly every case, it is the officer who decides the proper course of action. Moreover, there is no reliable way of knowing whether the decision made by the officer was the best possible one that could have been made. The police officer must rely upon gut instincts, past experience, and intuition when confronted with situations for which there are no clear-cut solutions.

Police patrol is conducted in a world of contrasts, ranging from the skid-row alleys of major cities to the quiet, tree-lined streets of suburbia. The physical settings in which patrol officers work help to shape their patterns of behavior. In some settings, and under some conditions, officers must be tough and uncompromising. They must be able to make quick decisions and be able to back them up; they must be prepared to use force as a last resort, if necessary. In other situations, and under different circumstances, patrol officers can afford to be more compassionate and flexible. They may be able to carefully assess a given situation and choose an option that offers the best possible outcome. Such circumstances afford officers time to demonstrate care and concern for the people they serve.

Patrol officers usually work alone and with little direct supervision. Except in the case of arrest or other official action, the decisions they make and what they do is rarely subject to review by higher authority. They enjoy great discretion in what they choose to do or not to do. Even though they may work under general policy limitations (for example, one hour of patrol time will be devoted to selective traffic enforcement), they have considerable leeway in the application of those policies.

Patrol work can be challenging, exciting, emotionally stimulating, and very often quite rewarding. At other times it can be monotonous, routine, and boring. Depending upon the location and the time of day, a patrol officer may speed from one call to another, knowing that the paper work will probably not get done until after the shift has ended. In other places, patrol officers have so much time on their hands that they must invent activities to keep them occupied. Some devote their attention to traffic-enforcement duties, while others concentrate on preventive patrol and look for that open door or broken window that might lead to an arrest.

Police patrol officers must possess a range of skills and abilities which set them apart from members of any other profession. They must learn to expect the unexpected, respond calmly in the face of danger or provocation, keep cool in the heat of battle, lend assurance to those who are in fear or have been injured, endure undeserved criticism, remain polite and courteous in the face of abuse and criticism, offer comfort and counsel to those in need, show compassion and mercy to victims of injustice, and remain objective and detached when dealing with emotional crises.

Information is the lifeblood of a police agency. The patrol officer learns early that success depends on obtaining, synthesizing, processing, interpreting, and (in some cases) disseminating information. Informants—anyone willing, for one reason or another, to pass on information to a police officer—become important allies. In some cases, this leads to the development of an unusual relationship between the police officer and those who lurk on the fringes of the law.

The world of the police patrol officer changes frequently. In most cases, this change is an advantage, since it helps to relieve the monotony and boredom that is often associated with the job. On the other hand, change can have an adverse impact upon the officer's physical and mental health. Some police departments maintain a system of shift rotation, for example, which requires officers to change shifts as often as every week or every twenty-eight days. Such frequent adjustments to the daily cycle of work, sleep, rest, and relaxation can place great demands on the officer's social life and physical well-being. On the other hand, changing shifts allows the patrol officer to experience a variety of conditions and people.

One of the unique features of patrol work is that the people and occurrences a police officer is likely to encounter differ considerably according to the assigned shift. In fact, it is often the case that the style of policing and the attitudes and behavior of the officers differ from one shift to the next.

The Day Shift

The day shift is usually, in most police departments, a shift devoted to service activities. The people that a police officer meets during the day shift are usually law abiding and create few problems. Handling service calls, dealing with traffic and parking problems in the downtown area, looking into youth-related problems at the local school or community park are common activities of the day shift. Some crimes do occur, of course, and they can be quite serious. Most bank robberies, for example, occur on the day shift. Quite a few cars are stolen during the day from parking lots at shopping centers and malls.

On the weekends, officers assigned to the day shift handle traffic accidents and respond to rescue calls involving serious injuries and illness. They often concentrate their patrol efforts around parks and recreation areas, particularly during summer months. Traffic enforcement is especially important in areas with a high rate of tourism. Police officers assigned to the day shift take a lot of reports of property stolen during the evening or early morning hours but not discovered until several hours later. Since officers assigned to the day shift do not do a lot of hard-core crime fighting, they tend to be more service oriented. In those departments with permanent shifts, the day-shift tends to attract those more interested in providing a service than in regulating human conduct.

Day-shift officers perform a great variety of duties. They drive patrol cars to the municipal garage or to the local car dealer for routine maintenance or repairs; make bank deposits for the city clerk's office, when no one else is available; deliver agenda packets for city council meetings; and transport prisoners to and from court. Of course, in larger cities, many of these functions are performed by nonuniformed personnel or officers not assigned to uniformed patrol. However, in the vast majority of small and medium-sized police departments, the patrol officer on the beat is often the only one available to perform these duties.

The Afternoon Shift

The afternoon shift—3:00 PM to 11:00 PM—is usually the busiest and offers the greatest variety of activity. Officers on the afternoon shift encounter a broad range of activities and people that make the job interesting and challenging. The afternoon shift combines the service orientation of the day shift with the criminal-apprehension functions of the midnight shift. Officers assigned to the afternoon shift will find themselves responding to house fires; multiple-injury traffic accidents; reports of lost and runaway children; tavern fights; domestic disputes; incidents of vandalism; thefts and burglaries; aggravated assaults; and an infinite variety of calls involving criminal and noncriminal activity.

The officer assigned to the afternoon shift is more likely than his fellow officers to respond to one call after another without time out for a coffee break or dinner. Dinner, it if comes at all, may consist of a cup of black coffee and a sandwich or doughnut eaten on the way to a call or during an infrequent break between calls. More often than not, between 40 and 50 percent of all the calls handled by the police department occur on the afternoon shift.

Officers assigned to the afternoon shift tend to make more arrests than their fellow officers and spend more time in court. They are assigned to a lot of traffic accidents and thus become somewhat traffic oriented. They will usually write more tickets than officers on either of the two other shifts, simply because of the nature of traffic flow during the later afternoon and evening hours.

The people encountered by an afternoon-shift officer are different from those encountered on the day or midnight shifts. Runaways, drunken persons, and vagrants are often confronted on the afternoon shift, usually after dark. Petty criminals, such as peeping Toms and sex offenders, can often be found lurking in the shadows of apartment buildings. Juvenile auto thieves, often on probation for other offenses, will frequently be found driving a car stolen from a nearby town.

Officers assigned to the afternoon shift will become accustomed to responding to burglar alarms at the high school and at various businesses in town. Most of these will turn out to be false alarms, but the patrol officer will get to know the names of the cleaning people and janitors at these places and learn to recognize the cars they drive.

Police officers assigned to the afternoon shift tend to be those in the middle ranks of experience within the department. They have probably been with the department for five years or more and are still young enough to enjoy the excitement of the afternoon shift. They must be agile enough to wrestle a drunken person or chase a juvenile attempting to steal a car from a parking lot. The afternoon-shift officer likes to work traffic and knows that each call brings with it something new and different. The officer on the afternoon shift usually expects to make police work a career.

The Midnight Shift

The midnight shift has been called the "dog watch," which suggests it is the penalty box of police work. It has also been called the "graveyard shift" to denote the lack of activity on the shift. The midnight shift is typically the least desirable of all police shifts and it is for this reason that, due to seniority provisions in union contracts, the youngest and least experienced officers often end up on this shift. This is unfortunate for several reasons. First, officers on this shift are exposed to a great deal of potential danger. Rookie officers have not yet learned how to protect themselves and are often too eager in the face of potentially dangerous situations. Second, officers fresh out of the academy need to be assigned to work with more seasoned officers who can serve as role models and teach them the fundamentals of being a good patrol officer. They do not have the chance to do this on the midnight shift.

There are, though, some officers who prefer to work the midnight shift, either because they like the kind of activity (or lack of it) commonly associated with the shift, or because they like the freedom derived from not having regular contact with "the brass." As already pointed out, rookie officers often end up on the midnight shift because they lack the seniority to be assigned to a more preferred shift. Those who work the midnight shift because they prefer it are a different breed of patrol officer. They tend to be more independent and, among themselves, they form a very close-knit group. Perhaps the fact that they work when most people sleep tends to give them a distorted view of the world around them.

> An officer steps into the dark night with a psychology... For night brings shadows, furtiveness, gloom. Night seems to accentuate crime, lawbreaking and human suffering. Cops, doctors, and clergy all agree. Speedsters are speedier, drunks are drunker, assaults fiercer, noises are noisier, gunfire is flashier, smells are smellier.
>
> The policeman feels his gun more often in the dark. He grips his truncheon more tightly. He thinks in terms of ambush. His senses are honed for action. He listens carefully for unfamiliar sounds, like the jungle animals. He is concerned with what is behind him. Everything looks suspicious. Night inks out the differences. Identification is difficult.[2]

The type of activity found on the midnight shift is different, too. Since there is very little activity, midnight-shift officers get to know their assigned areas very well and eventually develop a unique ability to detect the unusual. "Suspicious activity" often turns out to be nothing at all, but to the probing, alert, and aggressive patrol officer, anything out of the ordinary bears checking out.

> Probing suspicious circumstances means recognizing departures from a norm. A police officer familiar with an area comes to know the activities and behavior patterns of its residents. In commercial areas, for example, he comes to know where people are working at night, how different facilities are secured, and how bank deposits are made. In

[2] "Patrol, Patrol, Patrol," *Spring 3100* (July-August 1960), in Samuel G. Chapman, *Police Patrol Readings* (Springfield, Ill.: Charles C. Thomas, 1970) pp. 60-61.

> residential areas he learns which people gather regularly on street corners ... and the patterns with which people use and park their vehicles. Departures from these patterns alert police to the possibility that a criminal act has occurred.[3]

Danger is always potentially present on the midnight shift. Even though the patrol officer may encounter fewer people as a matter of course during this shift, the probability of encountering violence, either to the officer or to someone else, is much higher. Sexual assaults, aggravated assaults, robberies, muggings, and other crimes of violence seem to occur with greater frequency after midnight than during any other time of the day. Drunken, transient, and homeless people are also commonly encountered. Traffic accidents, when they occur, tend to be bad ones involving either serious injury or death. Burglar alarms are taken more seriously on the midnight shift, since there is less probability that they occur as a result of human error.

One problem on the midnight shift is the range of extremes between nothing occurring and everything happening at once. Since there are fewer calls on the midnight shift, most police departments tend (properly so) to assign fewer officers to this shift than to either the day or afternoon shift. But when something does occur on the midnight shift, it will often require every officer on the shift. This is particularly true in smaller departments where maybe only one, two, or three officers may be assigned to the midnight shift. A serious traffic accident can easily tie up all three officers for an hour or more. The search of a commercial building for a possible burglary in progress will almost always require two or three officers. Thus, the world of the midnight shift officer is often one of either very intense activity or one of no activity whatsoever. Officers assigned to this shift need to have the personal resourcefulness to sustain them over these peaks and valleys of activity.

COMMUNITY INFLUENCES

Community residents have a vested interest in the nature of police services provided by their local departments. Police administrators have an obligation to be aware of and respond to local community concerns when establishing departmental policies and priorities. At the same time, the police executive needs to be able to distinguish legitimate community concerns from isolated incidents of nit-picking by people looking out only for themselves.

Local political and community leaders are often not aware of police capabilities and thus need professional police administrators to aid them in identifying legitimate goals and objectives for the police. Thus, the police executive must play a leadership role in the community.

Communities, like people, have their own personalities. No two communities are exactly alike, just as no two people are identical. Many factors shape the "person-

[3]Herman Goldstein, *Policing a Free Society* (Cambridge, Mass.: Ballinger Publishing Co., 1977), p. 53.

ality" of a community: size, demographics, political leadership, form of government, cultural and ethnic differences, income levels, education, wealth... the list goes on and on. Each of these factors contributes to the personality of the community and makes it different from all others.

A successful police officer must understand the personality of the community and adapt a style of police behavior to suit that particular community. This is, of course, an important responsibility of the police executive. As a highly visible and active member of the community, the police executive must be able to translate community values and expectations into police policies and programs. In order to succeed today, a police chief cannot be isolated from the community.

Understanding the community is no less important to the patrol officer on the street. Even though the patrol officer is not expected to set departmental policy nor to maintain a high community profile (although some do, in smaller communities), the patrol officer does learn to translate community attitudes, values, and expectations into enforcement strategies. One of the purposes of assigning officers to patrol beats or districts is to make them accountable for the events that occur in their beat during their tour of duty. This does not mean, of course, that the patrol officer is held accountable for every single crime that occurs in his beat during his watch, but rather than he is expected to be accountable for his actions when those crimes occur.

For three reasons, the individual patrol officer should be familiar with all aspects of an assigned shift in an assigned patrol area. In many cases, officers may be routinely assigned to the same patrol beat for an indefinite period of time. In those cases, the patrol beat becomes the officer's "community," affording an opportunity to get to know some of the people and how they think. This occurs more frequently on the day shift and less frequently on the midnight shift, although officers on all shifts soon get to know the people they encounter on a regular basis.

The day-shift officer, for example, quickly learns the names of the business people, gas station attendants, bank tellers, waiters, waitresses, gardeners, and mail carriers frequently seen on his or her tour of duty. Similarly, the officer on the afternoon shift gets to know the tavern keepers, waiters, waitresses, nurses, doctors, desk clerks, ambulance drivers, tow-truck operators, and taxi drivers seen on a regular basis. The officer assigned to the midnight shift has fewer contacts, but forms closer relationships. Those who work in the early morning hours share a sense of camaraderie that creates a special bond. The midnight-shift officer gets to know the taxi drivers, the cleanup crew at the bank, the fry cook at the all-night eatery, the security guard at the plastics plant, and the newspaper-delivery driver.

Police chief executives should set the example for their officers by being active community leaders. They need to be more outspoken on public policy matters that affect their organizations. "The ultimate decision on the type of service provided by the police should be made by the community being served, but the police need to be much more active in placing choices before the community."[4]

[4]Herman Goldstein, *Problem-Oriented Policing* (Philadelphia, Pa.: Temple University Press, 1990), No. 4, 8, 19, p. 46.

Understanding the community and the local activity that it produces is just as important for the patrol officer as it is for the police executive. The patrol officer interacts with more people on a regular basis, and under much more difficult circumstances, than the police executive normally will.

In order to enhance a patrol officer's understanding of the community and to increase the officer's social interaction with its members, some communities have imposed residency requirements on their officers. Requirements that police officers live in the communities where they are employed have usually been upheld by the courts, and have certain advantages. Police officers who have little contact with the people of the community outside the scope of their official responsibilities may find it difficult to understand or accept local community values and attitudes. Thus, their style of enforcement or service orientation may be totally out of step with prevailing community values. In many cases, police officers seem to be out of touch with what the people in the community think and their enforcement practices may create conflict rather than solve problems.

> The police in American society today live and work in a world of their own. They are both occupationally and socially isolated from the overall community structure. Policemen share common values and standards of living with other policemen. Their frame of reference is an ideology common to all other policemen.[5]

Notwithstanding the need to have police officers intimately familiar with and involved in the community, they must also be able to remain objective and impartial in the performance of their duties. Officers must be a part of, rather than apart from, the community they serve. This is the reason that many police administrators have encouraged their officers to become actively involved in community and civic organizations. It is the same reason that many police departments have developed or supported police athletic leagues, youth soccer teams, scouting posts, and other programs designed to allow the police officers to get involved with members of the community. These are not just "public relations gimmicks," but have a very fundamental and worthwhile purpose.

Observers of police behavior have long recognized both the need for the police to become involved in community activities, and the dangers of the isolationism that has characterized some police departments. William Westley, in his classic study of police behavior, observed as follows:

> A police department is intimately related to the community in which it is located and which it serves. The personnel of the department are drawn from the community and have a personal involvement in the life and values of the community The men on the job are responsible to the public definitions of behavior. The nature of the community

[5]Charles D. Hale, *The Isolated Police Community: An Occupational Analysis* (Master's Thesis, California State University at Long Beach, 1972), p. 3.

> determines many of the problems that the police must meet. The political structure of the community may have an important influence on the department and the areas of enforcement that it emphasizes.[6]

The police should be, in a sense, a reflection of the community itself. That is, the police should reflect community values in its enforcement actions, operating style, and management philosophy. James Q. Wilson identified three basic "styles of policing—watchman, legalistic, and service style." These three basic styles could be further divided into dozens of substyles, each reflecting a somewhat different behavior type. In theory, at least, these subtle differences in policing style should parallel community differences.

Often, the "personality" of a police department is a reflection of the chief of police. Police chiefs set the tone of the organization through personal example and through the administrative policies and procedures they implement. A management style is revealed through the manner in which a chief deals with subordinates, encourages or discourages internal communications, listens to what others have to say, supports or suppresses free thinking and employee participation in the decision-making process, acts upon employee grievances, and implements new programs and policing strategies. The successful police chief is one who can mold and shape a police department to be what the community wants it to be.

Too often, a police chief thinks of a department in a possessive sense. A chief may speak of "my department" and "my people," when in fact they do not belong to the chief at all. A police chief, regardless of tenure, is but a temporary steward. As such, a chief should have no sense of ownership, but recognize accountability to the public. It is, after all, their police department. The chief is responsible to the community, not the other way around. The motto of the Reno, Nevada, Police Department, for example, is "Your Department, Our Community." This seems to reflect quite well the type of relationship that should exist between a police department and the community it serves.

The task of developing a positive, understanding, and mutually supportive relationship between the police and the community is considerably more difficult in large cities than it is in smaller towns and villages. The personality of a small town is relatively easy to read, and it is relatively easy to get to know the people who live there and to understand their values. Often, the police chief and most of the officers in a small town grew up in the area and have roots in the community. Smaller towns and cities usually have more stable and homogeneous populations, where values and norms are commonly accepted. As a result, the job of policing is usually more clearly defined in small towns and cities, since the police better understand what is expected of them.

In large cities, however, populations tend to be more culturally, socially, and racially diverse. Community values are more diffuse and the personality of the com-

[6]William A. Westley, *Violence and the Police: A Sociological Study of Law, Custom, and Morality* (Boston, Mass.: Massachusetts Institute of Technology, 1970), p. 19.

munity is more elusive. Understanding how people think, what they want and expect, and what they believe in is more difficult. The task of policing a larger city is more difficult because the "community" is difficult to define. Often, there are dozens of different communities and within those there are subcommunities, subcultures, areas, and neighborhoods, each with its own peculiar value system and interests.

> The class structure of the city probably affects the extent to which the officer feels isolated or alien. In a small, middle-class residential suburb, where standards of appropriate conduct are widely shared, where the privacy of the residents is assured and where there are no illegal enterprises, the police may exist largely to protect the community from outsiders and to offer assistance to citizens who have lost their dogs, stalled their cars or drunk too much at parties. In such communities, police-public relationships may be very congenial. In bigger cities with a larger lower-class population, the situation is altogether different. Standards of conduct are not widely shared and resentment of authority is greater.[7]

The job of understanding community values and expectations and translating them into police programs and styles of police conduct extends to the lowest level in the police organization. Whether assigned to a specific patrol beat or to the entire town or city, the patrol officer needs to become familiar with the people, places, and events in that area. In the Problem-Oriented Policing strategy discussed in Chapter 6, the patrol officer plays a central role in identifying problems and developing and implementing problem-solving strategies within the community, neighborhood, or patrol area.

In many cities and towns, the patrol beat is the patrol officer's community. It is, for the tour of duty to which the officer is assigned, his or her world. What happens in that beat is the officer's responsibility. To the people who live, work, or travel through that area, the officer is the police department. The boundaries of the patrol beat may remain constant, or may be changed from time to time. They will often correspond to geographic features such as rivers, major highways, or railroad tracks. In larger cities, patrol beats may also correspond to neighborhood boundaries. The social, cultural, political and economic characteristics of one patrol beat may be altogether different from another one. These conditions will not only affect the role of the police officer on the beat, but may affect the style of policing and the choices made by the officer.

> The degree of enforcement and the method of application will vary with each neighborhood and community. There are no set rules, or even general guides, to the policy to be applied. Each policeman must, in a sense, determine the standard to be set in the area for which he is responsible.[8]

[7]James Q. Wilson, "The Police and Their Problems," *Public Policy,* 12 (1963), p. 196.

[8]Bruce Smith, *Police Systems in the United States,* 2nd ed. (New York: Harper & Row, 1960), p. 19.

The diversity of cultural norms, community problems, and local citizen attitudes toward the police make it impossible to develop a standard of police conduct that can be adopted in all cases. This can be both a disadvantage as well as an advantage. The disadvantage is that it sometimes forces the officer to make quick decisions based upon perception and the urgency of the situation. Even local policies and procedures, when they exist at all, do not cover the myriad of circumstances and situations that may arise in the field, nor do they indicate how they may be applied or interpreted differently in different situations. In many cases, policies provide only broad or general guidelines which leave much to the discretion of the officer.

The very thing that makes this a disadvantage for the officer can also make it an advantage. Even though patrol officers need general guidelines to help them make decisions on the street, they must also have the ability to adapt those guidelines to the prevailing circumstances. The exercise of discretion by an officer is, in the long run, an advantage. Patrol officers, if they are to mature and develop professionally, must:

- Make decisions based upon what is needed and appropriate at the moment.
- Tailor their actions to the needs of the community.
- Recognize that an enforcement standard considered acceptable in one community, or patrol area, may not be acceptable in another.
- Accept responsibility for police work that is responsive to local expectations and conditions.

POLITICAL INFLUENCES

Decisions regarding the level and scope of police services in a community are, in a very real sense, political ones. Politics, after all, is the process by which priorities are established within community programs, local needs identified, funds allocated, and needs fulfilled. Often, the word *politics* has negative connotations because we associate it with vote getting, patronage, partisan influences, vested interests, corruption, and other negative images. Probably for good reasons, the person who criticizes city hall for problems relating to local government services is frequently tempted to lay the blame on "politics."

Politics is often blamed when a popular program is discontinued, or when a decision to raise taxes or cut back on spending is made. Although these decisions are made on the basis of political considerations, they are not necessarily "political" in the negative sense. Politics is not merely the work of politicians, but is the foundation of the governmental process. As one textbook states:

> Governments and individuals who act politically are involved in the process of determining the distribution of benefits of public resources, setting collective goals and controlling deviant behavior.[9]

[9]Jay A. Sigler and Robert S. Getz, *Contemporary American Government: Problems and Prospects* (New York: Van Nostrand and Richmond Co., 1972). p. 3.

In a very broad sense, then, decisions that affect the level, scope, and quality of police services are political in nature. Public policy is the formal byproduct of the political process. At the very practical level, it is through the political process that the municipal budget is developed and adopted. That budget establishes programs and priorities and allocates funds for municipal departments, including the police. A police administrator today needs to understand the political process and to play an active (but appropriate) role in that process if the police department is to receive the level of funding it needs to do what is expected of it.

A police department is but one element of local government. Its needs must be weighed against those of other units of government (for example, fire, public works, recreation, library) for limited local resources. This is in contrast to the 1960s and 1970s, when rising crime rates, civil disturbances, and a fear of "crime in the streets" motivated local elected officials to allow police budgets to grow considerably. Too often, unfortunately, they did not realize a dollar-for-dollar return on their investment. During the 1980s, citizens across the country became increasingly concerned about rising property taxes and unrestrained spending by local government. This led to property tax revolts in several states (California and Massachusetts, in particular) which resulted in massive cutbacks in spending by local government. As a result, police administrators and their colleagues in local government had to find ways to maintain services with reduced staffing levels.

Professional police chiefs try to be apolitical. They maintain that they must remain aloof from politics so their organizations can operate with the degree of neutrality and objectivity needed to provide professional police services. Although they are correct in this point of view, they often fail to recognize that even the decisions that they themselves make have political overtones and consequences. The decision to enforce or not enforce parking regulations in the downtown area is a political decision, even though it may be made for nonpolitical reasons (for example, lack of manpower, too many complaints, inadequate off-street parking). The decision to place additional emphasis on traffic enforcement around the high school may be based on sound police experience (for example, residential growth in the area has created additional traffic and the potential for a greater number of accidents and injuries), but that decision could have political overtones.

On the other hand, blatant partisan politics has no place in a police department, nor in the day-to-day administration of the city. Local politicians have every right and responsibility to determine what priorities are attached to which programs, how local government funds will be allocated, and how local resources should be utilized. They do not, however, have a legitimate right to become involved in the operation of individual departments.

Elected officials take great interest in police departments. Because the police maintain a high level of visibility and exercise enormous power, this interest in their operations is understandable. A mayor, for example, can exert a great deal of influence over local business people if allowed to directly appoint (and remove) building inspectors. A public safety commissioner who controls the intelligence unit can find out a great deal about the opposition during an election campaign. A member of a municipal council—able to control the level or scope of criminal enforcement in a particular part of town—is in a very good position to do favors for friends and vested interest groups.

These are all examples of the manner in which politics can be misused in the administration of a police department. Unfortunately, they are all examples of actual situations that either exist today or that have existed in the not-too-distant past. When these kinds of scenarios ultimately come to the attention of the public, they cast a negative image on the police even though the police are often not the culprits and often have little power to resist or deny such partisan political influences. It is just these kinds of misuse of the political process that prompt police officials to endeavor to "keep the police out of politics."

Unfortunately, the history of the American police system has had more than its share of abuse of police power by partisan political power. This was particularly true during the late nineteenth and early twentieth centuries. In many cities during this period, the police were little more than tools of local politicians who used police power to serve their own interests and the interests of their constituents. Those parts of the politician's wards which were solidly behind him and which could be counted on to "get out the vote" on election day could count on receiving favorable treatment at the hands of the police. If an officer was not inclined to "go along with the program," he would soon find himself looking for another job. Appointments to the police department, promotions, and assignment to choice positions were made on the basis of bribes and political favors rather than merit. Police chiefs had little power to control these abuses, since they themselves were subject to political appointment,

A reform movement was initiated during the latter part of the nineteenth century. Civil-service commissions were created at the state and local level as a means of protecting police and other government workers from political influences. By controlling the means and methods by which appointments were made within police departments, and by insisting that merit and qualifications be the sole determinants of such promotions and appointments, it was believed that political influences could be kept out of a police department.

For the most part, these civil-service systems achieved their intended purpose, and civil service remains a viable aspect of local government today, although the manner in which it is practiced varies considerably among the fifty states. A very few states, such as Massachusetts, have adopted state-wide civil-service systems which regulate and control hiring and promotional processes in most (but not all) departments in the state. In several midwestern states, including Illinois, Iowa, and Wisconsin, local civil-service commissions regulate hiring, firing, and promotional decisions with varying degrees of autonomy.

The effort to remove politics from police departments has had adverse consequences in some cases. While it is necessary for the police to remain free from partisan political influences, they must nevertheless remain responsive to local community demands. These demands are expressed by local officials, elected to look after the needs of their constituents and to ensure that government serves the needs of the people. In this respect, local elected officials have a legitimate right and responsibility to exercise some degree of control over governmental operations, including the police.

Too often, in their zeal to remain independent of political control and interference, the police regard themselves as an independent and autonomous branch of government. Local officials, on the other hand, fearful that they may be charged with meddling in police affairs, are sometimes reluctant to exercise legitimate and necessary supervision of the police. This erodes the system of accountability that must exist if the police are to be sensitive and responsive to public demands. Without this system of accountability, citizens might make legitimate requests of the police only to have them ignored. They might lack the political "clout" or influence necessary to ensure police response to their needs.

The police sometimes react negatively to suggestions that they be held politically accountable for their actions, claiming that this tends to weaken the objectivity and impartiality necessary for them to perform their duties in a professional manner. For example, suggestions to create civilian review boards to exercise broader community control over investigations into police misconduct have been met with great resistance by the police.The police tend to resist and resent any attempt by "outsiders" to control their actions, monitor their performance, or offer criticism of their methods. In reality, however, the police are subject to the same level of supervision and control—no more and no less—to which other units of government are subjected.

The police are determined to establish themselves as professionals and they make the point that self-regulation and internal control are among the characteristics of a "profession." How, they ask, can the police ever acquire professional status if they must continually answer to others for their actions? They fail to recognize, however, that the police have traditionally not done a very good job of self-regulation. When serious abuses of police authority are uncovered, the natural tendency of the police is to protect the alleged offender(s) from criticism in order to protect the reputation of the department. Sometimes they go to great lengths to protect an officer who has abused the authority of office because the officer has an otherwise good record or was acting under pressure or provocation.

Since the 1960s, the police have come under increased pressure to observe certain constitutionally protected freedoms when dealing with criminal suspects. To the average observer, such freedoms would seem fundamental to the democratic process. Unfortunately, police training programs of the past did not place enough emphasis on safeguarding the constitutional rights of citizens. Consequently, deprivation of those rights by the police eventually forced the courts to take action. Looking back on some police practices, the modern observer would probably wonder why the courts did not act sooner. In some police departments, police behavior toward criminal suspects bordered on the barbaric.

> It is one thing to talk quietly to a suspect without his counsel and artfully, perhaps by deceit, persuade him to incriminate himself; it is quite another to hang a suspect out of a

> window by his heels from a great height or to beat a confession out of him by putting a telephone book on his head and pounding the book with a blackjack so it does not leave marks.[10]

Failure by police executives to hold their officers accountable for their actions and resistance by the police to attempts to exercise control over them create nagging doubts in the minds of the public as to the commitment by the police to the spirit of public service. This tends to erode public confidence in the police which, in turn, can create further alienation between the police and the public. Police administrators need to recognize that the public has a legitimate right to expect police to be accountable for their actions.

> No longer is it acceptable for the police chief to remain isolated from the social and political forces of the communities.[11]

The extent to which police officers resent the involvement of local officials in the development of police policies, goals, and priorities was clearly indicated in a survey of police-officer attitudes conducted in eight small and medium-sized police departments located in six different states (Slovak). In this survey, police officers were asked to rank various individuals or groups (for example, city manager, city council, police chief) according to their importance in police policy making. Consistently, police executives were ranked as most important while elected and appointed officials were ranked as least important. As the author of this study states:

> The issue of who controls the police is one that has been hotly debated and yet remains unsolved. In fact, various reform movements of the present and the recent past have posited solutions which move in opposite directions. The movement toward professionalism... has often implied more internal police control of themselves; self-discipline is, after all, one of the hallmarks of a profession. By the same token, the community control/civilian review movement of the 1960's and early 1970's has generally taken the reverse approach, emphasizing the "public servant" role of the police officer and postulating the importance of the role of those served in controlling the actions of the police.[12]

To what extent should local officials, elected or appointed, as well as the general citizenry, exert control over the police? To whom are the police accountable? There are no easy answers to these questions. Growing public skepticism concerning the quality of government services, as well as the tax-reform movement that is still

[10]Jerome H. Skolnick, "Democratic Order and the Rule of Law," in Thomas J. Sweeney and William Ellingsworth, *Issues in Police Patrol: A Book of Readings* (Kansas City, Mo.: Kansas City Police Department, 1973), p. 6.

[11]Robert Wasserman, "The Governmental Setting," in *Local Government Police Management,* ed. Bernard L. Garmire, (Washington, D.C.: International City Management Association, 1982), p. 31.

[12]Jeffrey S. Slovak, *Police and Their Perceptions* (Chicago: Public Administration Service, 1977), pp. 29-30.

alive and well today suggest that the community is coming to expect and demand more from agencies of local government, but is not willing to pay more for those services. In addition, the police can no longer be the sole judge of what constitutes fair and efficient police service. Public involvement in police policy making is inevitable if the police are to be fully responsive to public concerns. While the police must continue to resist undesirable political influences which may affect their independence and objectivity, they must nevertheless understand the role they play in the political process as well as the political consequences of their actions.

LOCAL GOVERNMENT INFLUENCES

Although police agencies of various kinds and sizes exist at all levels of government, the municipal police department is the most common form of police agency. The municipal police department is a creature of local government. That is, it was created by and operates under the authority of the local legislative body.

The form of government under which a police department operates may have a direct influence on how that police department is organized and operates. The three most common forms of government are mayor-council, council-manager, and commission. Within these three types, there are a number of different variations. Each type may have a direct or indirect influence on police operations in the community.

The Mayor-Council Form of Government

Under this form of government, the mayor serves as the chief executive officer of the city and provides administrative direction and control over all municipal departments. The mayor also presides at meetings of the city council or local legislative body and may or may not have voting rights during such meetings. Under the strong-mayor plan, the powers of the mayor are quite broad and the mayor is usually seen as the chief administrative officer of the city as well. In some cases, the mayor may appoint an executive assistant or administrative assistant to assist in managing the affairs of the city. Depending upon the size of the city, past history and practices, and the personalities of the persons involved, the mayor's administrative assistant may have far-ranging authority over municipal operations. The mayor may also delegate to the administrative assistant limited authority over certain operating departments.

Depending upon the local charter, the mayor under this form of government may have the authority to appoint department heads, including the chief of police, subject to the approval of the city council or local legislative body. In some cases, the chief of police may be appointed by an independent board or commission. In Wisconsin, for example, all police and fire chiefs are appointed by a Police and Fire Commission. However, the chief of police reports to the mayor for day-to-day operational and administrative purposes.

Under this form of government, there must be a good working relationship between the chief of police and the mayor. This is not always easy, since mayors, like other elected officials, tend to be political and may not understand the basic difference between politics and administration. As a result, a chief of police may be asked or directed to do something which, in the eyes of the chief, may seem like a violation of professional ethics. When this occurs, it is necessary for the chief to point out to the mayor the reasons why the action requested or directed is improper or unwise. In the final analysis, the chief of police may be forced to make a choice between keeping the job and upholding professional principles. In most cases, however, some compromise can usually be reached that will be agreeable to both sides. Elected officials, through their years of experience on the political scene, generally recognize and respect the advice given to them by their professional department heads.

Some police officials dislike the mayor-council form of government because it makes them feel too dependent on elected officials who may be motivated by self-serving interests. They may also believe that elected officials, lacking the professional education and experience in police administration, may make administrative or policy decisions which are not conducive to effective law enforcement, but which are politically popular. Police chiefs also recognize that, under this form of government, they are often subject to removal from office at the discretion of the mayor and a majority of the legislative body. Unless they are protected by civil service (and very few are), or have some other protection against unwarranted removal from office, police chiefs may feel threatened by partisan political influences.

In the vast majority of cases, however, this is not the case. Police chiefs have worked under the mayor-council form of government for over a century and, for the most part, have survived without compromising their professional ethics. Some of the most professional, progressive, and competent police administrators in the last fifty years have worked under the mayor-council form of government without adverse consequences. While there have, of course, been cases of good police chiefs being forced out of office for political reasons, these are the exceptions rather than the rule.

It is clear, nevertheless, that there needs to be not just a good working relationship between the chief of police and the chief executive officer of the city, but also a clear understanding of authority and responsibility. The chief of police needs to know what is expected of the police and what the police can expect from the mayor; the mayor needs to make it clear to the chief of police the limits of the chief's authority. For example, who is to handle press releases at the scene of a major crime or natural disaster? Often, the mayor will reserve this privilege for himself and will ask that the chief of police defer all such requests to the mayor's office. Understanding and accepting the ground rules helps to establish a more effective and mutually enjoyable working relationship.

The Council-Manager Form of Government

The council-manager form of government has gained increasing popularity since its development in the early part of the twentieth century. The reason for its popularity is the growing recognition that elected officials cannot, due to their limited training and experience in government, provide the kind of professional administration needed for the complex affairs of a municipality. Under this form of government, a trained manager or administrator is appointed by the mayor and the council to serve as the chief administrative officer of the city and to supervise and coordinate day-to-day municipal operations. Typically, all department heads report to the manager or administrator, who in turn reports to the mayor and council. Owing to the growing complexity of municipal government today and the tremendous diversity of administrative skills needed to manage municipal affairs, the council-manager form of government seems like a logical choice for all but the smallest towns and villages where administrative responsibilities can still be performed by part-time officials.

Depending upon the manner in which the position was created and the authority vested in the position, the city manager or administrator may have authority to appoint department heads, including the chief of police. Under a "true" city manager form, the manager appoints all department heads without the concurrence of the legislative body. Even in such cases, the manager will usually keep the council fully informed of such decisions and will attempt to make appointments that will not arouse the opposition of the legislative body. In still other instances, appointments are made by the manager or administrator, subject to the approval of the mayor and council.

Many police chiefs prefer to work under the council-manager form of government for two basic reasons. First, they consider themselves professionals and prefer to work for people who are professionals. City managers and administrators typically have distinguished themselves through years of academic training, professional development, and experience in lesser positions in local government. Not only have they "worked their way up" in their profession, but they adhere to a comprehensive set of professional ethics which have been developed over the years by the International City Management Association (ICMA).

Second, police chiefs feel that they are more insulated from political influences when they work under a city manager or administrator. This is particularly true if they are appointed by, and may only be removed by, the manager or administrator. Since the manager or administrator is a professional and, at least in theory, operates according to a high set of professional standards, police chiefs feel that they can perform their jobs without worrying about the political consequences of their actions. For the most part, but not in all cases, they are correct in this assumption.

Unfortunately, not all city managers or administrators are professionals, and some may be less professional than others. Some may have much stronger convictions than others. Those with weak convictions or professional ethics may subject the chief of police to political influence, or may direct the chief of police to take actions which the chief considers politically irresponsible or inappropriate. Fortunately, this is not usually the case. Most professional managers and administrators are willing to accept the professional judgment of the police chief as long as the chief's decisions comply with the city's policies and administrative regulations.

It is important for the chief of police to have a good working relationship with the city manager or administrator and to have a very clear understanding of what is expected of him by the manager or administrator. Even though the manager or administrator may be a professional, he or she may not be the easiest person to get along with. Indeed, because they got where they are on the basis of personal influence, popularity, and "people skills," mayors are often much easier to get along with than managers or administrators. Managers and administrators are often very technically competent, but lack the interpersonal skills necessary to develop effective working relationships with either department heads or elected officials. This may be one reason for the relatively short tenure (three to five years) of managers and administrators.

The Commission Form of Government

The Commission form of government has dwindled in popularity over the years, although it is still found in some areas of the country today. The commission form of government is unique in that elected officials serve as both policy makers and administrators. In effect, a person elected to the commission becomes a department head over one or more operating departments of the city. For example, the Public Safety Commissioner would usually supervise the police and fire departments, while the Public Works Commissioner would supervise the streets, sanitation, and water departments. In some cases, commissioners may be elected directly to specific posts, while in other cases their assignments are made at the beginning of their term of office during their first organizational meeting. In most cases, one of the commissioners will serve as mayor and may also supervise one or more administrative departments in the city.

The obvious disadvantage of the commission form of government is that a person elected to manage a municipal department may have no skills or training in the field. The consequences of such a situation can be disastrous and can have a very damaging effect on operational efficiency and public confidence. There was a case, for example, in which a patrol officer was elected Commissioner of Public Safety and promptly fired the Chief of Police who had refused to promote him due to lack of job proficiency. The Public Safety Commissioner then appointed one of his cronies as Chief of Police for political purposes. During the remainder of the commissioner's term in office, police morale declined as did public confidence in the police department. The Commissioner of Public Safety was defeated at the next election and went back to his job as a patrol officer in the police department.

If it could be assured that persons elected to office had the necessary credentials and training to competently manage the departments to which they are assigned, the commission form of government would be much more acceptable than it is. If there could be some assurance that an elected commissioner could only remove a department head for justifiable cause, or could only appoint a department head on the basis of merit or qualification, then the commission form of government would pose much less of a threat to effective and professional administration. This is not to say, of course, that those cities which operate under the commission form of government are not properly managed, for some of them clearly are. There are, however, limitations which exist under this form of government.

Independent Boards and Commissions

Over the years, a number of different kinds of independent boards and commissions have been created to minimize the impact of politics on the police, to regulate the hiring and promotion of personnel, to establish operating policies and procedures, and to hear and adjudicate allegations of unprofessional conduct by police personnel. Most of these independent boards and commissions are regulatory, administrative, or judicial in nature and can have a very direct impact on the management and administration of police operations. In any event, they are forces to be reckoned with by the police administrator.

These independent boards and commissions have been set up by local legislative bodies either as a means of insulating the police from political influences or as a means of ensuring independent and impartial control of police operations. It is difficult to say whether these attempts have been successful since there are so many examples of successes and failures on record. Whether these independent boards and commissions serve as positive or negative influences on the administration and operation of a police department to a large extent depends upon the personalities and philosophies of those appointed to them, the nature of the board or commission, and the political environment.

Perhaps the most common type of independent board or commission is the local civil service commission, which is appointed to oversee the hiring, firing, promotion, and discipline of members of the police department. Members of this board are usually local citizens who are appointed by the chief executive officer or legislative board of the city. Members of the civil service board, merit commission, or police and fire commission are responsible for seeing to it that persons are appointed to the police department solely on the basis of merit and qualifications. They are also responsible for setting up criteria and developing promotional programs which ensure that only the most qualified persons are advanced in rank. Similarly, they are charged with reviewing disciplinary actions taken by the chief of police and conducting hearings on all cases involving dismissal of an officer.

The greatest source of conflict between civil service boards and the chief of police seems to come about as a result of the board's attitude that, in order to remain impartial and independent, it must keep itself isolated from the chief and his staff. As a result, some civil service boards do not even consult the chief of police when entry-level testing or promotional examinations are being designed and considered, or when new appointments to the department are being planned. Since the chief of police must ultimately supervise and work with those who are appointed to and promoted within the department, it seems only fair and appropriate that the chief should have some voice in these decisions. Unfortunately, in their zeal to remain independent, some civil service boards do not even bother to communicate with the chief on such matters.

In some states and in some cities, police commissions have been set up to oversee the administration and operation of the police department. Where such commissions do exist, the chief of police is directly responsible to the board and must answer to it on virtually all aspects of police operations. Such commissions, for example, may approve the police budget, establish operating policies and procedures, establish staffing levels and deployment plans, and administer discipline within the department. In effect, these commissions become the head of the police department and the chief of police becomes a figurehead.

The rationale for such commissions today is difficult to understand. Clearly, the persons appointed to such boards and commissions, despite their dedication and commitment, are not qualified by education or training to administer a police department. Decisions made by such commissions may be made on what they consider to be broader community interests but may have devastating effects on the police department. While it may be desirable to have a commission which can be advisory in nature to the chief of police, it should not have the authority to inject itself into the policy- or decision-making processes of the department.

A third type of independent board or commission that has been created over the years to "watchdog" the operations of police departments is the police review board. Police review boards became popular during the 1960s and early 1970s when urban riots, civil unrest, and civil-rights demonstrations brought police tactics and behavior under close scrutiny. Charges of police brutality and discriminatory treatment by police officers raised questions about the ability and willingness of police officials to objectively and impartially investigate allegations of police abuse. Police review boards were created in a number of cities to review police conduct and to impose disciplinary actions against officers.

The rationale behind an independent civilian review board is that the police are either incapable or unwilling to conduct thorough and impartial investigations of allegations of misconduct by their own officers. Clearly, it is not a matter of lack of capability, since conducting major criminal investigations is one of the things that the police do best. It is thus a question of their willingness to undertake such an investigation and to, if necessary, take disciplinary actions against an officer. Although there may be some justification for doubting the sincerity or integrity of police agencies in such investigations, the fact of the matter is that the police have done a reasonably good job of "keeping their house in order."

The police, after all, have a vested interest in implementing and maintaining high standards of professional conduct among their members. They need public support if they are to be successful in accomplishing their mission. Moreover, the police like to be regarded by others as "professional," and one of the hallmarks of a profession is the ability of that occupation to monitor and regulate the conduct of its own members. Certainly, there have been some lapses by the police in some very notable situations, but by and large, they have shown a sincere and aggressive commitment to investigate the actions of their officers in an impartial and objective manner and to take appropriate action when abuses have been discovered. Indeed, the system of internal discipline found within a police department, it may be argued, is weighed much more heavily in favor of the public than is the system of criminal justice itself.

Perhaps for this reason, the popularity of civilian review boards has waned in recent years. Perhaps this is because the public has gained increased confidence in the ability of the police to police themselves, or simply because the issues that originally brought the need for police review boards to the public consciousness are no longer issues. Even where they do exist, however, it is difficult to assess what impact, if any, they play upon the level or quality of police services or upon the performance of officers in the field.

If they do nothing else, civilian police review boards create a feeling in the minds of the police that no one trusts them. This adds to the feeling of alienation and isolation that many police officers feel toward the "outside world." As Berkley has noted, a civilian police review board

> ...makes the policeman feel that he has been singled out for special scrutiny and control. It tends to increase his feeling of group solidarity. The existence of a civilian review board may well widen the gap between "us" and "them."[13]

Whether a police officer will be more careful in dealing with the public on the street, deciding which laws to enforce or which person to arrest, or in making any other decision based on the existence of a police review board is doubtful. Chances are, the officer will make the same decision in a given situation without considering subsequent evaluation. By and large, officers make decisions based upon the circumstances that exist at the moment. It is only later, if at all, that they consider the long-term implications of those decisions on their careers. Indeed, we would have good reason to question the legitimacy and the quality of decisions made solely or primarily upon consideration of the impact they might have upon an officer's career.

THE POLICE ORGANIZATION

At a glance, the typical police organization may appear similar to many other organizations. It is, after all, a collection of individuals working together toward a common purpose. Most police organizations, like other organizations, have a defined structure of authority and responsibility, and its workers are assigned tasks according to a prescribed plan. A hierarchy of rank and authority exists for the purposes of disseminating information, ensuring that orders are given and carried out, and that decisions are made at the appropriate level. In these respects, the police organization is not unlike many other kinds of organizations.

All organizations have values. It is these values—ideas, beliefs, principles, and customs—that shape and influence organizational action and that make each organization distinct from all others. Moreover, the value system of the organization identifies the underlying philosophy or point of view for which the organization stands.

[13]George E. Berkley, *The Democratic Policeman* (Boston: Beacon Press, 1969), p. 146.

> Values are the beliefs that guide an organization, and the behavior of its employees. The most important beliefs are those that set forth the ultimate purposes of the organization.[14]

In the private sector, organizational values help to form public opinion for or against an organization and its products, services, and policies, and can be a powerful force in motivating consumers. Public-sector organizations, while not driven by the profit motive, have no less need to aspire to favorable public opinion of their services.

Values are used to define a code of conduct for members of the organization. To accomplish this purpose, values must be shared by members in all levels of the organization. When leader or members fail to support or stand behind the values of the organization, the potential for conflict is created and there may be a disparity between the values of the organization and the conduct of its members.

> ...only when the formal values of the organization match those acted out by the rank and file can the organization be considered 'high performing.'[15]

The police organization has its own personality and helps to shape the attitudes and perceptions of those who work in it. Some police agencies appear to be much more open than others and their officers and employees reflect this characteristic. Leadership, political environment, and the community are factors which influence the development of values in each police agency. Organizational values directly affect individual values in the work place, the way in which employees approach their work, and the level of satisfaction they receive from the work they do. The task of management is to inculcate certain desirable values into the organization and stress the importance of these values to their subordinates. Studies have shown that there is a relationship between organizational cohesion, individual commitment to the values of the organization, and job satisfaction among employees.[16]

There are several characteristics of the police organization which set it apart from other types of organizations. These include militarylike reliance upon a rigid system of discipline, rank structure, written directives and standard operating procedures; its closed system of promotion from within; and the existence of a unique subculture within the police organization.

Militarylike Characteristics

Police departments operate very much like military organizations. This is partly due to the similarity between the kind of work they do (maintaining order, regulating con-

[14]Robert Wasserman and Mark M. Moore, "Values in Policing," National Institute of Justice, *Perspective on Policing* (November, 1988), p. 1.

[15]Wasserman and Moore, "Values in Policing", p. 7.

[16]W. Randy Boxx, Randall Y. Odom, and Mark G. Dunn, "Organizational Values and Value Congruency and their Impact on Satisfaction, Commitment, and Cohesion: An Empirical Examination Within the Public Sector," *Public Personnel Management* (Spring, 1991), pp. 195-205.

duct, enforcing the peace) as well as the means by which they do it (force or threat of force). Police departments must rely upon a strong sense of discipline because officers who make most of the decisions work without any direct supervision. They must therefore be motivated (self-disciplined) to act within certain prescribed standards of performance and conduct. For the same reason, written regulations and standard operating procedures are issued to anticipate a variety of situations in which officers may find themselves and to provide guidelines for them to follow in such situations. These guidelines help reduce the need for officers to rely solely upon their own judgment or individual instincts.

It is also very possible that the militarylike organizational model of the police affects the officer's outlook on life and attitudes toward others. Military organizations are conservative in nature and tend to promote conservative behavior by their members. Police departments, being military in nature, also tend to promote conservative attitudes among their members. These conservative attitudes may very well influence how an officer interacts with others.[17]

The reliance of police agencies on a militarylike rank structure is not as easy to explain or understand. It is not clear why a first-line supervisor in a police department could not be called a foreman or simply a supervisor as opposed to a sergeant or corporal. Nor is it clear why a deputy chief could not be called a director or a manager, or why the person in charge of a patrol division or investigative bureau could not be called a commander. Indeed, civilian titles have been adopted by some police agencies, but most still cling to the comfortable custom of military-type rank designations.

Closed System of Selection

Police departments are unique (comparable, perhaps, only to fire departments) in that virtually all promotions and assignments are made from within. With a few rare exceptions, all promotions and assignments are made from within the organization. Only at the highest levels of the department (chief, deputy or assistant chief, and sometimes commander) are persons from outside the organization allowed to apply.

This kind of restriction affords certain obvious advantages to those within the organization, since it limits the field in which they must compete for promotion. To the organization, however, the pact is largely negative, since it limits the field of persons who may be considered for promotion and thus imposes restrictions upon the quality of the applicants themselves. Few businesses could survive in the highly competitive world in which they operate if they were forced to choose their managers and executive officers from the same persons who entered their organizations as property clerks and bookkeepers. In the years ahead, we may very well see this closed system of selection begin to change in favor of more open methods.

[17]Jerome H. Skolnick, "Democratic Order and the Rule of Law," in Thomas J. Sweeney and William Ellingsworth, *Issues in Police Patrol: A Book of Readings* (Kansas City, Mo.: Kansas City Police Department, 1973), p. 13.

The Police Subculture

The police comprise a unique subculture within society. A highly regulated and intense selection process produces a subculture dominated by certain kinds of personality types, value systems, and behavioral styles. A well-defined code of conduct requires intense loyalty to the organization. In addition, all members of the organization are expected to subscribe to a code of secrecy.

By the very nature of its work, a police department is not an open organization and must guard against the premature or unauthorized release of information available to it. But the aura of secrecy that pervades the typical police agency goes beyond the necessity of protecting confidential information from premature or unnecessary public disclosure. The police are often suspicious of any attempt by external forces to inquire into their activities, fearing apparently that such information might ultimately be used against them. Police officers have a tendency to withdraw behind what Goldstein has referred to as the "blue curtain" in order to shield themselves from public criticism.[18]

Police officers often develop an "us versus them" attitude which, in the long run, can be very harmful to relations between the police and the public. Police officers often believe (sometimes with reason) that no one else understands their problems and that their job is made more difficult by an apathetic, uncaring, unsupportive, antagonistic, and sometimes hostile public.

> He sees the public as a threat. He seldom meets it at its best and it seldom welcomes him. In spite of his ostensible function as a protector, he usually meets only those he is protecting them from, and for him *they* have no love... To them, he is the law, the interfering one, dangerous and a source of fear... To him they are the public, an unpleasant job, a threat, the bad ones, unpleasant and unwilling, self-concerned, uncooperative, and unjust.[19]

Under such circumstances, it is easy to understand how police officers develop their own identity, code of conduct, and value system. It is also no wonder that the police organization becomes more to the police officer than just a place to go to work. It is a home away from home and in some cases takes on the characteristics of an extended family. Officers see the police organization as a refuge from the unfriendliness of the outside world. It is a sanctuary where officers can share their experiences with the only people who will appreciate and understand them—their fellow officers. The police organization becomes the officer's world and supersedes all other external influences on the officer's behavior.

[18]Herman Goldstein, *Policing a Free Society* (Cambridge, Mass.: Ballinger Publishing Co., 1977), pp. 165-166.

[19]William A. Westley, *A Sociological Study of Law, Custom, and Morality* (Boston, Mass.: Massachusetts Institute of Technology, 1970), p. 49.

THE CRIMINAL JUSTICE SYSTEM

The police are also unique in that, although they are a creation of local government, they are expected to enforce laws enacted at the state and federal level. In addition, they are part of a larger system of criminal justice which extends beyond the authority of the municipal government of which they are a part. They are, therefore, a part of both a local service-delivery system and a broader system of criminal justice.

There is little confusion over the jurisdictional boundaries between the police, prosecution, the courts, and correctional agencies within the criminal justice system. For the most part, the responsibilities and authority of each of these agencies is fairly well defined and understood. The system of criminal justice does not generally suffer from a lack of cooperation among the agencies within that system, although there are some obvious exceptions to this statement.

To some extent, the agencies of criminal justice have competing and conflicting interests. The role of the police, for example, is to take people who are suspected of crimes off the street, while the function of the courts is to determine guilt or innocence. The function of correctional officials and institutions is to either incarcerate or rehabilitate, depending upon the sentence of the court, the nature of the offense, and the characteristics of the offender.

Of all the agencies of the criminal justice system, however, the police have the most direct impact since the process of criminal justice begins with the actions of the officer on the street. This is not to say that the police are the most important element of the criminal justice system, but rather that it is the actions of the individual police officer that most directly affect the outcome of the system. In this respect, then, the role of the prosecution, the courts, and the corrections authorities is secondary to that of the police. This obviously places a very heavy responsibility on the police, and especially on the individual patrol officer.

The decisions made by patrol officers on the street, although removed from public view, may have lasting consequences upon individual freedoms as well as upon the quality of justice. Nearly every major court decision concerning criminal procedure during the last fifty years has been based either entirely, or in part, upon the actions of police officers either prior to, during, or after an arrest, apprehension, or interrogation.

Regardless of how these court decisions are viewed by legal scholars, police administrators, or other agencies within the criminal justice system, the fact is that they have had a major impact upon the ways in which the police and other agencies of criminal justice go about their work. It is therefore important that the police understand their role in the criminal justice system and how their work and their individual actions and decisions ultimately impact upon and are reflected by that system.

INTERAGENCY COORDINATION

The police must interact daily with a variety of other municipal departments (fire, public works, recreation), as well as with local community and civic groups, religious organizations, and educational institutions. These organizations play a key role in determining the quality of life in the community. The police interact with these organizations almost on a daily basis and must form constructive relationships with them if the members of these agencies are to understand and support police goals and objectives. Indeed, one of the principal roles of the police executive is to foster and maintain effective working relationships with community groups.

In addition, there are a multitude of governmental structures at the local level which sometimes have confusing and conflicting jurisdiction and authority. There are towns, villages, cities, boroughs, districts, townships, and counties, each with different levels of authority and responsibility. Many of these governmental forms have their own police department, with the result that there may be six or seven police agencies in a small area of less than ten square miles. One of the results is that more and more governmental agencies compete for fewer tax dollars. Consolidation of agencies in such areas could result in substantial tax savings and an improved level of service to the public. Unfortunately, consolidation of services is often not a viable alternative from a political standpoint.

Confusion over jurisdictional boundaries at the local government level makes the task of policing more difficult. In large cities, for example, responsibility for providing police services may be divided among housing-authority police, transit-authority police, parks police, city police, and one or more other special-district police agency. Each of these police agencies has its own organizational structure, legislative body or governing board, and operational agenda. This further complicates coordination and communication at all levels, leads to overlapping jurisdiction, and duplication of efforts. Unfortunately, this seems to have always been the case in the United States, and the situation has not significantly improved since Bruce Smith made the following observation in 1940:

> It is at the local levels of government that the greatest variety of police units begins to appear. This is chiefly due to variety in the types of local governments themselves, to their overlapping jurisdictions and their occasionally complex interrelationships; but it is also attributable to the increasing tendency of local government units to set up rival police agencies—some of them with restricted police authority—within a single local area.[20]

SUMMARY

The police operate in a dynamic environment and are affected by a variety of conditions, organizations, and environmental influences. Too often, they try to isolate themselves from that environment rather than integrate themselves into it. The police are an essential part of the criminal justice system that plays such an important role

[20]Bruce Smith, *Police Systems in the United States,* 2nd. ed. (New York: Harper & Row, 1960), p. 20.

in the quality of life in our communities. The police cannot afford to be apart from, but must rather be a part of, the society in which they operate.

As society changes, so must the police adapt themselves to changing conditions around them. They cannot continue to rely upon custom and tradition, but rather they must incorporate new thinking, experimentation, and creativity into their operating practices. The brunt of these changes will be felt by the patrol officer more than others in the police organization. It is the patrol officer who interacts directly with the public and who will be impacted most directly by changing social conditions. The patrol officer, therefore, must be able to cope with those changing conditions.

It is an exciting time for the police, but one that offers many challenges. Police managers of the future will need to determine whether those challenges will become obstacles or opportunities.

REVIEW QUESTIONS

1. Discuss some of the ways in which the environment in which the police work affects the manner in which they perform their jobs.
2. What kind of information is important to the police patrol officer? Give several examples.
3. Discuss the ways in which the three patrol shifts differ and how these differences might affect the role of the patrol officer on each shift.
4. Explain the difference between legitimate and illegitimate influences that local political leaders can bring to bear on the police.
5. Why is it important for a patrol officer to understand the political system of the community in which he or she works?
6. Explain why a police department should be a reflection of the community it serves.
7. In what way do community values affect the mission and role of the police?
8. To what extent should local government officials, political leaders, and community representatives control the police?
9. List the advantages and disadvantages of creating independent boards and commissions to oversee police department operations.
10. Discuss organizational values and how they may affect the operations of a police department.

REVIEW EXERCISES

The following exercises are designed to reinforce or amplify comprehension of the material contained in this chapter.

1. Prepare a graphic model characterizing and comparing the three patrol shifts described in this chapter. Use such characteristics as physical hazards, typical problems encountered, clientele, etc.
2. Prepare a list of the various physical hazards that may confront patrol officers and how they might be minimized or avoided altogether.
3. In small group sessions, compare and contrast the strengths and weaknesses of the various forms of municipal government and their positive or negative affect on police job performance.
4. In class, discuss the legitimate and illegitimate influences that may be placed on the police by local political forces.

REFERENCES

Berkley, George E., *The Democratic Policeman*. Boston: Beacon Press, 1969.

Boxx, W. Randy, Randall Y. Odom, and Mark G. Dunn, "Organizational Values and Value Congruency and their Impact on Satisfaction, Commitment, and Cohesion: An Empirical Examination Within the Public Sector," *Public Personnel Management* (Spring, 1991), pp. 195-205.

Goldstein, Herman, *Policing a Free Society*. Cambridge, Mass.: Ballinger Publishing Co., 1977.

Hale, Charles D., *The Isolated Police Community: An Occupational Analysis*. Master's Thesis, California State University at Long Beach, 1972.

"Patrol, Patrol, Patrol," *Spring 3100* (July-August, 1960), in Samuel G. Chapman, *Police Patrol Readings*. Springfield, Ill.: Charles C. Thomas, 1970, pp. 60-61.

Siglerm Jay A., and Robert S. Getz, *Contemporary American Government: Problems and Prospects*. New York: Van Nostrand and Richmond Co., 1972.

Skolnick, Jerome H., and David H. Bayley, *Community Policing: Issues and Practices Around the World*. Washington, D.C.: National Institute of Justice, 1988.

Skolnick, Jerome H., "Democratic Order and the Rule of Law," in Thomas J. Sweeney and William Ellingsworth, *Issues in Police Patrol: A Book of Readings*. Kansas City, Mo.: Kansas City Police Department, 1973, pp. 3-24.

Slovak, Jeffrey, *Police and Their Perceptions*. Chicago: Public Administration Service, 1977.

Smith, Bruce, *Police Systems in the United States* (2nd. ed.). New York: Harper & Row, 1960.

Wasserman, Robert, "The Governmental Setting," in *Local Government Police Management*, ed. Bernard Garmire. Washington, D.C.: International City Management Association, 1982, pp. 30-51.

Wasserman, Robert and Mark M. Moore, "Values in Policing," National Institute of Justice, *Perspectives on Policing* November, 1988.

Westley, William A., *Violence and the Police: A Sociological Study of Law, Custom, and Morality*. Boston, Mass.: Massachusetts Institute of Technology, 1970.

Wilson, James Q., "The Police and Their Problems," *Public Policy*, 12 (1963).

CHAPTER 4
POLICE PATROL HAZARDS

CHAPTER OUTLINE

[1]Jeffrey T. Mitchell, "Development and Functions of a Critical Incident Stress Debriefing Team," *JEMS*, December 1988, pp. 43-46.

LEARNING OBJECTIVES

After completing this chapter, the student should be able to

1. Name the hazards that are frequently associated with police patrol duty.
2. List measures that can be taken to identify, neutralize, eliminate, or avoid direct confrontation with the hazards of police patrol work.
3. Describe the effect of organizational stress on the job performance of the police patrol officer.
4. Explain what critical incident stress is and describe the role it plays in police patrol work.
5. Define *police corruption*. Describe how it develops, its effect on police patrol work, and what steps can be taken to identify and eliminate it.

In this chapter, the phrase *patrol hazards* refers not just to the rather obvious physical hazards that threaten the safety of a patrol officer on a regular basis, but also the psychological and environmental hazards that can ultimately take a toll on the patrol officer. These include burnout, fatigue, stress, medical problems, poor motivation, and a variety of other factors that can have an adverse impact on the patrol officer.

Although a majority of police patrol encounters do not involve violence, many of them are potentially hazardous. Whenever a patrol officer intervenes in a family dispute, attempts to remove a drunken or disorderly person, stops a speeding motorist, checks an open door, checks out a suspicious person in a darkened alley, or attempts to make an arrest, the potential for violent conflict is always present.

How dangerous is police work? Strangely, it is not the most hazardous occupation. Mine workers, firefighters, and others suffer more job-related injuries and deaths. Nevertheless, police work can be extremely dangerous. Chapman, for example, estimates that one million assaults on police officers occurred between 1960, when the FBI began to keep records of such incidents, and mid-1985. During this same period of time, some 2,129 police officers were murdered and another 328,000 were injured.[2]

The traditional wisdom is that police officers can predict with some degree of certainty those types of situations which are potentially more dangerous than others. The fact of the matter, though, is that any situation, no matter how seemingly innocent or harmless, is potentially hazardous for the officer. The adage that there is no such thing as a "routine traffic stop" can be expanded to include all police encounters.

Certainly, there are certain types of situations that, by definition, are dangerous. Armed robberies, hostage takings, high speed chases, riots, and family disturbances

[2]Samuel G. Chapman, *Cops, Killers and Staying Alive* (Springfield, Ill.: Charles C. Thomas, 1986), p. xiii.

are all situations which, by their very nature, alert the police officer to the fact that the incident may erupt in violence. Experience and forethought help officers prepare themselves mentally and physically for what may occur. Thus, if violence does occur, the officer may be better able to deal with it.

Other situations, in which violence and danger to the officer are not as likely to occur, present much greater potential harm to the officer. Situations in which all circumstances appear normal, which have occurred hundreds of times before with nothing extraordinary occurring, and which seem to be completely innocent on their surface are the ones which may cause the greatest harm to the officer. When an officer does not expect anything to happen, he or she is often not prepared for it when it does.

Even those situations that do not result in physical violence can take a toll upon the physical and mental well-being of the officer. Psychological trauma and job stress are common in police work. Unfortunately, they are rarely recognized and often inadequately treated. Only in recent years has the full magnitude of job stress and its effects upon police officers come to be recognized. Even though the number of police officers who die in the line of duty each year is alarmingly high, there are even more who die, suffer crippling injuries, or leave their jobs prematurely due to poor mental and physical health.

So long as the police are empowered to impose their authority over others, to control human conduct, to quell disturbances, and to track down, arrest, and incarcerate lawbreakers, the potential for violence will always be with them. There are ways, however, to improve management procedures, training programs, equipment and technology, and working conditions of police officers to reduce or minimize the risks associated with police work. For example, police administrators and local government officials need to give more consideration to the impact of physical and emotional stress on police officers.

To accomplish this, training programs for police supervisors and middle managers should be developed to help them recognize and identify the physical and emotional signs of job stress on police officers. Training programs can also be developed to help patrol officers recognize and avoid the kinds of field encounters which produce the greatest potential for violence directed toward them. Improved selection methods and standards may also help to weed out persons who may later prove unable to deal with the physical and emotional hazards of police work.

It is not the purpose of this chapter to overstate these negative factors, nor to suggest that a patrol officer's career is not a rewarding one, for just the opposite is true. It is rather the purpose of this chapter to identify the various negative factors that can affect the patrol officer in order to lessen their potential impact.

PSYCHOLOGICAL HAZARDS

The job of the patrol officer is packed with psychological hazards that contribute to frustration and job stress. As discussed in Chapter 3, police officers often feel isolated and emotionally divorced from the society in which they work. They seem to have very little in common with the very people they are expected to serve. They often feel that their efforts are not appreciated by the public, that they are blamed for rising crime rates and other societal ills over which they have no real control. When they ask for more resources to do a job that seems at times impossible, they must compete with other agencies which may be looked on with greater approval by local citizens.

Police officers also operate in an organizational climate that does not adequately recognize or reward superior performance. To a great extent, the organizational climate in which a patrol officer works promotes feelings of mistrust, cynicism and alienation. Within the police organization are found small groups and subcultures organized along lines of race, religion, neighborhood associations, political persuasion, age, and gender. These organizational subcultures create an informal communication network which is often more effective than official communication channels in disseminating important information about the organization.

Because of the intense psychological pressures experienced by police officers, applicants for police service are required to undergo rigorous screening programs. Character checks, psychological and physical examinations, background investigations, lie-detector tests, credit and criminal-history checks, and other methods can be used to assess a candidate's mental and emotional stability and suitability for police work. Unfortunately, the methods used by many police agencies to screen applicants are still not adequate.

Insufficient funds are allocated to the personnel departments, civil service boards, commissions, and others responsible for conducting employment testing. Those who are responsible for overseeing the testing process are themselves often not qualified by virtue of training, education, or experience to reliably determine proper testing methods and procedures. Civil-service boards and commissions are often too defensive to delegate testing responsibility to others who are more qualified.

Even in those departments that have established good screening procedures, it is virtually impossible to successfully predict whether an individual will be able to cope effectively with the mental, physical, and emotional rigors of police work. Psychological and job aptitude tests are helpful, but limited. Additional screening measures take place during recruit training at the police academy; these sometimes result in "washing out" a few people who cannot take the pressure. Further evaluation and screening takes place during the on-the-job training program. A carefully designed and well-managed Field Training Officer Program can yield very satisfactory results in terms of identifying individuals likely to become problems later on in their police careers. Regular, systematic, and intense performance evaluations during the probationary period can provide justification for eliminating persons whose subsequent performance on the job can be expected to be only marginal.

Unfortunately, few departments have established procedures for periodically evaluating the emotional or mental health of their personnel once they have reached tenure on the job. Given the extensive evaluation and testing that is done prior to hiring, it is surprising that virtually none is done after the person has become a permanent employee. It appears that an assumption has been made that, once an officer is hired, he or she will pose no further problems. All too often, this assumption false. Moreover, rigid civil-service procedures in some states make it virtually impossible to terminate an officer unless substantial evidence of unfitness for duty can be provided. Clearly, the law and the philosophy of the courts protects the officer from the department. The community itself often suffers the consequences of this misplacement of values.

OCCUPATIONAL STRESS

Police work has been identified as being one of the most stressful of all occupations.[3] Stress in police work can come in a number of forms and from several different sources, including (a) the nature of police work, (b) the police organization, (c) external factors, and (d) individual factors. Job stress can result in

> ...high blood pressure, cardiovascular disease, chronic headaches, and gastric ulcers. It can lead to severe depression, alcohol and drug abuse, aggression, and suicide. Stress affects officers' alertness, their physical stamina, and their ability to work effectively. ...Excessive absenteeism, disability and premature retirement compensation, and high replacement costs for disabled officers all direct resources away from effective crime prevention and law enforcement activities.[4]

Job stress in police work can come from a number of sources (see Figure 4.1). These include inadequate or inferior supervision; lack of administrative support for the individual officer, lack of recognition or support for a job well done, inconsistent application of discipline; emphasis on quantity of work rather than quality; inadequate, ambiguous, or unclear written directives; policies and procedures which leave officers unsure of where they stand or what is expected of them.

Job stress can result in a wide range of organic disorders that, if severe enough, can lead to temporary or permanent disability. Problems can include headaches, backaches, muscle cramps, asthma, hyperventilation, high blood pressure, heartburn, ulcers, and thyroid disorders.[5]

It should be kept in mind that stress is not an indication of mental imbalance or mental disease, nor does stress only affect persons who have "something wrong with them." Stress affects—-to a greater or lesser degree—-all persons, but in different ways, for different reasons, and under different conditions. A person does not have to have "something wrong" to feel stress, nor should someone feel that experiencing stress is an indication of some underlying mental or emotional problem.

[3]Gail A. Goolkasian, Ronald W. Geddes, and William DeJong, "Coping With Police Stress," in Robert G. Dunham and Geoffrey P. Alpert, *Critical Issues in Policing: Contemporary Readings* (Prospect Heights, Ill.: Waveland Press, 1989), p. 498.

[4]*Ibid.*, p. 498.

[5]*Ibid.*, p. 503.

1. *External Stressors*
    - Frustration with the American judicial system
    - Lack of consideration by courts in scheduling officers for court
    - Lack of public support
    - Negative or distorted media coverage
    - Officers' dislike of administrative decisions
2. *Internal Stressors*
    - Policies and procedures that are offensive
    - Poor or inadequate training and inadequate career development opportunities
    - Lack of identity and recognition
    - Poor economic benefits and working conditions
    - Excessive paperwork
    - Inconsistent discipline
    - Perceived favoritism
3. *Stressors in Law Enforcement Work*
    - Rigors of shift work
    - Role conflict
    - Frequent exposure to life's miseries
    - Boredom
    - Fear
    - Responsibility for protecting other people
    - Fragmented nature of the job
    - Work overload
4. *Stressors Confronting the Individual Officer*
    - Fears regarding job competence
    - Necessity to conform
    - Necessity to take a second job
    - Altered social status in the community

Source: Richard M. Ayres, *Preventing Law Enforcement Stress: The Organization's Role* (Alexandria, Va.: National Sheriff's Association, 1990), pp. 4-5.

Figure 4.1 Sources of Law Enforcement Stress

Even the most well-adjusted person can experience significant levels of stress from problems on the job, with the kids, with his or her spouse, or with difficult clients or bosses.[6]

The unpredictability of activity on a given shift, beat, or day of the week can be a source of stress to the patrol officer. There is no way to accurately predict busy times and slow times (although general patterns can usually be found). An officer may face periods of high activity sandwiched between periods of inactivity, during which boredom can be the single greatest opponent. Lack of activity can be just as fatiguing and injurious to the officer as too much activity.[7]

[6]Lynn Hunt Monihan and Richard E. Farmer, *Stress and the Police: A Manual for Prevention* (Pacific Palisades, California: Palisades Publishers, 1980), p. 2.

[7]William H. Kroes, *Society's Victim—The Policeman: An Analysis of Job Stress in Policing* (Springfield, Ill.: Charles C. Thomas, 1976), pp. 23-25.

Stress is likely to occur in situations in which there is imminent danger (for example, a high-speed car chase on a rain-slicked highway, or when entering a building occupied by armed intruders). Stress can also come from those situations in which there is no physical danger to the officer, but rather a high state of emotional excitement (for example, a serious traffic accident involving one or more fatalities; investigating the disappearance, beating, or death of a small child). These situations clearly take a toll on the officer that may not become apparent until much later.[8]

Reducing stress in the police organization is in the interest of everyone, but is the ultimate responsibility of the administration. Stress-reduction programs need not be expensive, terribly elaborate, or difficult to administer, even in small agencies, nor do they necessarily require the services of a trained psychologist or psychiatrist. Stress-reduction programs must begin with the identification of the most common stress-producing factors in the organization and the development of strategies for eliminating these factors or minimizing their negative effects on employees.

Many people work under stressful conditions. The police are not alone in this regard. Some of the most routine jobs can be stressful at times. Clerical workers can be placed under great stress when given impossible deadlines or tasks beyond their capabilities. Sales representatives can be placed under great job stress if they continually fail to meet their sales quotas. Midmanagement and executive officials in major corporations work under some of the most stressful conditions in the world and they compete in a highly volatile and competitive environment.

Clearly, the police do not have a corner on job stress. One study, for example, examined the objective indicators of mental health disorders among a number of occupations by examining the occupational records of persons admitted to mental health hospitals or clinics. It was found that persons with the greatest admission rates were employed as waitresses, nurses, musicians, and general laborers. Police officers, by comparison, ranked seventieth among the occupational categories studied.[9]

The job of the police officer is unique in one respect, however. Police officers are among the many classes of workers whose jobs and associated stressors have a direct impact upon their family lives. For many police officers, it is very difficult to separate their jobs from other aspects of their lives. Too often, police officers become personally involved with the people they come into contact with on a regular basis. Although they are often reminded to "leave the job at the door" when they leave each day, it is easier said than done. Though they may work only eight (sometimes more) hours a day, they are cops twenty-four hours a day, seven days a week. Although they are not always in uniform, department policies usually require them to carry their badges and identification (sometimes their service revolvers), with them everywhere they go.

Some police officers make a conscious effort to separate their job from their social lives, but it is not easy to do. Some even try to hide the fact that they are police

[8]*Ibid,* p. 64.

[9]Robert Kotulak, "Is Your Job Driving You Crazy?" *Chicago Tribune*, September 18, 1977, pp. 1, 5, citing study by Michael J. Culligan and others, "Occupational Incidence Rates for Mental Health Disorders," *Journal of Human Stress,* 3 (September, 1977), pp. 34-39.

officers, apparently fearing that this could affect their social status or opportunity to be treated like others in the same social strata.

> I try not to bring my work home with me, and that includes my social life. I like the men I work with, but I think it's better that my family doesn't become a police family. I try to put my police work into the background, and try not to let people know I'm a policeman. Once you do, you can't have normal relations with them.[10]

Having "normal" relations with other people becomes very difficult once a man or woman has joined the police organization. The code of loyalty to the badge, and to those who wear it, is strong and surpasses all other moral commitments. The job becomes more than just a job. It consumes the thoughts and emotions of those who are part of it.

The police see themselves as somehow different from the people they police. By virtue of their formal authority, they are placed in a supervisory position over others in society. They possess a power held by no other occupational group. Yet, on a social level, they often see themselves as inferior to these same people.

> In adapting to this situation, the policeman comes to be governed by norms different from, and sometimes in conflict with, those which govern the persons with whom he comes in contact.[11]

Police officers see other people differently from the way they did before they learned about the world of policing. They become more suspicious of things, people, and events that to others appear innocent or normal. Small things that might once have gone unnoticed now become perceptual impulses. The police officer must be alert for anything unusual that might mean trouble. A door in a darkened alley that has been checked a thousand times before might be found ajar the next time it is checked. The bank alarm that regularly goes off because new employees can't seem to understand the shut-off procedure may be signaling a real emergency the next time it sounds.

Police officers develop a set of instincts that set them apart from others, and these differences are not left in their lockers at the end of the day. Although they may try to leave the job behind, they cannot leave behind behavioral patterns, attitudes, or conditioned reflexes they develop as they become assimilated into the police culture. The attitudes they develop toward certain groups of people on the job are carried with them away from the job. As they develop into more mature police officers, they begin to lose touch with high school and college friends and neighborhood acquaintances.

[10]Jerome H. Skolnick, *Justice Without Trial: Law Enforcement in Democratic Society* (New York: John Wiley & Sons, Inc., 1967), p. 51.

[11]James Q. Wilson, "The Police and Their Problems: A Theory," in Arthur Niederhoffer and Abraham S. Blumberg, *The Ambivalent Force: Perspectives on the Police* (Waltham, Mass.: Ginn & Co., 1970), p. 294.

Another irony of police work is that most people—and police officers are no different—like to be liked. They take pleasure in feeling that their work is appreciated and that what they are doing counts for something. When asked why they want to be police officers, police applicants most frequently respond that they want to be of service to the public. They want to help others and to make a contribution to the quality of life in the community. Unfortunately, they may later find that their efforts are not always appreciated. Indeed, members of no other occupational group, except perhaps baseball umpires, receive so little respect or praise for the work they do.

One of the difficulties associated with police work is that police officers have very few acceptable ways in which to release the pressures brought on them as a result of the job. They are expected to remain cool, unemotional, and professional at all times, even under the most adverse circumstances. They cannot afford to show any sign of emotion that might indicate weakness on their part. They must remain impassive in the face of threats or hostility. They must be calm and cool when counseling an emotionally distraught mother whose small child has been abducted. They must be able to handle fatal car crashes, drownings, suicides, stabbings, and an assortment of brutal, bloody, and revolting incidents with total objectivity and a lack of emotionalism.

Despite what is expected of police officers, they have the same emotions as everyone else. They, however, must learn to control, hide, and disguise them so that they will not interfere with the job they are trying to do. Later, long after the situation has been cleared up, and well out of public view, the emotion will take its hold on the officer. In some cases, the release is quite apparent and dramatic. In other cases, it is much less obvious, and may not be visible at all. These cases may have the most adverse impact upon the officer later on.

Police officers must also be prepared to respond to critical situations immediately, knowing that a single wrong move may place themselves or others at risk. Not only are they expected to protect the lives and property of others, but they are expected to do it in a manner that will protect the constitutional rights of all concerned, even those who have broken the law. Failure to respect these rights, even under the most difficult and trying conditions, places the officer in jeopardy of being sued, or having an entire career ruined.

Fortunately, most police officers do not consider these possibilities as they go about their duties. If they seriously considered the many adverse implications of their job, they might very well choose another line of work. At the very least, they might change the way in which they perform their duties, which could have a negative impact on the job performance.

Despite all their hard work, the results of the police officer's efforts are often disappointing. An officer may risk everything to apprehend a dangerous criminal only to have the criminal released on low bail and be back on the street in a few hours. The work of the police officer prior to, during, and after an arrest may be scrutinized over and over again by the courts. In all too many cases, technical violations are discovered which eventually result in the arrest being thrown out and the criminal returned to society.

The police officer is also aware that most persons who commit crimes are never apprehended and that there is little truth to the saying that "crime doesn't pay." Indeed, for some clever criminals, crime pays very well. Even those who do get caught—those in the minority—can usually afford a skilled attorney who is almost better prepared than the public prosecutor. It is not unusual, therefore, for police officers to see themselves as part of the losing team.

It is not hard to understand how and why police officers develop a cynical view of the world in which they work. All too often, this cynicism inhibits the ability to derive a sense of pleasure and accomplishment from the work they do. Moreover, the disappointment and dissatisfaction experienced on the job may carry into the officer's personal life. Divorce, suicide, alcoholism, child abuse, spouse abuse, and self abuse may be the unpleasant results of the emotional distress of the job.

While it is difficult to assess with any degree of certainty the effects of job stress on police officers, anyone familiar with police work need only imagine the many adverse consequences and deleterious effects that emotional strain can have upon the officer's performance. In a violent confrontation, for example, overreaction on the part of the officer could conceivably result in an escalation of the incident, the injury or death of the officer or one or more of the participants, or a civil suit against the officer at some later time. The officer's failure to take appropriate action can have just as many adverse results.

The cause-and-effect relationship of job stress apparently varies from one occupation to another. In one study, nearly 1500 workers were classified according to the amount of stress normally encountered in their jobs. The workers were then rated (from above average to below average) according to ten "strain indicators" that were considered likely to affect their job performance. High job stress was found to be frequently associated with above-average depression and below-average self-esteem, life satisfaction, and motivation to work. Thus, it appears that those who are least satisfied with their jobs are also those most likely to experience job stress. This finding has important implications in terms of the working conditions and management of police agencies.

Surveys of police officer attitudes frequently result in findings of high job satisfaction and feelings that the job they are doing is important, juxtaposed with attitudes expressing a lack of confidence in elected officials, dissatisfaction with working conditions, and feelings that they are neither appreciated nor properly compensated for the job they do.[12] Such attitudes tend to reinforce the "us versus them" syndrome that seems to permeate many police organizations and indicates the lack of trust and confidence police officers have in those for whom they work.

Another study examined the effects of job stress on individual performance in

[12]The "politics" that is often found in police organizations, consisting of "playing favorites," application of double standards, "back stabbing" and other episodes of favoritism and lack of fairness in the enforcement of policies and procedures contributes to stress among police offers. See Lynn Hunt Monihan and Richard E. Farmer, *Stress and the Police: A Manual for Prevention* (Pacific Palisades, California: Palisades Publishers, 1980), p. 13.

the Miami, Florida Police Department. In the Miami study, it was discovered that very few police encounters could be classified as stressful; in only about 1.5 percent of all encounters was the officer's performance significantly affected by stress. The average patrol officer nevertheless suffers from a considerable amount of tension and anxiety as a result of the nature of the police function. The researchers further concluded that

> The stress of police work, like most life stress, derives not from a large number of stressful events, but from the anxiety provoked by relatively infrequent but very real and serious stress. Although the frequency of stressful events appears to be low, it should be borne in mind that even a small percentage is enough, presumably, to keep a patrolman worried as he goes about his job.[13]

Much of the stress from which police officers suffer derives from the authority they are expected to exercise. In the exercise of authority, police officers become involved in many troublesome situations in which their presence and actions are perceived by others as potentially harmful and threatening. While the authority of the police officer may be sufficient to quell most situations of this kind, the officer never knows when there will be an open and hostile opposition to his authority which may lead to violence toward himself or others.

There may be no reasonable way to reduce the stress associated with the police role, especially in critical incidents. It should be possible, though, to identify those types of encounters that are most likely to produce job stress, to analyze the dynamics of each, and to begin to develop remedial programs tailored to specific stressful episodes. Clearly, more scientific research is needed to analyze the full spectrum of stressful events that police officers encounter on a routine basis, to assess their impact upon the officer's mental and psychological health, and to design programs to alleviate negative stressors. The work by Monihan and Farmer provides useful and practical guidelines that can be used to identify the sources of stress and problems associated with them, as well as remedial steps that can be taken to eliminate the sources of stress.[14]

In addition, there need to be more sophisticated early warning programs in police departments to recognize and deal with the symptoms of job stress and burnout before they become major problems. Employee assistance programs, which have been introduced into many municipalities in the country, are good examples of what can be done to deal with these kinds of problems. Unfortunately, some of these programs are quite limited in scope and employees and their supervisors are often not fully aware of what they have to offer nor of how to avail themselves of their services. Supervisory personnel, in particular, need to become thoroughly familiar with such programs and how and when they can or should be used.

[13]Daniel Cruse and Jesse Rubin, "Police Behavior, Part I," *The Journal of Psychiatry and Law,* 1 (September 1973), p. 193.

[14]Lynn Hunt Monihan and Richard E. Farmer, *Stress and the Police: A Manual for Prevention* (Pacific Palisades, California: Palisades Publishers, 1980).

Occupational stress affects not only the police officer, but his or her family as well. Police work is not the kind of work one can leave at the office. The day's events, regardless of what they are, are carried home to the wife, husband, or children.

> The family often bears the brunt of the negative effects of stress for it is one of the few places where we can legitimately and, more importantly, safely express our feelings.[15]

As a result, programs designed to deal with the effects of job stress should take into consideration the effects which are felt by the officer's family too.[16]

Supervisors also need to be better educated and trained to recognize the problems of job stress that affect their employees. Just as they are expected to monitor the performance of their officers and to detect and deal with performance deficiencies, they need to be sensitive to the possible reasons for a decline in performance. In particular, they need to understand the dynamics of job stress, how it affects each person differently, and to recognize the symptoms.

Critical Incident Stress

One topic that is currently getting much attention in police departments is **critical incident stress.** This type of stress is qualitatively different from the stress we discussed earlier. Although it produces a similar physiological response, it is much more intense. It overwhelms the normal coping mechanism so that the officer is not able to resist its effects.

An officer may feel tense if the sergeant criticizes his or her handling of a traffic stop, or angry if a report has to be redone, but these stressful situations are typical of the daily routine. This type of stress can be successfully managed through exercise, or by asking the sergeant what could be done differently next time.

But how does an officer work off his tension level after he discharges a weapon and kills an armed felon? What happens when an officer is first on the scene and discovers the mangled body of a three-year-old boy (who happens to be the same age as the officer's son)? How does the officer cope with getting shot and wounded?

These incidents produce physical and mental reactions so intense that not even strenuous exercise or sleeping twelve continuous hours are sufficient to reduce stress to a comfortable level. Concentration, problem solving, memory, and judgment skills are all affected.

In critical incidents, more than 50 percent of officers experience a moderate to intense stress reaction.[17] Some know they just feel differently. They are easily irritated, can't sleep for more than a few hours at a time, and have difficulty making simple

[15]Lynn Hunt Monihan and Richard E. Farmer, *Stress and the Police: A Manual for Prevention* (Pacific Palisades, California: Palisades Publishers, 1980), p. 13.

[16]Ellen Scrivner, "Helping Police Families Cope With Stress." *Law Enforcement News* (June 15/30, 1991), pp. 6, ff.

[17]Roger Solomon, "Police Shooting Trauma," *The Police Chief* (October 1988), pp. 121-123.

decisions. Other officers deny they feel any differently and report that the incident did not bother them. All part of the job, they say. Only a very small percentage of officers are truly not affected by a shooting or dealing with the dead bodies of young children.

It is important that supervisors recognize these reactions as normal under the circumstances and prepare officers for what to expect. The stress reaction is a normal reaction to an abnormal situation. What follows is a sampling of the signs and symptoms of a stress reaction:

Physical

Officers may experience fatigue, nausea, rapid heart rate, headaches, thirst, dizziness, vomiting, weakness, and profuse sweating.

Cognitive

Officers may experience confusion, memory problems, inability to make good decisions or solve problems, difficulty accepting responsibility, concentration problems, difficulty identifying familiar objects or people, disturbed thinking and hypervigilance.

Emotional

Officers may experience anxiety, denial, fear, guilt, feeling overwhelmed, uncertainty, depression, intense anger, loss of emotional control, and agitation.

Behavioral

Officers may experience withdrawal, emotional outbursts, suspiciousness, inability to rest or excessive sleeping, alcohol consumption, hypersensitivity, intensifying of startle reflex, change in sexual functioning, change in speech pattern or communication, and loss or increase of appetite.

Reactions to stress may appear immediately following the trauma or may occur several days later. In some cases stress reaction may occur weeks or months later. The signs and symptoms of stress reaction may last for a very short while or for several weeks following the event. Stress reactions depend on both the nature and severity of the stressful incident.

Police officers with no prior experience in dealing with a critical incident are most vulnerable to feeling highly distressed and may come to believe they are "going nuts." Nightmares, flashbacks, and intrusive thoughts are among the more common symptoms. Not knowing what to expect may actually increase the probability of these responses. Most victims of critical incident stress benefit from the attention of a psychologist or social worker. Sometimes only one session is needed to assist an officer regain emotional stability.

Stress Debriefing[18]

The individual or group debriefing goes a long way toward reassuring the police officer that his or her reactions are normal. The purpose of the debriefing is to accelerate the recovery process so the officer can return to duty as soon as possible, particularly after a shooting. The debriefing should take place within 24 hours of a shooting and within 72 hours of other incidents.

The process is short-term and educational. Police officers' reactions and thoughts related to the trauma are discussed. Specifically, they are asked to describe in detail what they remember doing during the incident and how they reacted after it was over. Officers are told what to expect in the way of a stress reaction over the next few days or weeks. Further details are given regarding stress-management techniques and interactions with family and friends.

A collective sign is often heard near the end of a group debriefing, signifying that the participants know they will be okay and that the worst of the experience is over. By listening to experiences similar to their own, they come to realize that their feelings are normal for the circumstances. This knowledge has a tremendous calming and restorative effect on participants involved in the process. It also satisfies their curiosity about where the other officers were and what they were doing as the situation unfolded.

The Lauri Dann school shooting that took place in Winnetka, Illinois in 1988 illustrates the value of group debriefings. Six debriefings were held to accommodate all the rescue personnel involved. Since fire, police, and medical personnel all interacted on the scene, they were debriefed together. Two debriefings were also offered to spouses.[19]

A debriefing should be mandatory for all involved personnel following a shooting, line-of-duty death, or suicide. The reality in most departments is that attendance for a critical incident stress debriefing (CISD) is voluntary. Unfortunately, some officers who need the process the most do not attend. It helps to have command-level attendance, but only those officers directly involved in the critical incident are asked to attend. Certainly, where commanding officers are directly involved, as in the Lauri Dann shooting, the police chief, deputy chief, and lieutenants attended.

Supervisors should pay attention to those officers who do not attend a debriefing, but were directly involved in the critical incident or close associates of an officer who died. Changes in behavior resulting from a stress reaction may not become evident until weeks or months later. Many police officers—especially those with less than seven years experience—keep their grief or anger submerged believing that is

[18]Jeffrey T. Mitchell, "Development and Functions of a Critical Incident Stress Debriefing Team," *JEMS* (December 1988), pp. 43-46.

[19]Debriefings were conducted by the Northern Illinois Critical Incident Stress Debriefing Teams, as part of a network of several hundred volunteer teams throughout the United States. Contact American Critical Incident Stress Foundation at P.O. Box 204, Ellicott City, MD. 21041, (410) 750-0856.

what is expected. This "John Wayne Syndrome" can result in changes in their work performance, alcohol intake, and amount of complaining.

It is the younger officer who has the more intense stress reaction. Younger officers lack the exposure and experience with critical incident stress. They find that denial, or the belief that "I can handle this event," is woefully inadequate to prepare them for their reactions. A seasoned officer who has been through a similar event should be assigned to talk to the "rookie" about what to expect.

Incorporating stress debriefings as routine procedure in police departments following certain kinds of trauma will help to prolong the lives and careers of its police officers. This approach accelerates emotional recovery, is cost effective in reducing turnover, and demonstrates to front line personnel that the administration cares for their welfare.

JOB- RELATED HAZARDS

Police work entails many hazards unlike those found in other occupations. This does not necessarily means that police work is more hazardous, but rather that the hazards associated with police work are, in many instances, unique. It is important for the police administrator and patrol supervisor to understand the nature and scope of these hazards so they can deal with them as effectively as possible.

Organizational Frustration

One constant source of irritation to many police officers is the traditional style of management still practiced in many police organizations today. Traditional management practices, which are becoming less acceptable today, rely heavily on authority, discipline, and obedience to rules to gain compliance by members of the organization.

In many police departments, there is considerable pressure put upon the individual officer "to produce." Production may be measured in terms of miles patrolled, arrests made, citations issued, doors and windows checked, traffic stops made, persons interviewed, property recovered, and so forth. Patrol officers often are required to complete daily activity logs which contain numerous categories of activities for which the officer may be given credit and account for time spent. Too often, the emphasis is on the quantitative aspects of police work rather than the qualitative dimensions of it. There is more concern about how many arrests were made than there is whether the arrests were "good" (that is, necessary and legitimate). As long as an officer's numbers are up for the month or the quarter, he or she probably will not have to worry about being called in by the sergeant to discuss job performance.

Because of the kind of work they do, police officers work in organizational structures that place a great deal of emphasis on order, discipline, and rigid adherence to rules. One of the characteristics of a well-managed police department is a set of comprehensive, well-written rules, regulations, and general orders which establish the policies and procedures of the department. These directives provide guidelines within which the police officers are expected to conduct themselves. Deviations from these directives can result in disciplinary action. Thus, within a democratic society that prides itself on individual freedom, police officers work in an environment in which freedom of choice is quite limited. This presents an interesting paradox that is unique to police work.

Since the late 1960s, there has been a continued emphasis on raising the educational level of police officers. Many cities offer tuition-reimbursement programs and cash incentives for two-year, four-year, and graduate degrees. Some police departments have included degree requirements for original appointment or promotion in the department. As a result, there are probably more college-educated police officers today than at any other time in history. This emphasis on college education has had an interesting impact on the police organization and has made management of the police organization more difficult.

Educated persons, including educated police officers, are taught to question the world around them, to take nothing for granted, to ask why things are the way the are, and to reject custom and tradition as a panacea to all problems. Most police organizations, however, do not operate in a democratic mode and are not set up to tolerate, let alone encourage, critical thinking and questioning of the status quo. Few police managers encourage creative thinking by their subordinates, and still fewer allow traditional procedures to be challenged by the lowest-ranking members of the department.

Police officers with college degrees, therefore, often find themselves in an organization in which they are expected to conform to established principles and procedures rather than to think for themselves. They often become frustrated at this lack of individual freedom and sometimes rebel against the organization and its management. They may eventually become targets of the disciplinary system or seek other careers in which the ability to think critically and to question the status quo will be more tolerated.

It is interesting to note that there was a time, not so long ago, that a great many of the men (there were few women in the police service then) who entered the police service had prior military service. Indeed, in the 1960s and before, prior military service was almost expected of those entering police work, even though it was not a job requirement. After having served two or more years in a military organization, most of these persons were accustomed to taking orders and working under a very rigid code of conduct. Thus, the regimentation found within the police organization was nothing new to them. Adapting to this kind of work seemed to be easy.

In more recent years, however, the percentage of those entering police work with prior military service has declined considerably. These men and women may be high school graduates, but more than likely have at least some college education. For many, the police department is the first "real job" they have had other than temporary jobs in fast-food restaurants, gas stations, or as sales clerks. As a result, their orientation to work is much different; they do not have the same concept of discipline that those who have served in the military have. They react differently to the bureaucratic regimentation found in many police agencies and often find themselves out of step with the norms of the organization.

If authoritarian management techniques were once acceptable in police agencies, they are much less so today. One major reason for this is the significant advances made by collective bargaining in the public sector during the last twenty years or so. There is a great deal of truth to the adage that poor management creates strong unions. While some states still do not allow collective bargaining for public-safety employees, the vast majority do.

Collective bargaining has become a reality in the management of police agencies, and has created additional stressors within the police organization. Labor-management relations in many police departments often are reduced to a power struggle between the leadership of the department and the leadership of the union. Officers may be torn between their loyalty for one or the other. Blue-flu epidemics, work slowdowns, and even more visible demonstrations of union solidarity have escalated tensions within police departments. Unhappiness with management polices have led to votes of "no confidence" against the chief of police. Many of these protests have resulted in the replacement of the chief of police.

More enlightened police chiefs work with union leaders to achieve organizational as well as individual goals. Participatory management techniques have been developed to encourage members of the police organization to contribute their ideas and efforts to the achievement of organizational objectives. Police chiefs have learned that the foundation of a healthy organization is a motivated and satisfied employee, the things that once motivated workers are no longer as important, and that job enrichment and self-fulfillment are important determinants of individual job satisfaction.

> Job tenure and a livable wage are no longer enough to keep the young policeman happy. Like others in our society who are upwardly mobile, when survival and security needs have been met, he wants to be in on the organizational action. He wants to know the reasons for actions affecting him, and to feel that he has some say in the decision-making process.[20]

Other sources of organizational frustration in police work include the methods and procedures surrounding promotion, assignment, and discipline. Many objective and job-related advances have been made in the design and development of promotional examinations in recent years. Nevertheless, the manner in which they are implemented often gives rise to charges of favoritism and bias on the part of the administration. In many cases, it appears that those loyal to the chief are promoted over those who are equally, if not more, qualified. On the other hand, it is important that the chief be able to trust the people promoted to positions of responsibility. There is, after all, no vested right to promotion in a police department. While it is

[20]Martin Reiser, "Some Organizational Stresses on Policemen," *Journal of Police Science and Administration*, 2 (June 1974), p. 157.

important that only qualified persons be promoted, the chief of police should have some voice in the selection of persons who will help carry out the mission of the department.

Traditional methods of police management—which concentrate decision-making authority in the hands of a few, do not provide for discussion of important issues among employees, and fail to recognize or consider the consequences of organizational policies upon individual members—are no longer suitable. While the ultimate responsibility for managing the police organization cannot be delegated, managers who share their authority and who seek input from all members of the organization before making decisions affecting them will usually be more successful than those who rely upon traditional or authoritarian management methods. There is a very real need to "humanize" the management of police organizations, to decentralize the decision-making process, and to recognize the importance of the individual within the organization.

Active involvement of police officers in the mainstream of organizational management can yield significant dividends. Police officers who are encouraged to think for themselves, to be creative, and to participate actively in the decision-making process will develop a great commitment to the organization and to its leadership. The greater their involvement in the decision-making process, the more they will support the decisions that are rendered by it. Even though a police organization is not, in reality, a democratic organization, it can support democratic principles in management.

The new breed of police officer, who is generally better educated than his or her predecessors, and who has been raised in a generation of critical thinkers and questioners, can, if allowed, contribute much to expanding the capacities of the police organization. Unfortunately, many police chief executives, particularly those who are unsure of their own abilities, are often unwilling or unable to delegate authority to subordinates. Failure to delegate is one of the surest signs of managerial incompetence and can have devastating effects upon the morale of the members of the police organization.

Police administrators frequently fail to recognize or appreciate the tremendous contribution that rank-and-file officers can make to the organization. Few police organizations and management systems are designed to allow, let alone encourage, participation by patrol officers in decisions that affect the way the organization operates or even how it performs its basic responsibilities. This is indeed ironic, since it is the patrol officer on the beat who is in the best position to know what works and what doesn't in terms of fulfilling the basic mission of the police. Such management strategies are myopic at best, since they fail to make the best use of the resources available to the organization. Moreover, they produce a climate which stifles individual creativity and undermines employee morale.

> Officers quickly learn, under these conditions, that the rewards go to those who conform to expectations—that nonthinking compliance is valued. [21]

[21]Herman Goldstein, *Problem-Oriented Policing* (Philadelphia, Pa.: Temple University Press, 1990), p. 27.

Role Conflict

Another source of irritation in police work is the frustration officers suffer from the conflict that is inherent in the police role itself. This is due in part to the discrepancy regarding what the public expects of the police, how the police themselves define their role, and the day-to-day realities of police work. The police have inherited a myriad of tasks. Society expects much of the police and shows little gratitude for what the police do. They expect the police to solve problems which develop from causes and conditions over which the police have no control. Society expects the police to "hold the line" on crime, but is unwilling to provide the police with the resources necessary to work toward that goal.

Police officers often see themselves as crime fighters, when in fact they spend very little time actually fighting crime or apprehending criminals. While there is no question that this is an important aspect of the job, it is clearly not the most important part of what they do. Because they see themselves as crime fighters, police officers sometimes develop an adversarial attitude toward the people with whom they come into contact. They regard some people as potentially dangerous or possessing criminal tendencies on the basis of stereotypes and generalizations derived from past experience. They tend to think in terms of good or evil, right or wrong, legal or illegal.

While this way of thinking may simplify the way the police see the world around them, it distorts reality and contributes to role conflict. Improved police training programs can help to ease the impact of role conflict on the police officer. These training programs should stress the conflict inherent in the police function and should illustrate the service orientation of the police role over the crime fighting aspect of it. In addition, police policy statements need to emphasize the diverse and complex nature of the police role in society. Police selection programs should be designed to screen out those individuals who are likely to develop stereotypical attitudes toward people on the basis of limited experience. The ability to see the entire range of police duties is equally important and needs to be understood by all who choose to enter police work.

Moral Conflict

Police officers, by virtue of their position of trust, authority, and responsibility, are expected to maintain high standards of character and integrity. At the same time, they are exposed to many situations in which they may be tempted to violate ethical standards in order to further their own self-interests. A few police officers do succumb to these temptations and the results are often disastrous for the officer and for the entire police organization.

Unfortunately, there are no practical means to remove temptations from the working environment of the police patrol officer. Those who have power over others will always be subject to temptation to use that power to serve their own purposes. Such is human nature. Moreover, it is impossible to predict with any degree of certainty whether and under what conditions an officer who is otherwise honest might be tempted to accept a bribe, protect a drug dealer, perjure himself, or take property belonging to others.

Circumstances as well as the officer's state of mind at the moment can make the difference between the officer upholding the oath of office or surrendering to moral temptation. It should be possible, however, through better training programs, improved supervision, and better counseling methods to assist officers in avoiding such temptations. In addition, much more study and scientific analysis needs to be devoted to identifying the conditions under which police corruption is likely to occur, the causes of it, and the means by which it can be avoided.

Police administrators must acknowledge that such temptations exist. They must also send a clear message to their subordinates and to the general public, through word and example, that absolute integrity is expected of each member of the police organization. Those who seek to undermine that integrity, by creating or tolerating corruptive influences, should be prosecuted. Those few officers who succumb to temptation should be severely punished, dismissed from the force and in extreme cases, criminally prosecuted.

Some police agencies have created police chaplain programs, in which ministers of different faiths become actively involved in the police department and are available to discuss personal problems with officers. These discussions are strictly confidential and may result in the officer being referred to a counselor, psychologist, or some other individual or organization for further assistance. A number of communities have created employee assistance programs that are available free to all city employees. These programs offer a wide variety of services to employees who suffer job-related problems including drug abuse, alcoholism, job stress, and emotional duress. These programs are relatively inexpensive to operate, particularly in view of the considerable investment the city has in its employees.

While these programs are not specifically designed to deal with the issue of moral temptation and corruptive influence in police work, they could certainly be modified to serve this purpose. It is likely that those same persons who allow themselves to fall prey to corruptive influences may benefit from the same kind of counseling methods or remedial programs that are helpful to others suffering from job-related disabilities.

Corruptive Influences

When discussing the many hazards associated with police work and the patrol function specifically, the critical importance of corruptive influences on the police cannot be overlooked. Corruption in police work can be even more devastating than other types of hazards since its effects are likely to be felt throughout the entire organization, to say nothing of the effect on the community. Unfortunately, many police officials often attempt to deny the existence of police corruption and, by doing so, only compound the problem. This type of denial instills a sense of permissiveness that permits the problem to become even more widespread. It also sends a message to the outside world that corruption is tolerated in the police profession, which further erodes public confidence in the police.

The police are subject to a number of corruptive influences that permeate their working environment.[22] Because of the power they possess, many persons are anxious to grant the police favors and rewards in return for small tokens of gratitude which they hope the police will be able to bestow upon them. Many of these situations are innocent enough, but have adverse consequences. A local merchant, for example, may distribute bottles of liquor to patrol officers in return for protection of his store. A more common example is the restaurant owner who offers police officers half-price meals and free coffee as an inducement to come in frequently. In either case, the merchant is purchasing the favors and attention of the police. It also places the police officer who accepts such favors in a compromising position in the event enforcement action may later be needed against the merchant.

Most police officers do not see the acceptance of such gifts and gratuities as a serious problem. They do not foresee that such practices may be used against them at some point in the future. In addition, some police officers justify the acceptance of favors, since the merchant is, in fact, getting something for his money—more attention—from the police. As a result, they seem to believe that the equation balances out. The truth of the matter, however, is that such practices are demeaning to the police and tend to place them in the position of compromising themselves and denying their ethical principles.

Such actions tend to characterize police officers as "moochers" and people tend to view this practice as proof that police officers are always looking for a free meal or handout. As a result, it is not difficult to understand why many people honestly believe that police officers in general are willing to look the other way as long as there is something in it for them. Not only are such actions by police officers demeaning, they tend to discredit the police agency the officers represent. Acceptance of gifts is a clear violation of the Law Enforcement Code of Ethics, to which nearly all police agencies—and therefore their members—subscribe. This code of ethics, included in the Appendix, sets forth very clear guidelines concerning the conduct of law enforcement personnel.

Police corruption may exist in many forms and attempts to formulate a generally acceptable definition of police corruption have not proven successful. Goldstein defines police corruption as "the misuse of authority by a police officer in a manner designed to produce personal gain for the officer or for others."[23] Goldstein's definition is not entirely acceptable in that it fails to include situations in which the officer is merely the passive recipient of gifts or favors and where no misuse of official authority is clearly evident. Perhaps a better definition of police corruption might be

[22]V. A. Leonard and Harry W. Moe, *Police Organization and Management,* 7th ed. (Mineola, New York: The Foundation Press, Inc., 1987), pp. 135-138.

[23]Herman Goldstein, *Policing a Free Society* (Cambridge, Mass.: Ballinger Publishing Co., 1977), p. 188.

any act (or failure to act) by a police officer designed to place the interests of the officer, or other persons, over the law or the common good.

Even if a commonly acceptable definition of police corruption cannot be reached, it is likely that most people can agree on what does or does not constitute corrupt behavior by a police officer. By the standards of most people, the acceptance of a free cup of coffee or a half-price meal by a police officer is not corrupt behavior, even though it may fall within the rather broad definition of corruption offered by some. Can the same be said of free movie passes, discounts at local clothing or department stores, free passes to sporting events, and other similar favors? At what point do these small gifts and gratuities begin to add up to subtle attempts to gain the favor of the police officer? At what point do these acts or policies become inducements to influence the official acts or decisions of the officer? At what point does the officer begin to place more value on the friendships and generosity of persons who are in a position to reward him for his help than on the course of action dictated by the rule of law and by professional ethics?

Police administrators often question whether some relatively innocent behavior, such as accepting discounts on merchandise or free passes to movies, really constitutes corruption. Clearly, in nearly every case, there is no intent on the part of the officer to grant special favors to those who provide these kinds of gifts and gratuities. And, in most cases, those who offer these gifts and gratuities to the police have no ill intent or purpose in mind. In most cases, they see these attempts to cultivate good will with the police as being good for business and a legitimate business expense.

In the vast majority of cases, the offering of free gifts, discounts, and other gratuities to the police will not result in any harmful behavior. However, the potential is clearly evident. The image created by such actions is a negative one and does not enhance the status of the police in the eyes of the public. The term moocher most often comes to mind when a citizen sees a police officer receive a benefit to which no one else is entitled.

Professional police administrators reject the idea that there is any such thing as "minor" corruption, and they make every effort to discourage and eliminate even the most innocent kind of behavior that might be labeled or misinterpreted as corrupt. These police administrators will not turn their backs on this problem, but will confront it openly and will challenge those who insist that nothing can be done about it. They will not hesitate to issue policy statements prohibiting the acceptance of gifts and gratuities by police officers and will back them up by appropriate disciplinary action when violations of these policies are discovered.

No police executive will be so naive as to believe that even the strongest statements prohibiting improper conduct by their officers will ever eliminate such behavior altogether. However, they can, through example and disciplinary action, ensure that such conduct is the exception rather than the rule. Not only will such actions by the police chief executive send a clear message to the members of the department, but it will serve notice on the general public that any attempts to influence police conduct or to curry favor with police officers by unethical or unlawful means will not be tolerated.

If the police are to deal effectively with the problem of police corruption, they must recognize and understand what it is and how it happens. While there are a number of factors that contribute to police corruption, the following are the most important:

1. **The Nature of Police Work**. The authority of the police and the environment in which they work make them likely targets of corruptive influences. The police, for example, recognize that they cannot enforce all the laws equally and must exercise considerable discretion in their law enforcement role, thus making it easier for them to rationalize particular courses of action that may ultimately work to their own benefit.
2. **The Police Officer's Clientele.** The police officer comes into contact with many unsavory people who will do anything they can to protect themselves from the consequences of their own misdeeds. They do not share the same values of honor and integrity with the rest of society and they place their own interests above all others. Some police officers develop personal relationships with these people and, if they are not careful, may end up sharing their value systems. When this happens, the police officer's sense of integrity and character can easily be compromised.
3. **Public Attitudes.** Police officers often believe, sometimes for good reason, that the general public is apathetic about them and their job. Public apathy and tolerance of unethical conduct play a key role in creating a climate conducive to police corruption. Indeed, citizens often take the lead in initiating situations leading to corrupt behavior by a police officer. The problem of police corruption cannot be solved unless and until citizens take a strong and united stand against it and demonstrate to their elected and appointed officials that they will not condone such conduct.
4. **Leadership.** The professional climate in a police organization is determined by the examples set by the police executive and the command staff. Strong leadership results in a strong and healthy organizational climate. There is probably no more important influence on performance in a police organization than the character and integrity of its leadership. It is important that police executives demonstrate, through their own acts and words, their intolerance of police corruption.

There are no simple solutions to the problem of police corruption. Corruption may be thought of as a failure of individual integrity and it is a natural consequence of human weakness. In a broader sense, though, it represents a failure of the police organization and its management. It is difficult to understand how a single officer could be corrupted in an organizational climate that strongly denounces corrupt behavior. It is difficult to understand how a police officer could succumb to corrupt influences in a police organization with strong administrative controls, performance review procedures, and disciplinary processes. It is difficult to understand how a single officer could be corrupted in an organization where peer pressures and supervisory controls encourage and reward professional conduct and denounce self-serving actions by individuals.

Although it may be possible for one or two officers to commit corrupt acts without the knowledge of their peers, it is very unlikely that such incidents could go undetected for a long time. Indeed, such systematic or regular pattern of corruption by even a single officer could not occur without the knowledge ar collusion of the officer's peers.[24]

There needs to be a strong spirit of intolerance toward corruption in police organization. This spirit of intolerance must begin at the top of the organization and extend to the rank-and-file members. If officers understand that their peers will actively seek out and expose corruption in their midst, they will be less susceptible to corruptive influences. This means that the code of silence that permeates the police organization and which police officers are expected to uphold, may sometimes be ignored so that guilty officers may be discovered and ejected from the police organization. An officer's oath of office is, after all, more important than his or her loyalty to an officer who has violated the integrity of the badge.

PHYSICAL HAZARDS

There are a great many physical hazards associated with police work and they should not be ignored. Due to the nature of police work, these hazards cannot be eliminated altogether. However, it may be possible to find ways to minimize their frequency and severity.

Physical Conditioning

One of the contradictions of police work is that, while police officers must always stay in top physical condition, most police work requires limited physical exertion. Many police officers spend a great deal of their on-duty time sitting at a desk or behind the wheel of a patrol car. They must nonetheless be prepared, on the spur of the moment, to outrun a fleeing robber or chase a young purse snatcher through backyards and over fences.

While police officers must undergo a thorough medical and physical examination before being appointed to the force, there is no requirement (in most departments) that they remain in good physical condition. In addition, much of police work is sedentary in nature. Few police agencies require their officers to undergo annual or periodic physical examinations, nor do they provide any incentives for police officers to stay in good physical condition. As a result, they are often not physically prepared

[24]William Ker Muir, Jr., "Skidrow at Night," in Carl Klockars, *Thinking About Police: Contemporary Readings* (New York: McGraw-Hill Book Co., 1983), pp. 241-242.

to handle emergency situations requiring strenuous effort. Their lack of good physical conditioning often results in needless physical injuries and permanent disabilities.

The importance of good physical conditioning among police officers should not be underestimated. A police officer represents a substantial investment on the part of the local municipality. Police officers who must be prematurely retired or who take long periods of sick leave to recover from on-duty injuries sap the strength of the agency. It is important that public officials and police administrators recognize the importance of good physical conditioning and take the lead in implementing programs designed to upgrade the physical condition of police officers.

Coronary heart disease is the leading cause of death in the United States. It is the leading killer of men in their forties and women in their sixties. Health-maintenance programs designed especially for police officers can result in reduced risk of developing premature heart disease, stroke, cancer, and diabetes.

In the New York City Police Department, a team of health and medical professionals assigned to the Hypertension Screening Unit of the Health Services Division is responsible for conducting cardiovascular risk assessment of police officers. The unit follows the guidelines of the National Cholesterol and High Blood Pressure Education Program begun in the 1980s. The unit has also developed a nutrition awareness program for members of the department.

> Just as cars and equipment need preventive maintenance, people need programs to prevent disease. Health promotion programs also result in improvements in the health and appearance of the police officer and cost savings for the department. Health benefits include reduced blood pressure and cholesterol levels, weight control and possible prevention of heart disease, stroke, cancer and diabetes. Cost savings include a reduction in days lost due to illness and a decrease in injuries and their attendant cost. Health promotion programs may also result in improvements in staff morale, productivity, and job satisfaction.[25]

Police departments need to give more attention to the physical as well as the mental health and condition of their employees, both sworn and civilian. Seriously overweight or out-of-shape people pose high health risks and are candidates for on-the-job injury, workers compensation claims, and disability retirement. Nor can people who are not in good physical shape work at their optimum level of efficiency. As Chapman has noted, "Officers who are not in good physical condition endanger not only their own lives but those of other police. Out-of-shape people lose some of their physical and mental keenness in crisis situations, too."[26]

It is ironic that most police agencies require officer candidates to be in top physical shape, but very few departments have any requirements that specify physical requirements for tenured employees. Even if employees do not take pride in

[25]Rebecca M. Arlies, "Healthy Hearts for New York City Cops," *The Police Chief* (July 1991), p. 22.

[26]Samuel G.Chapman, *Cops, Killers and Staying Alive* (Springfield, Ill.: Charles C.Thomas, 1986), p. 74.

themselves and do not care enough to look after their own physical conditioning, police agencies have an obligation to look after the physical well-being of their employees. Physical conditioning programs, diet and nutrition programs, stress-reduction programs and other health maintenance strategies can and should be used by municipalities to ensure the health and well-being of their employees.

Many modern police agencies have set aside rooms in their facilities for physical conditioning equipment, encouraging their officers to work out during their off-duty time. Other departments have enrolled their officers in local health clubs, with the membership fees paid by the municipality for those officers who wish to participate. These programs are solid investments in the physical condition of the officers as well as in the vitality of the police organization itself.

More police agencies need to develop formal physical-conditioning programs for officers. These programs should be based on the same concept as the annual or semiannual performance evaluation of the officer. Physical conditioning should be considered a job requirement for all personnel and officers should be given annual physical examinations to ensure their fitness for duty. Those who fail the annual physical should be placed on a remedial program designed to bring them up to minimum standards within a prescribed period of time. Failure on the part of the officer to achieve minimum physical standards should be viewed in the same way as failure to maintain minimum job proficiency. In extreme cases, disciplinary action, and perhaps even dismissal, might be required.

As a nation, we are becoming much more conscious of our health and physical fitness than at any previous time in our history. Regulations against smoking in airplanes and in other public places have increased dramatically in recent years. Many major corporations have banned smoking altogether in their headquarters and other offices, and a large number of municipal governments have followed this example. Some cities have taken the lead by making abstinence from smoking a job requirement. Other cities have prohibited police officers and firefighters from smoking on the job. Some police departments have designated certain patrol cars as no-smoking cars, so that non-smoking officers are not exposed to the odors left by cigarette smoking.

A majority of police officers working today have either given up smoking or never tried it. The percentage of smokers in police departments will decline even more dramatically in the future. In the not-too-distant future, the smoking of cigarettes, cigars, and pipes will be the indulgence of a few nonconforming individuals who cannot resist the urge to engage in conduct which others find unhealthy or distasteful.

The tendency toward no smoking is only one indication of the growing popularity of physical fitness and health maintenance among American workers. Police officers are among the large number of workers who recognize that a healthy body and good physical condition are directly related to good on-the-job performance. A study conducted in the Dallas Police Department several years ago confirmed that physically fit officers generally perform better than officers who are in poorer physical condition.

In the Dallas study, volunteers were divided into control and experimental groups. The group of experimental officers was enrolled in a rigorous physical-fitness program for six weeks. Officers in the control group maintained their normal level of physical activity. At the end of the experiment, officers in the experimental group were found to surpass members of the control group in terms of attitude, enthusiasm, self-control, and human-relations skills. In addition, members of the experimental group demonstrated an overall decrease in sick leave during the six-month period compared with an average increase in sick leave for the control group.[27]

We are becoming a more health-conscious society, and more attention is being given to increasing the physical fitness of employees in nearly all occupations. This is especially important in police work, where the nature of the basic police function will always require a patrol officer to perform strenuous physical activity at a moment's notice. Continued emphasis needs to be given to physical fitness for police officers; municipalities must continue to provide opportunity and encouragement for officers to stay in shape. The beneficiaries of such efforts will be the officers themselves, the police organization, and the public.

Fatigue

Job fatigue, like stress, is not unique to police work, but its effect on police performance is profound. Police officers, often lulled into inaction by the routine nature of the job, must be prepared to respond swiftly to emergencies in full command of their mental and physical faculties. Diminished readiness or physical impairment on the part of the officer may result in serious injury or death. Moreover, an officer suffering from job fatigue cannot be expected to devote full attention to assigned duties and may prove to be a serious liability in an emergency.

It is generally assumed that overwork can cause job fatigue, and this is no doubt true in many instances. In police work, however, one important factor that has been found to contribute to job fatigue is inactivity. While this may seem strange to those who are unfamiliar with police work, it is nonetheless true. Police officers assigned to quiet shifts or districts in which nothing much ever happens are just as likely, if not more likely, to experience job fatigue as are officers assigned to high-activity shifts or districts. This was one of the findings of a study of job stress among police officers in Miami, Florida.[28]

In the Miami study, Cruse and Rubin compared incidents of job fatigue with several variables and discovered that the most significant correlation was with the average number of activities handled by the officers. They found that an increase in job fatigue was positively associated with those shifts, such as the early morning

[27]See Donald A. Byrd, "Impact of Physical Fitness on Police Performance," The Police Chief (December 1979), pp. 30-32.

[28]Cruse and Rubin, "Police Behavior."

watch, in which relatively few incidents occur. The researchers thus concluded that job fatigue among patrol officers is largely a product of inactivity.[29]

While some police authorities may reject the notion that inactivity among patrol officers may create job fatigue, those who have accompanied patrol officers during long, quiet shifts characterized by inactivity and routine incidents should recognize the validity of the conclusions of the Miami study. It is a fact that patrol work is, in many instances, quite dull and boring, and patrol officers who come to work each day expecting hours of excitement and adventure will soon become disillusioned and frustrated. Hour upon hour of uneventful, routine patrol can be extremely tiring and can sap not only the strength but the alertness and enthusiasm of the officer. For police officers who are action oriented, this kind of inactivity can be demoralizing and detrimental to their job performance and attitude.

Obviously, it is not possible to eliminate the element of boredom that is so naturally associated with police work. The natural cycle of everyday living dictates that there will be periods of ebb and flow in human activity. It may be possible, however, to find ways to make the officer's tour of duty more stimulating and challenging during periods of inactivity. Officers could be given specific tasks related to crime prevention, for example, to perform during times when there is not much activity. The nature of assignments would, of course, depend upon the shift, the geographic area, and the type of criminal activity usually found at that time and location. Perhaps officers could be given greater discretion in the manner in which they handle certain calls: how much time they devote to them, the priority they attach to them, and the resources they utilize in handling them. This technique would allow them to make better use of their uncommitted patrol time by using their own creativity and initiative, thereby increasing their level of satisfaction.

It may also be advisable to periodically rotate officers from one shift to another, to change their beat assignments from time to time, and even to change the hours of their shift in order to reduce periods of monotony and boredom. For example, rather than beginning a shift at 4:00PM, start two hours later or earlier for a period of thirty days and then return to the original schedule for another sixty days. In summary, any change that can be made to reduce the monotony and boredom so often associated with patrol duty should be tried to see if it achieves the desired effect.

Physical Strain from Shift Rotation

Because police work is a 24-hour-a-day activity, it is customary to assign patrol officers to different shifts in order to provide around-the-clock coverage. There are a variety of shift arrangements used in law enforcement, some of which will be discussed later in the text. However, the eight-hour shift remains the most common arrangement in police agencies. Some police departments do rotate patrol personnel from one shift to another periodically. This is done for two reasons: the desire to

[29]*Ibid.*, p. 197.

expose patrol officers to a variety of shift assignments and the need to assign officers to shifts on an equitable basis.

The subject of shift rotation is a controversial one. Some authorities have suggested that shift assignments should never be changed so that officers may become thoroughly familiar with the unique characteristics of the shift to which they are assigned, and thus more adept in dealing with the conditions and behaviors prevalent on that shift. Others feel that shift assignments should be made on the basis of seniority, as a reward to officers with more time on the job. Still others believe that periodic shift rotation is good for the organization and for the individual. The proponents of this position feel that a periodic change in assignment keeps the officer fresh and precludes the possibility of the officer becoming "stale."

While some shift rotation is probably more desirable than none, caution should be taken to ensure that officers are not rotated more frequently than is necessary. A shift-rotation schedule of three to six months is probably most desirable. More frequent change in shift assignments may have adverse affects on the individual officer.

Monthly or periodic shift rotation affects the body's natural rhythm and takes a toll on an officer's ability to perform at optimum capacity.[30] It is not uncommon for police departments to rotate shifts every twenty-eight days, and some rotate as often as every week. The more frequent the rotation, the more significant are the consequences to the individual. It usually takes several days to adjust the eating, working, and sleeping cycle to a new schedule. In addition, police departments may very well discover that rotating shift assignments makes it difficult to proportionately distribute the patrol force according to workload. There are other distinct advantages of fixed shifts:[31]

1. efficient use of available personnel
2. fewer adverse physiological problems
3. shift preference for senior officers
4. most officers seem to prefer fixed shifts
5. court scheduling is made easier
6. more effective supervision and evaluation of personnel

The scheduling practices of police agencies vary widely and there is probably no such thing as a "typical" or "standard" method of scheduling police personnel.

[30]Ronald W. Geddes and William DeJong, "Coping With Police Stress," in Robert G. Dunham and Geoffrey P. Alpert, *Critical Issues in Policing: Contemporary Readings* (Prospect Heights, Ill.: Waveland Press, 1989), pp. 498-507. See also William H. Kroes, *Society's Victim—The Policeman: An Analysis of Job Stress in Policing* (Springfield, Ill.: Charles C. Thomas, 1976), pp. 29-33.

[31]William W. Stenzel and R. Michael Buren, *Police Work Scheduling: Management Issues and Practices* (Washington, D.C.: U.S. Department of Justice, 1983), p. 12.

However, it is generally acknowledged that effective scheduling in a police agency can achieve a number of purposes, including[32]

1. reduced sick leave
2. more efficient use of personnel and equipment
3. reduced overtime
4. improved service levels
5. improved morale

Dealing with Communicable Diseases

The escalating proliferation of communicable diseases during the last ten years has forced police agencies to adopt policies designed to educate officers regarding preventive measures. The AIDS (Acquired Immune Deficiency Syndrome) epidemic, in particular, has increased awareness among police administrators of the need for preventive measures to protect law enforcement personnel.

Of all the infectious diseases with which law enforcement officers come into contact, none poses a greater threat, both physically and psychologically, than the Human Immunodeficiency Virus (HIV)—the AIDS virus. Unfortunately, what we don't know about this killer disease causes more harm than what we do know.[33]

There are basically only three known ways in which to contract the AIDS virus: (1) through homosexual and heterosexual activity; (2) through contact between infected mother and infant; and (3) through blood-to-blood contact with an infected person. Of these, the third type of exposure is the one that poses the greatest on-the-job risk to law enforcement personnel.

Blood-to-blood contamination can occur in a variety of ways. Health-care professionals as well as police officers, firefighters, emergency medical technicians, and others who come into contact with injured or infected persons must take reasonable precautions. Basic preventive measures and hygienic procedures must be followed in any situations in which there is a risk of exposure to contamination. Gloves, masks, and protective eyewear should be worn anytime there is the likelihood of exposure to contaminated blood or other bodily fluids.[34]

Education and training help police officers overcome fear of infectious diseases. They need to know the ways in which disease can be transmitted as well as

[32]Stenzel and Buren, *Police Work Scheduling*, p.xvii.

[33]For a good summary of the facts about AIDS, see Theodore M. Hammett, "AIDS and the Law Enforcement Officer," National Institute of Justice, *NIJ Reports* (November-December 1987), pp. 2-7.

[34]Theodore M. Hammett and Walter Bond, "Risk of Infection with the AIDS Virus through Exposure to Blood," National Institute of Justice, *AIDS Bulletin* (October 1987), p. 3. See also Daniel B. Kennedy, Robert J. Homant, and George L. Emery, "AIDS and the Crime Scene Investigator," *The Police Chief* (December 1989), pp. 19, 22.

Issue/concern	Educational and action messages
Human bites	Person who bites usually receives the victim's blood; viral transmission through saliva is highly unlikely. If bitten by anyone, milk wound to make it bleed, wash the area thoroughly, and seek medical attention.
Spitting	Viral transmission through saliva is highly unlikely.
Urine/feces	Virus isolated in only very low concentrations in urine: not at all in feces: no cases of AIDS or AIDS virus infection associated with either urine or feces.
Cuts/puncture wounds	Use caution in handling sharp objects and searching areas hidden from view: needle-stick studies show risk of infection is very low.
CPR/first aid	To eliminate the already minimal risk associated with CPR, use masks/airways: avoid blood-to-blood contact by keeping open wounds covered and wearing gloves when in contact with bleeding wounds.
Body removal	Observe crime scene rule: Do not touch anything. Those who must come into contact with blood or other body fluids should wear gloves.
Casual contact	No cases of AIDS or AIDS virus infection attributed to casual contact.
Any contact with blood or body fluids	Wear gloves if contact with blood or body fluids is considered likely. If contact occurs, wash thoroughly with soap and water: clean up spills with nine parts water to one part household bleach.
Contact with dried blood	No cases of infection have been traced to exposure to dried blood. The drying process itself appears to inactivate the virus. Despite the low risk, however, caution dictates wearing gloves, a mask, and protective shoe coverings if exposure to dried blood particles is likely (that is, crime scene investigation).

Source: Theodore M. Hammet, "AIDS and the Law Enforcement Officer," National Institute of Justice, *NIJ Reports* (November-December 1987), pp. 2-7.

Figure 4.2 Responses to AIDS-Related Law Enforcement Concerns

basic preventive measures that can be taken to protect themselves. AIDS-awareness training should be part of the basic training programs as well as the annual in-service training for all officers.[35]

In addition, police agencies need to implement policies and procedures that provide reasonable and common-sense guidelines for officers to follow in situations where contact with AIDS or any other communicable disease is likely. Figure 4.2 provides a general outline of the basic procedures that should be covered in this regard.

The Mesa, Arizona Police Department adopted its infection control policy in 1987.[36] That policy is broken down into three distinct sections. The section on Prevention offers general recommendations for the use of gloves, masks, and other preventive devices based on universal precautions. Another section addresses disinfection and decontamination of equipment. Officers are encouraged to use extreme caution and available precautionary equipment in handling any potentially infectious materials. The third section deals with officer reporting requirements and establishes an in-house reporting system to monitor officer exposures for possible insurance claims. Officers are required to categorize each exposure on a range from direct blood-to-blood contact to merely being in the presence of someone with an infectious disease.

SUMMARY

Many of the hazards associated with police work are integral components of the nature of police work and cannot be eliminated. Others can be either eliminated altogether or their adverse impact minimized through research and a greater understanding of how police work affects human behavior. Police officers need to be physically, mentally, and emotionally strong in order to perform at their best. The patrol administrator needs to recognize which hazards inherent in police work can be properly managed to reduce the adverse impact on the patrol officer.

REVIEW QUESTIONS

1. Discuss and compare the relative dangers of physical hazards and emotional stress on the well-being of the patrol officer.
2. Describe some practical ways to minimize the threat of physical hazards to the police patrol officer.

[35]Martin Forst, Melinda Moore, and Graham Crowe, "AIDS Education for California Law Enforcement," *The Police Chief* (December 1989), pp. 25-28.

[36]Correspondence received from Guy Meeks, Chief of Police, Mesa Police Department, dated July 17, 1990.

3. What are some of the sources of job stress for the police officer and how might they be minimized or eliminated?
4. What can supervisors and middle managers do to minimize the impact of job stress on patrol officers?
5. What are critical incidents and how do they effect job stress among police officers?
6. How do debriefings help to minimize the effects of critical incident stress?
7. Explain how characteristics of the police organization may contribute to frustration with the job among police officers. To what extent are these characteristics unique to the police organization?
8. How do traditional methods of police management contribute to organizational frustration and what can be done to change this?
9. What are some of the kinds of role conflict that affect police officers the most? To what extent are such conflicts found in other occupations?
10. What can be done, if anything, to minimize or eliminate some of the corruptive influences that are a constant threat to police officers?

REVIEW EXERCISES

The following exercises are designed to reinforce or amplify comprehension of the material contained in this chapter.

1. Conduct library research on the subject of role conflict in other occupations. Compare and contrast the findings in the literature against what the text says about role conflict in law enforcement.
2. Prepare a paper (no more than five pages) describing the causes of organizational stress in law enforcement and what steps might be taken to eliminate or minimize them.
3. Conduct additional research on critical incident stress and relate the findings to the information contained in the text.
4. In small group sessions, discuss the various forms of police corruption. Develop a list of things that might be done in a police organization to minimize the threat of this problem.

REFERENCES

Arlies, Rebecca M., "Healthy Hearts for New York City Cops," *The Police Chief* (July 1991), pp. 16-22.

Byrd, Donald A., "Impact of Physical Fitness on Police Performance," *The Police Chief.* (December 1979), pp. 30-32.

Chapman, Samuel G., Cops, *Killers and Staying Alive.* Springfield, Ill.: Charles C. Thomas, 1986.

Cruse, Daniel, and Jesse Rubin, "Police Behavior, Part I," The *Journal of Psychiatry and Law,* 1 (September 1973).

Dunham, Robert G., and Geoffrey P. Alpert, *Critical Issues in Policing: Contemporary Readings.* Prospect Heights, Ill.: Waveland Press, 1989, pp. 498-507.

Forst, Martin, Melinda Moore, and Graham Crowe, "AIDS Education for California Law Enforcement," *The Police Chief* (December 1989), pp. 25-28.

Geddes, Ronald W., and William DeJong, "Coping With Police Stress," in Robert G. Dunham and Geoffrey P. Alpert, Critical *Issues in Policing: Contemporary Readings.* Prospect Heights, Ill.: Waveland Press, 1989, pp. 498-507.

Goldstein, Herman, *.Policing a Free Society.* Cambridge, Mass.: Ballinger Publishing Co., 1977.

Goldstein, Herman, *Problem-Oriented Policing.* Philadelphia, Pa.: Temple University Press, 1990.

Goolkasian, Gail A., Ronald W. Geddes, and William DeJong, "Coping With Police Stress," in Robert G. Dunham and Geoffrey P. Alpert, *Critical Issues in Policing: Contemporary Readings.* Prospect Heights, Ill.: Waveland Press, 1989, pp. 498-507.

Hammett, Theodore M., "AIDS and the Law Enforcement Officer," National Institute of Justice, *NIJ Reports* (November-December 1987), pp. 2-7.

Hammett, Theodore M., and Walter Bond, "Risk of Infection with the AIDS Virus through Exposure to Blood," National Institute of Justice, *AIDS Bulletin* (October 1987).

Kennedey, Daniel B., Robert J. Homant, and George L. Emery, "AIDS and the Crime Scene Investigator." *The Police Chief* (December 1989), pp. 19, 22.

Kotulak, Robert, "Is Your Job Driving You Crazy?" *Chicago Tribune,* September 18, 1977, pp. 1,5, citing study by Michael J. Culligan, and others, "Occupational Incidence Rates for Mental Health Disorders," *Journal of Human Stress,* 3 (September 1977), pp. 34-39.

Klockars, Carl, *Thinking About Police: Contemporary Readings.* New York: McGraw-Hill Book Co., 1983, pp. 239-251).

Kroes, William H., *Society's Victim – The Policeman: An Analysis of Job Stress in Policing.* Springfield, Ill.: Charles C. Thomas, 1976.

Leonard, V. A., and Harry W. More, *Police Organization and Management* (7th ed.). Mineola, New York: The Foundation Press, Inc., 1987.

Mitchell, Jeffrey T., "Development and Functions of a Critical Incident Stress Debriefing Team," *JEMS* (December 1988), pp. 43-46.

Muir, William Ker, Jr., "Skidrow at Night," in Carl Klockars, *Thinking About Police: Contemporary Readings.* New York: McGraw-Hill Book Co., 1983, pp. 239-251.

Monihan, Lynn Hunt, and Richard E. Farmer, *Stress and the Police: A Manual for Prevention.* Pacific Palisades, California: Palisades Publishers, 1980.

Niederhoffer, Arthur, and Abraham S. Blumberg, *The Ambivalent Force: Perspectives on the Police.* Waltham, Mass.: Ginn & Co., 1970, pp. 293-307.

Reiser, Martin, "Some Organizational Stresses on Policemen,"*Journal of Police Science and Administration,* 2 (June 1974).

Scrivner, Ellen, "Helping Police Facilities Cope With Stress," *Law Enforcement News* (June 15/30, 1991), pp. 6, 8, and 10)

Skolnick, Jerome H., *Justice Without Trial: Law Enforcement in Democratic Society.* New York: John Wiley & Sons, Inc., 1967.

Solomon, Roger, "Police Shooting Trauma," *The Police Chief* (October 1988), pp. 121-123.

Stenzel, William W., and R. Michael Buren, *Police Work Scheduling: Management Issues and Practices.* Washington, D.C.: U. S. Department of Justice, 1983.

The Four-Ten Plan – California Law Enforcement, California Commission on Peace Officer Standards and Training, 1981.

Wilson, James Q., "The Police and Their Problems: A Theory," in Arthur Niederhoffer and Abraham S. Blumberg, *The Ambivalent Force: Perspectives on the Police.* Waltham, Mass.: Ginn & Co., 1970, pp. 293-307.

CHAPTER 5
VIOLENCE AND THE POLICE

CHAPTER OUTLINE

LEARNING OBJECTIVES

After completing this chapter, the student should be able to

1. Explain why violence is often a byproduct of police patrol work.
2. List the legal liabilities and restrictions imposed on police use of deadly force.
3. Describe the hazards associated with high-speed chases by the police and the restrictions that may be reasonably imposed on them.
4. Outline the principles guiding the recommended use of firearms by police officers.

Violence is, unfortunately, an occupational hazard of the police. The regrettable fact is that the police are often called upon to use force as a means of completing an assignment, and they often are confronted with violent reactions from those upon whom they must impose their authority.

All too often, police officers are the recipients of the kind of violence that comes from the barrel of a gun, the end of a club, or the edge of a knifeblade.

One form of force the police use to apprehend lawbreakers is the high-speed chase or "police pursuit." Indeed, while the average police officer will retire without ever having fired a shot in the line of duty, high-speed chases occur with alarming frequency in police departments throughout the country.

Although it may at first seem odd to address the issues of the use of force and the high-speed chase in the same chapter, it is becoming more and more evident that they are closely related issues. The use of physical force and the high-speed chase can both result in violence and pose serious threats to the community. Moreover, both activities need to be carefully controlled and regulated by police policies and procedures. This chapter addresses these two very important aspects of police work.

PHYSICAL HAZARDS OF POLICE WORK

Police officers can be killed or injured in a variety of ways—some more predictable than others. For example, it is well-established that domestic disturbance calls are the most dangerous the police are asked to handle. Research has shown that more police officers die responding to domestic disturbance calls (22 percent) than to any other type of incident.[1]

In 1984, over 60,000 assaults on police were reported to the Federal Bureau of Investigation (FBI).[2] In 1990, nearly 72,000 city, county, and state police officers

[1]Joel Garner and Elizabeth Clemmer, "Danger to Police in Domestic Disturbance—A New Look," in Robert G. Dunham and Geoffrey P. Alpert, *Critical Issues in Policing: Contemporary Readings* (Prospect Heights, Ill.: Waveland Press, 1989), p. 156.

[2]*Ibid.*, p. 516.

were assaulted while performing their duties, and 132 police officers were killed in the line of duty.[3]

Ten times as many police officers are assaulted in disturbance calls than in responding to robberies-in-progress and burglaries-in-progress combined. Nearly one-third of all assaults on police officers occur in connection with disturbance calls. The next highest category of assaults on police officers is in attempting other arrests (20 percent).

In 1990, 26,031 law enforcement officers were injured as a result of being assaulted. The injury rate was 6 per 100 officers. Over 80 percent of the assaults on police officers committed in 1990 involved personal weapons (hands or feet) as opposed to clubs, knives, firearms, or other types of weapons. Eighty percent of the officers assaulted in 1990 were assigned to uniformed patrol. Assaults on police officers can be prevented through better training programs. For example, officers must be taught to recognize and deal more effectively with situations in which violence is likely to occur. In addition, police officers need to be taught to be more alert and less careless in the handling of what appear to be "routine" field encounters.

> It is because they are human that police officers make fatal errors. They will work hard for long hours at a job with great stress. They save lives, protect the innocent, and even those not so innocent. Then on occasion, they commit an error. They get preoccupied. They relax. An act of great, but unnecessary courage is diluted with carelessness; or after many years, like most people, they get apathetic. For police officers, these are fatal and deadly errors. For police officers, remaining constantly alert is not only a way of life, it is a necessity for survival on the street.[4]

Apathy, boredom, and monotony are among the worst threats to the police officer's safety, for they lead to carelessness and inattention which often result in just plain poor police work. In all too many cases, police officers are their own worst enemy. Complacency by a police officer is an open invitation to violence. Complacency seems to parallel job tenure. Senior officers seem more likely to let down their guard and unnecessarily expose themselves to violent attacks than do officers with less time on the job. Well-designed and effective training programs, coupled with better supervision, can help to reduce carelessness by police officers.

> Training is the single most important factor in preventing an officer from becoming a statistic. By far, the most important training is that which is geared to reducing officer

[3]The statistics cited here are taken from Federal Bureau of Investigation, *Uniform Crime Reports: Law Enforcement Officers Killed and Assaulted 1991* (Washington, D.C.: U.S. Department of Justice, 1991), pp. 41, 45.

[4]Pierce R. Brooks, "...Officer Down, Code Three" (Schiller Park, Ill,: Motorola Teleprograms, Inc., 1975), p. 95.

carelessness and complacency. Periodic retraining will minimize these mind sets, and casualties will be reduced.[5]

USE OF FORCE BY POLICE OFFICERS

Police officers must sometimes use force to impose their authority on others. In the vast majority of cases, the use of force by police—even deadly force—is warranted by the circumstances. On occasion, the use of force is not justified. When this occurs, it makes all police officers suspect.

There are many possible explanations as to why police officers sometimes overreact and use force unnecessarily or to a greater degree than is warranted by the circumstances. In some cases, police officers may simply be manifesting their own frustrations with a variety of factors that make their jobs as police officers more difficult and often less satisfying: public apathy, unsupportive public officials, insensitivity, and uncaring administrators and supervisors. In other cases, police officers may simply be trying to "show off" or prove their toughness to their colleagues.

> More than anything else, policemen—like most of the rest of us—tend to respond in kind to the actions of the people they meet. If they can learn that their position imposes on them special responsibilities, their use of force may become much less of a problem. Also, like most of the rest of us, the police appear to be concerned with maintaining their image in the eyes of their colleagues and the public. If they can learn to submerge short-term considerations of face in long-term considerations of professional prestige, they may find their use of force less controversial and their professional status on the rise.[6]

As Chapman has observed, the best police "come-along" is the spoken word.[7] Police officers can achieve much more through the art of persuasion than they can by resorting to force. With few exceptions, each time a police officer resorts to force to resolve a situation, it should be considered a defeat rather than a victory. Too often, police officers resort to action before words. What they often realize too late is that words are more effective tools for overcoming resistance and hostility.

This does not mean, however, that each time an officer becomes involved in a violent confrontation that it could have and should have been avoided, or that the officer was wrong for resorting to force. Clearly, force is sometimes the only remedy and the officer must be prepared to use it within appropriate limitations. However, each violent police-citizen confrontation should provide an opportunity for subse-

[5]Samuel G. Chapman, *Cops, Killers and Staying Alive* (Springfield, Ill.: Charles C. Thomas, 1986), p. 58.

[6]Robert J. Friedrich, "Police Use of Force: Individuals, Situations, and Organizations," in Klockars, *Thinking About Police,* p. 311

.[7]Chapman, *Cops, Killers,* p. 64.

quent review, examination, and analysis to determine whether mistakes were made and how they might be avoided in the future. In addition, police officers need to be taught and constantly reminded that the art of persuasion and negotiation is one of the most powerful tools available to any police officer and one that can generally accomplish a great deal more than force and aggression.

Police officers are unique among nearly all occupations, except the military, in that they possess the awesome authority to use **deadly force** in the performance of their duties. In this case, deadly force is defined as, "...such force readily capable of, or likely to cause, death or serious bodily harm."[8]

Despite the importance attached to the use of deadly force by the police, there are no reliable central repositories established to collect and analyze data concerning such events. Although the FBI does collect information on police use of force from contributing agencies, these figures are not published, apparently because the FBI does not consider them to be reliable. However, the number of police shootings which result in the death of a citizen is relatively low, Police shootings tend to occur most often in connection with robbery incidents, traffic stops, and disturbance calls.[9]

Civil litigation against the police for use of excessive force focuses on the reasonableness of the force used in light of the specific circumstances. However, defining what is reasonable can be a very complex and demanding task. Determining the reasonableness of a police officer's actions in using deadly force must be considered in the light of what the public expects of its police officers as well as what is considered reasonable to an average police officer. Unfortunately, the reasonableness of the officer's actions will often be evaluated long after the acts themselves occurred.

Obviously, the public expects to be protected from violent criminals and has given the police the authority as well as the means to use force to ensure that end. Police officers are trained in the use of force, including deadly force. However, their training must also include the limitations under which that force may be used. If anything, then, the police standards of reasonableness should be more stringent than those imposed by the general public.

Because they have been authorized to use force to achieve their objectives, the police sometimes assume that such force can be used indiscriminately. The use of excessive force by police seems to occur most often in those agencies where supervision is lax, training is inadequate, and leadership has failed to issue stringent policies against the unnecessary use of force. As one publication has indicated, the culture of the police organization, which is characterized by its set of value systems, may have a very direct impact on the frequency with which incidents of excessive force occur.

> The "culture" of a police department reflects what that department believes in as an organization. These beliefs are reflected in the department's recruiting and selection

[8]Geoffrey P. Alpert and Lorie A. Fridell, *Police Vehicles and Firearms: Instruments of Deadly Force* (Prospect Heights, Ill.: Waveland Press, 1992), p. 12

.[9]*Ibid.*, pp 35-41.

practices, policies and procedures, training and development, and ultimately, in the actions of its officers in delivering services.[10]

Clearly, the "organizational climate" of a police department can have a lot to do with the number of episodes in which violence is used by officers against citizens and how those incidents are treated. Organizations characterized by strong leadership, aggressive supervisory practices, strict-but-fair disciplinary systems, clear-cut policy guidelines, a positive self-image, and progressive training programs are less likely to experience or to permit incidents of violence by police officers against citizens. Such episodes are regarded as exceptions to normal conduct and are treated accordingly.

On the other hand, there are, unfortunately, a small number of police agencies where mistreatment of citizens, suspects, and arrestees is tolerated and even encouraged. In these police agencies, citizens are viewed as adversaries and officers think in terms of "us versus them." This kind of attitude is a throwback to the old days of "cops and robbers" and has no place in modern law enforcement. Progressive and enlightened police managers will work aggressively to eliminate these kinds of attitudes from their organizations and will take a firm stand against violence by officers against citizens and other forms of police misconduct.

On the other hand, we must not be too quick to condemn police officers who, in a moment of passion, anger, fear, or uncertainty, overreact to a potentially dangerous situation. Police officers are trained to react quickly to potentially dangerous situations and they know that their failure to act, or to act properly, may result in death or serious injury to themselves or to someone else.

How a police organization and its membership, including its leadership, views the use of violence by its officers against citizens may be related to the values of the organization. Values are an important and necessary part of the organization's culture and identity. Values tell us much about how the organization operates and what it sees as its central purpose. Values also tell us much about how the members of the organization behave and what is expected of them. Values can serve a number of important purposes for the police organization:[11]

1. They set forth its philosophy of policing (for example, crime prevention, social control, public service).
2. They serve as the basis for establishing policies and procedures.
3. They set parameters for organizational flexibility.
4. They provide the basis for organizational strategies.
5. They provide the framework for judging officer performance.
6. They serve as the framework by which the department itself can be evaluated.

[10]Community Relations Service, U.S. Department of Justice, "Principles of Good Policing: Avoiding Conflict Between Police and Citizens," in Robert G. Dunham and Geoffrey P. Alpert, *Critical Issues in Policing: Contemporary Readings* (Prospect Heights, Ill: Waveland Press, 1989), p. 164.

[11]*Ibid., p. 193.*

We provide police officers with the tools of violence, but expect them to use them only when absolutely necessary. We expect police officers to be law enforcers and peacekeepers, even when these two goals are themselves contradictory. We expect police officers to work under the most difficult, threatening, and demanding conditions and to do so with professionalism, humility, and a positive attitude. When they fail to perform with absolute perfection, we are disappointed and seek to know why.

Clearly, there are instances in which police officers use violence and force unnecessarily and under conditions in which they could have and should have known better. There are, without question, may instances in which the police overreact and bring shame to themselves and to all police agencies. The fact of the matter, however, is that these are the exceptions rather than the rule and must be acknowledged and treated as such. When they do occur, however, it is important that every lesson that can be learned from them be recorded and studied and that training programs be modified in order to teach others to avoid acting similarly.

The shooting of an unarmed burglar who is attempting to flee is clearly a violation of the principle of reasonableness, simply because the value of human life outweighs the public benefit associated with the apprehension of the suspect. On the other hand, a fleeing burglar who seizes a heavy board, pipe, or brick, and makes an overt attempt to harm the pursuing officer to make good his escape may well provide the officer with justification for the use of deadly force.

If the fleeing burglar simply has the means of harming the officer (that is, the suspect is known to be armed), but makes no overt attempt to use the weapon to effect his escape, then there can be no justification for using deadly force. In other words, simply because the officer knew that the suspect was armed would not justify the use of deadly force to effect an apprehension. The officer must show not only that the suspect was armed (or was capable of using deadly force himself), but that the suspect actually attempted to, or was about to, use that force against the officer or upon some other person to effect his escape.

In the landmark case of *Tennessee v. Garner,* 471 U.S. 1, 105 S. Ct. 1694, 85 L. Ed. 2d 1 (1985), the United States Supreme Court ruled "the use of deadly force to prevent the escape of an apparently unarmed fleeing felon may not be used, unless the officer has probable cause to believe that the suspect poses a significant threat of death or serious physical injury to the officer or others."

The court further stated, "Where the officer has probable cause to believe a suspect poses a threat of physical harm, either to the officer or to others, it is not constitutionally unreasonable to prevent escape by using deadly force. Thus, if a suspect threatens the officer with a weapon or there is probable cause to believe that he has committed a crime involving the infliction or threatened infliction of serious physical harm, deadly force may be used if necessary to prevent escape, and if, where feasible, some warning has been given." (Page 1701 of the Supreme Court decision.)

Many in the law enforcement community felt that the ramifications of the Garner decision would seriously impact the ability of the police to defend themselves when necessary. In actuality, many departments had already adopted a similar standard and had been operating effectively for years prior to Garner. The policy of "defense of life—your life or the life of another person," had been commonly stated in many department policies and the Supreme Court's ruling in the Garner case only affirmed what was already in place.

"When writing policies on the use of deadly force, some police departments become sidetracked.They attempt to define forcible felonies by providing guidelines to their personnel on specific instances involving deadly force. Sole reliance on specific felony situations without further elaboration on the circumstances or offender behavior should be avoided."

For example, burglary is defined as a forcible felony in some jurisdictions, yet under the guidelines of the Garner case, an unarmed burglar running from the scene of the crime could not be shot by a responding police officer for failing to stop when ordered to do so.

Equally as deficient are those policies which use felony as the only measuring stick for the justified use of deadly force. For example, in some jurisdictions the pointing of a firearm at another person is a misdemeanor, unless the gun is actually fired. Certainly a police officer encountering someone with a firearm pointed directly at him or her would reasonably feel in immediate danger of death or serious physical injury. Although the crime may be a misdemeanor, deadly force could be justified on the condition that the officer's life is in jeopardy. This is but one example of the problems encountered in writing policies on the use of deadly force.

In attempting to be specific about the department's policy on the use of force, including deadly force, a law enforcement agency should specify that officers will use only the amount of force necessary to effect lawful objectives. The use of deadly force should be considered a last resort, and the policy of "defense of life" should be adopted.

Finally, police agencies should have procedures in place to review all use-of-force incidents by their members. Review should begin at the first-line supervisory level and continue through the department's hierarchy. In reviewing use-of-force incidents, the agency might discover the need for training or new equipment. Such a review could reveal something of the abilities of police personnel. Those members with a higher use-of-force average than the majority may require additional training in crisis intervention, cultural awareness, or control tactics. They could also be undesirable officers incapable of achieving objectives without being authoritarian or heavy-handed.

The use of force by law enforcement officers appears to be one area where the public will be expecting a higher degree of accountability during the next decade. For this reason, agencies must ensure that their policies are updated and that personnel understand alternatives to the use of force. Officers must use restraint in selecting a level of force when persuasion or voluntary compliance methods fail.

There are four models that have served as guidelines for police use-of-force policies. These range in degree of restrictiveness from that which allows the police to use deadly force to apprehend any fleeing felon (the "Fleeing Felon Rule") to the "defense of life" standard which, as its name implies, permits a police officer to use deadly force only when necessary to defend his life or that of someone else. The

Model Penal Code, on the other hand, permits an officer to use deadly force when the officer believes (1) "the crime for which the arrest is made involved conduct including the use or threatened use of deadly force," or (2) "there is a substantial risk that the person to be arrested will cause death or serious bodily harm if his apprehension is delayed."[12]

The **forcible felony rule** is similar to the **model penal code,** and provides that "officers could use deadly force to effect the arrest of persons suspected of certain specified felonies which pose a risk of great bodily harm or death."[13]

> The value of human life is immeasurable in our society. Police officers have been delegated the awesome responsibility to protect life and property and apprehend criminal offenders. The apprehension of criminal offenders and protection must at all times be subservient to the protection of life.[14]

The police administrator must be able to develop a written policy on the use of deadly force by officers that provides a reasonable and practical balance between the safety of the community and the needs of officers to protect themselves and others in the line of duty. It is difficult to generalize as to what constitutes an effective policy on the use of deadly force and it is certainly true that no written policy, except one which permits almost no discretion by the officer, can satisfy all situations.

The Commission for Accreditation of Law Enforcement Agencies, Inc. (CALEA) has adopted a total of sixteen different recommendations dealing with the use of force by police personnel. These recommendations cover topics such as the type and frequency of training programs, the type and specification of weapons to be used, and the investigation of shooting incidents. More importantly, the standards set forth very clearly state the limits that should be placed on the use of deadly force:

> 1.3.2 A written directive states that an officer may use deadly force only when the officer reasonably believes that the action is in the defense of human life, including the officer's own life, or in the defense of any person in immediate danger of serious physical injury.[15]

Similarly, Standard 1.3.3 proves that "...the use of deadly force against a 'fleeing felon' must meet the conditions required by standard 1.3.2."

The more restrictive the policy, the greater will be the protection of the agency from civil or criminal liability. However, overly restrictive policies may impede an officer's legitimate right to use force to protect himself or another person. While

[12]Alpert and Fridell, *Police Vehicles,* p. 71

[13]*Ibid*[1]

[14]Kenneth Matulla, *A Balance of Force,* 2d ed. (Gaithersburg, Md.: International Association of Chiefs of Police, 1985), p. 68.

[15]Commission on Accreditation for Law Enforcement Agencies, Inc., *Standards for Law Enforcement Agencies* (Fairfax, Virginia, 1983), pp. 1-2..

overly restrictive policies should be avoided, the International Association of Chiefs of Police has recommended the following provisions be included in use-of-force policies:[16]

1. Warning shots should be prohibited.
2. There should be strict restrictions on shots at or from a moving vehicle.
3. There should be restrictions on the carrying of secondary (backup) weapons by officers.
4. The department should control the type of weapon and ammunition used by officers.
5. Officers involved in shootings should be placed on "administrative leave" following the incident.
6. Psychological counseling should be required for all officers involved in shootings.

While there is little empirical evidence on the subject, what little research that has been conducted on police use-of-force policies has found that police departments with more restrictive use-of-force policies record fewer officer-involved shootings. More importantly, researchers have found that the implementation of more restrictive use-of-force polices has not resulted in greater risk to the officers.[17]

As a case in point, in 1971 the New York City Police Department issued Temporary Operating Procedure 237 which placed new restrictions on police use of deadly force. A subsequent evaluation of the effects of this new policy conducted by Fyfe revealed that a considerable reduction in the frequency of police shooting accompanied the new policy. It was further suggested that

> ...an obvious consequence of the implementation of clear shooting guidelines and their stringent enforcement is a reduction of injuries and deaths sustained by suspects...
>
> Of equal significance to the police administrator is the fact that these shooting decreases were not accompanied by increased officer injury or death. Conversely, since both these phenomena appear to be associated with the frequency of shooting incidents and related citizen injury, both declined pursuant to T.O.P. 237.[18]

FIREARMS TRAINING AND POLICIES

For many years, good firearms training programs were lacking in a great many police departments. Although some improvement has been made in the last two decades, it

[16]Alpert and Fridell, *Police Vehicles,* 74.

[17]*Ibid* pp. 91-92.

[18]James J. Fyfe, "Administrative Interventions on Police Shooting Discretion: An Empirical Examination," *Journal of Criminal Justice* (Winter 1979), p. 332. .

is still true today that firearms training programs in many police departments in the United States leave much to be desired. Typically, they lack in terms of the frequency with which the training is provided, the instructional methods used, or in the development of proper attitudes.

Too many police training programs, for example, tend to emphasize the use of force and firepower as survival techniques, rather than encouraging officers to use their brain power and common sense as a means of avoiding situations in which force may be necessary. These training programs, and the attitudes they foster, ignore the simple fact that many violent confrontations between the police and citizens could be avoided if police officers were taught to seek out alternatives to the use of force.

> As important as having proficiency in the use of firearms is having the skills to make the appropriate decision to shoot or not to shoot in a practical situation, and having the skills to minimize the possibility of a shooting.[19]

Police training programs need to be expanded and upgraded to include training in tactical strategies and decision making with emphasis on recognizing potentially violent confrontations before they occur. Trainees should explore alternative courses of action which could resolve confrontations through nonviolent means.

Situational training programs, in which officers are exposed to situations designed to simulate real field conditions, are the most valuable type of training in defensive tactics. Classroom training should be limited to instruction in the principles of field tactics, human relations, firearms mechanics, and similar topics. But situational training can do much more to prepare the officer to deal with the "real world" of unexpected confrontations. It is the best way to help officers understand the dynamics of particular situations and to help them identify suitable ways of dealing with potentially harmful situations.[20]

According to Alpert and Fridell, firearms training programs in some police agencies have

> ...advanced from the use of a simple bull's-eye course emphasizing marksmanship to much more sophisticated programs which provide training in decision-making diffusing potentially violent confrontations.[21]

As a result of the 1985 United States Supreme Court decision in *Tennessee v. Garner,* police departments have been required to drastically change their policies and training programs on the use of deadly force. In that decision, the United State Supreme Court ruled that the use of deadly force against an unarmed and nondangerous fleeing felon is an illegal seizure of that person, under the Fourth Amendment. The court ruled that deadly force may not be used unless it is necessary to prevent escape *and*

[19]Alpert and Fridell, *Police Vehicles,* p. 95.

[20]See Chapman, *Cops, Killers,* pp. 60-63.

[21]Alpert and Fridell, *Police Vehicles,* p. 94.

the officer has probable cause to believe the suspect poses a significant threat of death or serious injury to the officer or others.

This decision caused many police departments to greatly reduce the number of situations in which officers were permitted to discharge their firearms in the line of duty. Many agencies adopted the position that the use of deadly force was solely limited to the defense of the officer's life or the life of another person. The "defense of life" policy is probably the most common guideline existing in police departments today, although some organizations go to great lengths to specify each and every situation where the use of deadly force may be justified.

Certainly, the use of warning shots, firing at or from moving vehicles, firing into crowds or through doors should be strictly prohibited. Although shooting from moving vehicles or at the tires of an escaping vehicle is entertaining and sensational on television programs and in motion pictures, in real life such practices are very dangerous and should be prohibited in department policies.

Instruction Methods

For years, police departments in many jurisdictions concentrated their firearms training efforts on bull's-eye shooting. It was felt by some instructors that accuracy was paramount and that skill would be the single most important factor for an officer involved in a shooting situation. This same school of instruction required police officers to collect their spent-shell casings after shooting and drop them into a coffee can or other container so the range could be kept neat and tidy.

Unfortunately, these same housekeeping habits were brought to the field by the officers, and in documented cases, officers involved in shootings attempted to reload their weapons with previously-fired cartridges during the stress of the situation. Firearms instructors must insist that shell casings be discarded without regard for housekeeping considerations. The trainee's attention must remain focused on the target while quickly reloading. Instructors should concentrate on teaching students "center mass" shooting techniques. This amounts to shooting for the chest and upper abdominal area, where most vital organs are located. Firearms training should stress the philosophy of "shooting to incapacitate." Center mass instruction is consistent with this approach.

The Commission on Accreditation of Law Enforcement Agencies requires officers carrying firearms to qualify at least annually with their duty firearms. While this may seen minimal, and it is, there are officers throughout the United States who have not qualified in years. Virtually any experienced range officer in a larger department can give accounts of officers who present themselves for qualification at the firing range with ammunition that is rusted or green. The same officers, who have "slipped through the cracks" and not been near the firing range for extended periods of time, often carry handguns that do not function, or don't even know how to disassemble their weapons for cleaning. All of these situations are indicative of a firearms training program in need of significant improvement.

In 1979, the case of *Popow v. City of Margate* (476 F. Supp 1237) was decided in Federal Court in New Jersey. This case had significant ramifications for the development of defensible police firearms training programs. In this case, a police officer was chasing a suspected fleeing kidnapper and shot an innocent bystander, Darwin Popow, who died as a result of the accidental shooting. In court, it was shown that the police department's training program was inadequate. There were no exercises involving low-light or night shooting situations, no instruction on shooting at moving targets, no instruction on shooting in residential or populated areas, and no simulations or judgment training to teach officers when to shoot or not shoot. The court, in awarding a summary judgment against the city of Margate, concluded that the training program was not sufficient absent the types of training just mentioned.

Indoor firing ranges offer convenience, but are somewhat limited. While lights can be dimmed and targets generally moved to and from the firing line, side-to-side movement or any type of evasive-action simulation is only found on the most sophisticated and expensive firing ranges. Outdoor ranges provide greater flexibility, but in some areas of the country their use is limited due to inclement weather considerations.

Moving targets, or "Hogan's Alley" type training programs are much easier to arrange with an outdoor range. Some training programs use both types of arrangements. Indoor firing with simulated night shooting is done 1-2 times a year. Shoots with barriers, cover, and concealment exercises are done on outdoor ranges. This combination of shooting conditions, locations, and arrangements provide a balanced and practical alternative to traditional shooting programs. Putting officers through firearms training consisting of the same course each time without variation or opportunity for new exercises is both boring and inappropriate.

If the agency requires qualification on a standard course of fire, then other exercises should be inserted after the qualification course. Most qualification courses require less than 30 minutes per officer. An extra half-hour at the firing range firing other exercises is well worth the extra salary, ammunition, and targets if the validity of the shooting program is ever questioned in a lawsuit.

Judgment Training

To address the need for judgment firearms training, many agencies now use movies or video to expose officers to various scenarios. The officers stand before a screen where the situation unfolds, then react as they would on the street. In some training programs, this can be done in a classroom using a videotape and monitor. The officer points a pistol loaded with blanks or caps at the screen, and fires when appropriate. An alternative is to stretch a disposable tablecloth across an actual pistol range and project a 16mm film onto the tablecloth. As the scenarios unfold, officers fire at the tablecloth. These methods are simple to use and often stimulate discussion regarding the law, department policy, and officer liability. Instructors in these programs must be certain to control the ammunition, weapons, and students to avoid injuries.

Another option for judgment firearms training is the use of the computerized scenario training available from a variety of sources. Portable units are brought to the department and officers are exposed to various situations and required to make a decision to shoot or not. These sophisticated machines register the time it takes the officer to react, where the shot struck, and tell whether the shooting reflected good judgment. A printout is often provided for the department's training records verifying the officer's attendance and score. Many academies now have these devices in place to meet the need for judgment firearms training.

Firearms Instructors

Many police officers are first introduced to firearms at the police academy. To a new recruit with no previous experience handling weapons, the academy firearms instructor can become somewhat of a guru. Unfortunately in some cases, academy firearms instructors are limited in their training assignments to just the firing range. Their personalities and teaching styles are geared toward the survivalist approach, and the focus of instruction centers purely on marksmanship and tactical skill building. Little or no emphasis is placed upon confrontational alternatives, concealment, cover, or negotiation. A firearms training program lacking discussion of these items as well as a block of instruction on the legal, moral, and ethical considerations of using deadly force is deficient at best.

Selecting suitable candidates as firearms instructors is critically important. Some agencies automatically locate the best shooters on the department and assign them as range officers. This does not guarantee quality of instruction, nor does it often work out. Departments should attempt to identify individuals with satisfactory shooting abilities who show some acumen for teaching and recruit them as range officers. The worst situation occurs when personnel assigned to the range are relegated to merely counting the number of times a shooter hits the target, without regard for improving judgment skills. With practice, anyone can score a target, but not just anyone can be an instructor.

Personnel assigned to the firing range as instructors should periodically have their blood level tested for the presence of lead. During recent years there has been increasing evidence that routine exposure to lead from police ranges is a serious health risk and may be fatal. A simple blood test on an annual basis can monitor an individual's lead level.

Firearms Selection

One of the most emotional debates in law enforcement circles is which firearm is the best for use by police officers. Some maintain that the reliable revolver with its six-shot capability has proven its value over the years and provides the best service. Others counter with the need for increased firepower and the ease of reloading offered by the semiautomatic pistol. During the past ten years, there was tremendous movement among law enforcement agencies to make the transition to semiautomatic pistols. As of 1991, approximately 50 percent of the law enforcement officers in the United States carried semiautomatic pistols. This represents a significant increase from just a decade ago.

Many believe that the ability to have 16 cartridges ready to fire in a semiautomatic pistol translates into more firepower. Unfortunately, this is not necessarily true. Firepower should actually be gauged by the ability of an officer to strike an intended target efficiently. An officer armed with a six-shot revolver who shoots and hits an armed assailant with the first shot actually has more firepower than an officer with a 16-shot semiautomatic pistol who fires the weapon 16 times and doesn't hit a thing. There is no doubt that a semiautomatic is easier to reload during stress than a revolver, but many revolvers continue to provide adequate service to the officers who carry them.

A relatively new addition to the law enforcement arsenal is the double-action-only handgun. These weapons cannot be cocked and each shot starts with the hammer at rest. Some departments have learned from experience that cocked handguns increase the likelihood of accidental discharges. There is no question that a cocked weapon goes off easier than one requiring a full trigger pull. The introduction of the double-action-only weapon into law enforcement appears to be more than just a passing fad.

In 1990, one major manufacturer of the double-action-only handguns sold 120,000 to officers in the United States. These weapons are useful in training situations because a longer trigger pull makes them more forgiving of human error than weapons capable of being cocked. A cocked weapon may require only 2-4 pounds of pressure to pull the trigger compared to 6-8 pounds for a double-action-only handgun. The double-action-only handgun provides the officer with a safety margin not available with a cocked weapon.

Annual Inspection of Weapons

Law enforcement agencies must ensure that the handguns carried by their officers are inspected periodically by a competent armorer. A minimum of one inspection per year of the internal parts is recommended. Range personnel should inspect the external parts during regular range attendance.

It is essential that police agencies regulate the customizing of handguns carried by their members. Items such as trigger shoes, which slide over the factory-installed trigger and are held on by set screws, should be prohibited. Although these trigger shoes provide a wider trigger which some officers feel makes the trigger pull smoother, they can slip down and get hung up on the trigger guard. When the trigger shoe is stuck against the trigger guard, the trigger cannot be pulled back. This makes the handgun useless.

Another dangerous byproduct of using a trigger shoe is the fact that some of these devices make the trigger wider than the trigger guard. If the wider trigger gets caught on the officer's holster while the gun is being replaced, the hammer will go back and the gun will discharge accidentally.

The customizing of handguns to suit individual officers should also be of concern. Just as many shooters automatically want to adjust the sights on their handguns to improve their accuracy, so too do others believe a smoother trigger will do the same. Although both of these situations can generally be corrected by increased range practice and familiarity with a handgun, making adjustments and modifications seems to provide the instant answer for impatient, inexperienced novices.

The problem arises when the officer takes a handgun to a friend who likes to "tinker." The friend may be an auto mechanic or machinist, and soon the solution to the problem boils down to cutting a few spring links on the gun or grinding down a few "rough edges." The result is that when the handgun comes back to the officer, it feels very smooth and seems much easier to fire. That's because it probably is. It also is more than likely unsafe.

If an officer is involved in litigation surrounding a shooting, one of the first actions taken by the attorney representing the person shot is generally to subpoena for inspection of the officer's handgun. The handgun is then taken to a gunsmith on retainer by the victim's attorney, and the weapon is subjected to a series of tests to see if any of the tolerances exceed manufacturer recommendations. If it can be shown that the handgun was altered beyond the specifications of the manufacturer, or that the weapon has a "hair trigger," the results will probably not be favorable to the officer. For this reason, it is recommended that police officers have their handguns inspected annually by the department armorer, and that officers be prohibited from making any modifications to their handguns without permission from the department.

HIGH-SPEED CHASES BY POLICE

Since the civil disturbances of the 1960s and 1970s, a great deal of attention has been paid to the use of deadly force by police. As a result, there has been significant reform in this area. However, the same level of inquiry has not been devoted to police pursuits which, in their own way, offer an even greater threat to public safety. This is surprising, in view of the alarming figures that are reflected in the literature on police pursuits. For example:

- One out of five pursuits ends in death.
- Five out of ten pursuits results in serious injuries.
- Seven out of ten pursuits end in accidents.
- Four out of five pursuits are for minor offenses.
- Pursuits result in 500 deaths each year; 20 are law enforcement officers.[22]

[22]Alpert and Fridell, *Police Vehicles,* p. 99.

These figures are no better than crude estimates, however, since there is no reliable, comprehensive source of information on police pursuits. Despite what are being increasingly recognized as the serious consequences of police pursuits, there has been no systematic or uniform effort to collect, analyze, and disseminate detailed information about the number, severity, causes, circumstances, and outcomes of police high-speed chases. Only one state—Minnesota—has begun to collect such information on a state-wide basis.[23]

In all too many cases, the hazards created by police pursuits are far greater than the offenses which precipitated them. When serious injury or death results from a police chase, the reason for initiating the chase will be an important factor when determining whether the officer's actions were justified.

Police officers seem predisposed to engage in high-speed chases, despite the obvious dangers they pose to themselves and others. This is especially true of the youthful, less-experienced officers who are ambitious and eager to prove their skill and courage to their peers. Engaging in a high-speed chase is seen as a personal challenge to the officer, whose standing among his or her peers will be influenced by how well the officer handles the pursuit. Successful completion of a high-speed chase will be looked upon as a mark of distinction. Officers who successfully engage in high-speed chases, particularly those that become quite harrowing, sometimes think that they have proven that they are worthy of their badge of office. This view is often shared by their peers and by some of their superiors.

On the other hand, a person who runs from the police, if only for a minor traffic violation, commits the ultimate sin of challenging the authority of the officer. In the police lingo, the offender is guilty of "contempt of cop." Fleeing from the police is seen as a challenge to the officer's authority and makes the contest between the offender and the officer a personal one. Some officers feel they must prove to themselves that they are equal to the task. The officer must match wits, skill, and courage with the offender. Because the offender has challenged the officer, the officer feels compelled to go to any length to bring the offender to justice.

The cold, hard reality, of course, is that the high-speed chase rarely accomplishes anything worth the risk it entails. "Almost without exception, high-speed police pursuits are no-win situations."[24]

In far too many cases, the offender is wanted for little more than a minor traffic violation. Even when apprehended, the offender's punishment is often minimal and no where near proportionate to the risk associated with bringing the offender to justice. It is for this reason that some police agencies have either placed serious restrictions on high-speed chases or have prohibited them altogether. More police departments are likely to follow this example in the future as the dangers associated with police pursuits become more widely publicized and the general public becomes more opposed to them.

[23]*Ibid.*, p. 112.

[24]James O'Keefe, "High-Speed Pursuits in Houston," *The Police Chief* (July 1989), p. 33

Contrary to what they think or what they are trained to believe, most police officers are not expert drivers and are not qualified by either training or experience to engage in high-speed chases. In addition, "...overconfidence, pride, a false sense of proficiency, and impatience on the part of the officer all contribute to negative outcomes in pursuit driving."[25]

When police officers engage in high-speed chases, they are attempting to effect the arrest and apprehension of a law violator. Their intent is essentially the same as when they use deadly force to make an arrest, and the circumstances are similar. In both cases, the officer is using deadly force—a police car or a service revolver—to effect the arrest of a person. The limitations imposed upon such actions are also similar. The same standard of reasonableness that applies to the use of deadly force must also be applied to high-speed chases.

In high-speed chase, however, the risk to the general public is often much greater than when a police officer discharges his weapon or uses some other means of force against a lone suspect. The fact that the risk to the public is greater, coupled with the fact that those pursued by the police are often petty traffic violators who do not otherwise represent a serious threat to public safety, exposes high-speed chases by the police to a much greater degree of public scrutiny and concern.

Police departments have an affirmative responsibility to adopt policies that place at least **minimal** limits on police pursuits. These policies may include restrictions governing, for example, (a) the speed at which officers may drive while pursuing a fleeing suspect, (b) the distance over which a police pursuit may travel before being terminated, (c) the number of vehicles which may participate in a pursuit, (d) the use of rolling or stationary roadblocks and other devices designed to terminate the pursuit, (e) the role of the field supervisor and his or her ability to terminate the pursuit at such time as it appears necessary to do so, (f) the use of unmarked cars in pursuits, (g) communications procedures, and (h) multiagency cooperation and coordination during pursuits.

In addition, police policies governing pursuits should include the following general principles:[26]

1. Be workable in real-world situations.
2. Be adaptable to training.
3. Be written in a positive manner.
4. Reflect community values.
5. Include input from officers at all levels.

The reasonableness of an officer's actions in deciding to initiate a high-speed chase, and in deciding whether to continue the chase, must be determined in terms of the impact on public safety. Will public safety be threatened more by allowing the

[25]Alpert and Fridell, *Police Vehicles*, p. 112.

[26]*Ibid.*, p. 119.

pursued driver to go free than by continuing the chase? This is not an easy question to answer, since it is impossible to predict exactly what will happen if the pursued driver is not apprehended. However, assuming that the pursued driver is wanted by the police only for a minor traffic violation, there is a reasonable certainty that his escape will not result in significant risk to the general public. On the other hand, a person being chased by the police who is known to be armed and who is wanted for questioning in connection with a series of violent crimes is quite another matter. The public will generally support the actions of the police in the latter case, even though they pose obvious risks to public safety. There is much less likelihood that the public will approve of using deadly force (a high-speed chase) to effect the arrest of a minor traffic offender.

Not only does a police officer have a duty to enforce the law and to arrest law violators, he or she has an equally important duty to protect public safety. When an officer engages in acts which directly or indirectly threaten public safety, the reasonableness of those actions will be balanced against the necessity of effecting an arrest at that time and under those circumstances.[27]

The policy of the St. Charles, Illinois Police Department (included in its entirety in Appendix B), is fairly restrictive and sets forth rather specific guidelines for officers to follow. For example, this policy provides that officers **shall** (that is, **must**) terminate a pursuit under *any* of the following events or conditions:[28]

1. The emergency equipment of the police vehicle ceases to function properly.
2. It becomes evident that the risks to life and property begin to outweigh the benefit derived from the immediate apprehension or continued pursuit of the offender.
3. Upon the order of a supervisory or command officer.
4. When the suspect's identity has been established to the point that later apprehension can be accomplished, and there is no longer any need for immediate apprehension.
5. The environmental conditions indicate the futility of continued pursuit.

The role of the police supervisor in monitoring, controlling, and supervising the conduct of high-speed chases by police officers must also be addressed. Police field supervisors have a legitimate right and a certain responsibility to carefully monitor all high-speed chases which occur while they are on duty. Just as the individual officer must know whether later review and investigation will justify the decisions made during the pursuit, the field supervisor must determine whether the pursuit is justified and whether it is being conducted with all reasonable care and caution.

[27]See Alpert and Fridell, *Police Vehicles,* p. 28.

[28]St. Charles, Illinois, Police Department General Order 87-06, dated April 23, 1987.

The policy of the St. Charles, Illinois Police Department illustrates how supervisors can be made responsible for monitoring and controlling pursuits conducted by their subordinates.:[29]

1. It shall be the responsibility of the supervisor in charge of the shift to review the facts given by the pursuing officer and to make an independent judgement if the pursuit be continued.
2. Based on all the information available, the ranking shift sergeant will order the termination of the pursuit if, in his opinion, the dangers caused by the pursuit outweigh the need for an immediate apprehension.
3. Continuous monitoring of the pursuit's progress by the shift supervisor will be made until the pursuit ends or is terminated, whether by the officer or by the shift supervisor.

Police administrators need to recognize that the same care and sensitivity that goes into developing guidelines on the use of deadly force should be applied to defining limitations on police pursuits. The liabilities associated with police pursuits are similar to those associated with the use of deadly force by the police.[30]

Police training programs must include training in the techniques of high-speed chases, but also training in recognizing the threat to public safety that such chases pose.[31] Failure to provide such training may expose the police agency and the governmental unit to severe liability.

> Failure to train officers in the consequences of their actions, including the increased risk to citizens created even when the officers drive safely, should be considered as indifference on the part of the government.[32]

Just as most police departments provide for some kind of post-incident review and critique of officer-involved shootings, they should provide for the same kind of detailed evaluation of high-speed chases, particularly those that result in death, serious injury, or significant property damage. These critiques can provide valuable feedback to officers, supervisors, and trainers that can be used to reduce the frequency and severity of pursuits in the future. It is important that these critiques be conducted in a positive manner rather than as disciplinary measures.

[29]*Ibid.*

[30]Geoffrey P. Alpert, Questioning Police Pursuits in Urban Areas," in Dunham and Alpert, *Critical Issues*, pp. 216-229.

[31]James O'Keefe, "High-Speed Pursuits in Houston," *The Police Chief* (July 1989), p. 34.

[32]Alpert and Fridell, *Police Vehicles*, p. 31.

> The utility of this feedback process lies in the hope that this mechanism will evolve to be the tool that alters the organizational culture from "Chase them until the wheels fall off" to "Why risk your life to chase some traffic violator?"[33]

Post-incident critiques can serve a number of useful purposes, including: (a) determining whether the pursuit was justified and whether it was conducted in accordance with departmental policies, (b) identifying alternative steps that might have been taken to either avoid the pursuit or minimize its adverse impact, (c) identifying training deficiencies, and (d) suggesting changes in departmental policy.[34]

The long-term objective of any pursuit-critique program should be to reduce the frequency and severity of high-speed chases and to minimize their adverse consequences in terms of death, serious injury, and damage to property. Moreover, post-incident review of high-speed chases will serve notice to police officers, as well as the general public, that such incidents are not to be treated as routine occurrences.

Clearly, police executives, public administrators, educators, and legislators must work together to find a solution to the problem of high-speed chases by police. Training programs, improved legislation, and public education efforts may be part of the answer. The impetus for change has already begun. In 1990, the Deseré Foundation was organized by G. W. LaCrosse, in response to a series of tragic deaths resulting from police pursuits. The Deseré' Foundation is dedicated to finding ways to reduce the frequency and severity of police pursuits through better legislation, improved police training programs, officer awareness, and better technology. The Deseré' Foundation is the initial catalyst for those interested in reform of pursuit driving. The number of concerned citizens and public officials appears to be growing. Reform in pursuit driving may follow the trend established in the social movement against drunk driving and police use of firearms.[35]

SUMMARY

Violence is, unfortunately, an inherent element in police work. We live in a society in which violence occurs all too frequently and in which violence seems to be taken for granted. Violence is a part of our history and our heritage, whether we like it or not. As much as we might like to change this fact, wishing will not make it happen.

The police must learn to live with violence directed toward them as well as toward others. They need to be able to avoid violence directed toward them whenever possible, and at the same time use only that level of force needed to perform their duties.

[33]O'Keefe, *op. cit.*, p. 37.

[34]Alpert and Fridell, *Police Vehicles*, pp. 120-121.

[35]*Ibid.*, p. 150.

Police officers are human though, and there will, regrettably, be times when they use unnecessary force, either deliberately or without thinking, in the line of duty. When this occurs, corrective actions need to be taken and those guilty of obvious transgressions need to be punished. Sound training programs and alert and conscientious supervisors are the key to avoiding unnecessary conflict between the police and the public.

REVIEW QUESTIONS

1. What type of call poses the greatest threat to police officers and why?
2. Officer error is a major factor contributing to serious injuries and fatal assaults on officers. Why is this the case and what, if anything, can be done about it?
3. What are some of the things that could cause a police officer to overreact in a stressful situation, thus leading to a violent confrontation with an offender? How can confrontations be controlled or avoided?
4. Even though police officers are taught that the best police "come along" is the spoken word, they often overlook this fact and resort to force. Why is this and what can be done to deter this kind of behavior?
5. Why is it so difficult to evaluate the "reasonableness" of an officer's actions in using force against an offender after the incident has taken place?
6. Explain what is meant by the "culture" of a police organization and how it effects the actions of its members.
7. What are organizational values and why are they important to a police agency?
8. What is the **forcible felony rule** and how does it compare or contrast with deadly-force standards issued by the Commission for the Accreditation of Law Enforcement Agencies (CALEA)?
9. Describe what would, in your view, be a sound firearms training program for a municipal police agency.
10. What are some of the factors that should be considered in choosing a firearm for use by police officers?
11. What are some of the arguments in favor of limiting high-speed chases by police officers?
12. What kind of guidelines, if any, should be placed on the conduct of high-speed chases by police officers?
13. What is the similarity, if any, between use of deadly force and high-speed chases by police officers?

REVIEW EXERCISES

The following exercises are designed to reinforce or amplify comprehension of the material contained in this chapter.

1.. Conduct a review of the leading cases in police use of deadly force or high-speed police chases. Prepare a paper summarizing the results of your research and describe any trends that you have discovered.
2.. Conduct a survey of local police agencies in your area to obtain copies of their use of deadly force or high-speed chase policies. Analyze and evaluate these policies based on the principles outlined in this chapter. Present the results of your research in classroom discussion.
3. Survey local police departments in your area to obtain information about their firearms training. Prepare a paper outlining the results and identifying any trends or common themes you may discover. Describe how, or if, the results compare with the principles outlined in this chapter.

REFERENCES

Alpert, Geoffrey P., "Questioning Police Pursuits in Urban Areas," in Dunham and Alpert, pp. 216-229.

Alpert, Geoffrey P., and Lorie A. Fridell, *Police Vehicles and Firearms: Instruments of Deadly Force (*Prospect Heights, Ill.: Waveland Press, 1992).

Brooks, Pierce R., *"...Officer Down, Code Three" (*Schiller Park, Ill.: Motorola Teleprograms, Inc., 1975), p. 95.

Chapman, Samuel G., *Cops, Killers and Staying Alive (*Springfield, Ill.: Charles C. Thomas, 1986).

Commission on Accreditation for Law Enforcement Agencies, Inc., *Standards for Law Enforcement Agencies (*Fairfax, Virginia: 1983).

Community Relations Service, U. S. Department of Justice, "Principles of Good Policing: Avoiding Conflict Between Police and Citizens," in Dunham and Alpert, pp. 164-204.

Dunham, Robert G., and Geoffrey P. Alpert, *Critical Issues in Policing: Contemporary Readings (*Prospect Heights, Ill.: Waveland Press, 1989).

Friedrich, Robert J., "Police Use of Force: Individuals, Situations, and Organizations," in Klockars, *Thinking About Police,* pp. 302-313.

Fyfe, James J., "Administrative Interventions on Police Shooting Discretion: An Empirical Examination," *Journal of Criminal Justice* (Winter 1979), pp. 309-323.

Garner, Joel, and Elizabeth Clemmer, "Danger to Police in Domestic Disturbance—A New Look," in Dunham and Alpert, pp. 516-530.

Klockars, Carl B., *Thinking About Police: Contemporary Readings (*New York: McGraw-Hill Book Co., 1983).

O'Keefe, James, "High-Speed Pursuits in Houston," *The Police Chief* (July 1989), pp. 32-34.

St. Charles, Illinois, Police Department General Order 87-06, dated April 23, 1987.

*Uniform Crime Reports: Law Enforcement Officers Killed and Assaulted 1991 (*Washington, D.C.: U. S. Department of Justice, 1991), Preface.

CHAPTER 6

BACK TO BASICS: COMMUNITY-ORIENTED AND PROBLEM-ORIENTED POLICING

CHAPTER OUTLINE

LEARNING OBJECTIVES

After completing this chapter, the student should be able to

1. List the basic principles of community-oriented policing and problem-oriented policing.
2. Describe the historical antecedents and evolution of what are now termed *community-oriented policing and problem-oriented policing.*
3. Outline the issues and problems involved in implementing problem-oriented policing or community-oriented policing as well as the results to be expected from these programs.
4. Gain an understanding of how the implementation of community-oriented policing or problem-policing will impact upon traditional principles of police organization, management, and administration.

It would seem that if all the lessons learned over the years about the successes and failures of policing could be assembled into one integrated, cohesive, and organized program, it would be possible to develop a model policing program. There are certainly enough lessons gained from the failures and successes of the past from which to draw. And yet, as American policing enters the second half of its second century of existence, it seems that new lessons continue to be learned about what works and what doesn't.

This does not mean that police administrators, practitioners, and researchers are slow learners, but simply that the evolution of policing has advanced slowly and deliberately. It also means that, throughout its history, policing has struggled to keep pace with the world around it, which has changed at a much higher rate of speed.[1]

The evolution of policing has been sporadic. In recent years, it has been aided by the rush of new technology that has made it easier to move from place to place, to communicate, to detect and apprehend criminals, to obtain and produce evidence, and to deal with some of the human problems encountered in managing any organization. Yet, despite these great accomplishments, the police seem to be still playing catch up with the world.

No one has yet figured out how to really make much of an impact on crime, and the crime rate continues to spiral upward. Despite modern advances in criminology and forensic science, most criminals are never apprehended, and a great many of those that are caught are either acquitted or get off with light sentences. Despite advances in training and education and the greater "people skills" of its officers, citizens continue to be killed and brutalized by police officers and relationships between the police and some community members are not good. Despite the fact that we have

[1]For a thoughtful discussion of how policing evolves in the future, see Malcolm K. Sparrow, Mark H. Moore, and David M. Kennedy, *Beyond 911: A New Era for Policing* (New York: Basic Books, 1990).

a bumper crop of enlightened leaders in responsible positions in police agencies around the country, we continue to be besieged by labor problems, poor morale, and instability in police organizations.

Although the advances in American policing may seem to be too few and far between, they do represent progress rather than regression. During the last three decades, significant insights have been gained about the complexity of the police role and about what works and what does not work.

For the most part, our way of policing today is much better than it was twenty, thirty, or fifty years ago. For the most part, our police officers are better educated, better trained, and more capable of dealing with the tremendous pressures of their jobs. For the most part, police departments have become much more sensitive to the problems of the community and have developed ways of cultivating harmonious relationships with community residents. For the most part, they are doing a better job of dealing with the variety of tasks that are assigned to them.

The history of American policing is marked by a number of important milestones that have, in one way or another, significantly impacted on the quality and scope of policing in this country. The advent of civil service in the late nineteenth and early twentieth centuries gradually removed the influence of partisan politics and corruptive influences from the operations and management of police departments. The development of the telephone, the radio, and the automobile had a tremendous impact on the manner in which police operations were conducted and improved police efficiency significantly.

In more recent times, the President's Commission on Law Enforcement and Administration of Justice, the National Advisory Commission on Criminal Justice Standards and Goals, and the Commission on Accreditation for Law Enforcement Agencies each have significantly affected the caliber of police services in this country. Each of these developments has helped to change the manner in which police services in this country are performed.

Perhaps the most noteworthy development in American policing in the last decade has been the advent of the **Community-Oriented Policing** and **Problem-Oriented Policing** philosophies. Community-Oriented policing and problem-oriented policing are not the same thing, although they employ similar strategies and ultimately serve the same purposes. Both philosophies mark a radical departure from the more traditional forms of policing, and it is too early to know how effective either philosophy will be or the extent to which they will significantly impact the evolution of policing in this country.

Both of these new policing philosophies offer a fresh approach to solving old problems, but neither should be seen as a panacea to all the problems that affect the cities, nor should they be viewed as the best way of delivering police services in all communities.

It is also a mistake to believe that the traditional, reactive mode of policing can be eliminated altogether in favor of proactive strategies. Police must continue to respond to citizen calls for service, investigate crimes, and attempt to identify and apprehend criminals. These functions, which are reactive in nature, must be carried out, regardless of what style of policing is chosen. Proactive strategies, which incorporate problem identification as well as resolution, must also be encouraged. Although the early results of the "new" policing strategies are promising, much more needs to be learned before it will be possible to fully assess their impact on police operations.[2]

THE EVOLUTION OF AMERICAN POLICING

To better understand why community-oriented policing and problem-oriented policing represent significant departures from traditional policing modes, it is necessary to briefly review the political development, control, and accountability of American police forces during the last century.

One of the more recent attempts to clarify our understanding of the development of the American police over the last 150 years is the evolutionary model developed by Kelling and Moore.[3] Kelling and Moore have portrayed the history of American policing as consisting of a series of models, each of which is characterized by differences in the police role in the community, relationships with political leaders and governmental officials, organizational design, management principles, and police programs and strategies. The three principal eras of American policing are shown in Table 6.1.[4]

In the early days of policing in this country, it was not unusual for city police forces to develop very close relationships with local politicians. This was often a symbiotic relationship in which each party benefited. The police were the instruments of local government and owed their power and authority to the politicians who controlled the machinery of government. Politicians, on the other hand, stayed in power only as long as they could provide needed services to their constituents. The police, having the unique powers of regulation and control, were an ideal instrument by which politicians could render services to their constituents. Thus, the close ties between local politicians and the police led to the wide range of public services that the police are called upon to perform today.[5]

Now, of course, the police perform these services because of their intrinsic value to the community and because the community has come to expect them. In modern times, it is generally recognized that there should be a clear distinction

[2]Heidi Reuter, "Not Everyone is Sold on Community Policing," *The Milwaukee Sentinel,* November 21, 1991, pp. 1A, 8A.

[3]George L. Kelling and Mark H. Moore, "The Evolving Strategy of Policing," in National Institute of Justice, *Perspectives on Policing* (November 1988).

[4]For good review of the Kelling and Moore model, see Camille Cates Barnett and Robert A. Bowers, "Community Policing: The New Model for the Way the Police Do Their Job," in *Public Management* (July 1990), pp. 2-6.

[5]George L. Kelling and Mark H. Moore, "The Evolving Strategy of Policing," in National Institute of Justice, *Perspectives on Policing* (November 1988), p. 3.

Table 6.1 The Evolution of American Policing

	Political Era	***Reform Era***	***Community Problem-Solving Era***
Legitimacy and authorization	Primarily political	Law and police professionalism	Community support (political) law, professionalism
The police function	Crime control, order\` maintenance, broad social services	Crime Control	Crime control, crime prevention, problem solving
Organizational design	Decentralized and geographical	Centralized, classical scientific management: division of labor and unity of control, bureaucratic	Decentralized, task forces, matrices
External relationships	Close and personal	Professionally remote	Consultative, police defend values of law and professionalism, but listen to community concerns
Demand management	Managed through links between politicians and precinct commanders and face-to-face contacts between citizens and foot-patrol officers	Channeled through central dispatching activities	Channeled through analysis underlying problems
Principal programs and technologies	Foot patrol, call boxes, and rudimentary investigations	Automotive preventative patrol calls for service, telephones, radios	Foot patrol, problem solving, team policing, crime watch groups
Measured outcomes	Maintaining citizen and political satisfaction with social order	Crime control (Uniformed Crime Reports)	Quality of life and citizen satisfaction

Adapted from: "The Evolving Strategy of Policing," Kelling and Moore

Source: Camille Cates Barrett and Robert A. Bowers, "Community Policing: The New Model for the Way the Police Do Their Job," *Public Management* (July 1990), p. 4.

between the political process and the administration of police organizations, even though both have the same goal of public service.

During the "reform era" of the American police, there developed a growing awareness that the close relationship that had existed for decades between the police and local politicians was contrary to the professional ethics that now were being urged upon the police. It became increasingly clear that the police had to remove themselves from local politicians and conduct themselves in an impartial and independent manner. This growing realization led to the formation of independent boards and commissions whose sole purpose was to insulate the police from local partisan politics. "Political influence of any kind in a police department came to be seen as not merely a failure of police leadership but as corruption in policing."[6]

In this way, the police became one of the most autonomous units of local government that has ever been known. Ironically, this move to isolate the police from political influences also led to the isolation of the police from the very people they were sworn to serve. Thus, while one problem was eliminated, another was created.

The Professional Model

The "professional model" of policing was an attempt by police administrators to eliminate the excesses and deficiencies in law enforcement that had been exposed during the reform era and to develop a style of policing that would be characterized by objectivity, fairness, and impartiality.

> During the era of reform policing, the new model demanded an impartial law enforcer who related to citizens in professionally neutral and distant terms.[7]

The "professional model" of policing tended to isolate the police from the community. This problem became readily apparent during the race riots of the 1960s.[8] This points out the need to change our notion of what "police professionalism" is and what it represents.[9]

One of the attributes of the professional model was that police administrators became the sole arbiters of police policies, standards, and procedures. Professional police administrators, after all, were the most qualified persons to decide how police departments should operate, what their goals and objectives should be, how their performance should be evaluated, and the manner and extent to which they should be held accountable. Lost in this equation were community attitudes, sentiments, and expectations.

[6]*Ibid.*, p. 5.

[7]*Ibid.*, p. 6.

[8]Herman Goldstein, *Problem-Oriented Policing* (Philadelphia, Pa.: Temple University Press, 1990), p. 22.

[9]Jerome H. Skolnick, *Justice Without Trial: Law Enforcement in Democratic Society* (New York: John Wiley and Sons, Inc., 1967), pp. 238-239.

The professional model also helped to create the myth that the police are the "thin blue line" of defense against crime and social disorder. They came to see themselves as the only defense against lawlessness and the other ills of society. In fact, however, although these responsibilities have been routinely delegated to the police, the police are neither wholly or even primarily responsible for controlling crime or maintaining order in our communities. Rather, these responsibilities rest with the family, the church, the schools, and other social and community institutions.[10]

The professional model of policing also helped to influence the way police viewed the role of citizens in crime control. Under the professional model, the police saw themselves as the final authority in how police problems should be solved. Like members of other professions such as doctors, lawyers, and teachers, the police strove to have full control over all matters pertaining to their field of endeavor. The police saw themselves as the best ones to determine how police problems should be solved. Citizens were not seen as important participants in the process of problem identification, analysis, and resolution.

Far too often, the police have hidden behind a cloak of professional detachment, objectivity, and neutrality that has defeated their ability to be sensitive to the thoughts, feelings, needs, and expectations of the people they are supposed to serve.[11]

In later decades, police administrators and researchers began looking for ways in which to restore public confidence in the police and to improve police-community relations. It was discovered that motorized patrol, which had succeeded in making the police officer more mobile and able to respond quickly to calls for service, had also succeeded in isolating the patrol officer from the community. Foot patrols, on the other hand, could help to bring the police back to the community.

Some critics would argue, perhaps quite convincingly, that community-oriented policing is nothing more than doing the police job in the manner in which it was meant to be done. After all, integrating police and community interests, bringing the police into closer contact with community groups, allowing community representatives to identify problems, and using resources beyond those traditionally available to the police, do not sound like particularly revolutionary ideas. There is no question that community-oriented policing is, in a very real sense, a "back to the basics" approach to policing.

On the other hand, community-oriented policing, as well as problem-oriented policing, also represent a further milestone in the evolution of policing. Moreover, they represent a dramatic departure from the style of policing that was common during the third quarter of the twentieth century. Whether they will remain as viable strategies into the twenty-first century remains to be seen.

Community-oriented policing and problem-oriented policing grew out of an increasing awareness of the inadequacies of traditional styles of policing, including:

.[10]George L. Kelling, "Police and Communities: the Quiet Revolution," National Institute of Justice, *Perspectives on Policing* (June 1988), p. 2.

[11]Oettmeier and Brown, p. 126.

1. Police departments spend too much time responding to incidents and too little time dealing with the problems that create the incidents.
2. Public sentiment will no longer allow the police to be the sole arbiters of what kind and level of police services are to be provided and the manner in which they shall be performed.
3. The patrol officer on the street is in the best position to identify community problems and is, if allowed, capable of formulating means by which those problems can be solved.
4. The traditional organization structure, which places all authority at the upper levels of the organization, may not be the best one to support decentralized decision-making.
5. Citizens will, if allowed, gladly accept the challenge of working with the police to identify and solve community problems.
6. Police performance needs to be evaluated on more than the number of crimes committed or the number of arrests made and convictions obtained. More direct measures of police impact on quality of life in the community need to be considered.

It should be recognized that community policing is not just another new public relations program, nor is it a specialized assignment within the police organization. As one of the leading proponents of community policing has observed, it is not "a particular program within a department, but instead should become the dominant philosophy throughout the department."[12]

The fundamental philosophy underlying these new "community policing" programs is that the police not only need public support in order to accomplish their goal, but that they have an obligation to involve the community in the process of deciding what they do and how they do it. This is the essence of the "new professionalism" described by Skolnick and Bayley in their review of successful policing programs in several American cities.

> The new professionalism implies that police serve, learn from, and are accountable to the community. Behind the new professionalism is a governing notion that the police and the public are co-producers on crime prevention.[13]

[12]Lee P. Brown, "Community Policing: A Practical Guide for Police Officials," National Institute of Justice, *Perspective on Policing* (September 1989), p. 1.

.[13]Jerome H. Skolnick and David H. Bayley, *The New Blue Line: Police Innovation in Six American Cities,* (New York: Basic Books, 1985), pp. 212-213.

COMMUNITY-ORIENTED POLICING

Community policing is not, as some would have us believe, a "new" policing philosophy.[14] It is, rather, a reaffirmation of the basic premise upon which policing is based. The notion that police officers working with local residents can cooperate in identifying and solving local community problems is also not new, but is fundamental to the basic purposes for which the police were created.

The race riots of the 1960s followed by student protests; peace demonstrations; and an emerging social consciousness about the plight of the poor, the homeless, ethnic and racial minorities; and a variety of other social problems helped focus attention on the role of police in society and the manner in which the police carried out their duties.

Police relationships with the community, and more specifically with certain elements within society, came under greater scrutiny. The cool detachment with which the police often viewed community problems was interpreted as a lack of sensitivity. What had once been accepted as common police practices were no longer taken for granted. The whole idea about how the police do their job came into question. Traditional beliefs about preventive patrol, response times, investigative efficiency, and a host of other issues became legitimate areas of inequity.

Community policing, like the "team policing" philosophy that emerged in the 1970s, grew out of a recognition that the police cannot afford to be isolated from the community, but rather that they are an integral element of the community service system. The emphasis is clearly one of trying to establish a harmonious relationship between the police and the community, not just so that they can "get along" but rather so that they can work together toward solving problems of mutual concern.

At the heart of any community-oriented policing program is the idea of identifying and utilizing a variety of resources to deal with and solve targeted problems. Typically, these include the resources of the police department, as well as other city departments and agencies, community organizations, neighborhood groups, civic clubs, fraternal organizations, educational institutions, and others. Too often, police officials are inclined to look no further than their own organizations when seeking resources to deal with new or unique problems. When this occurs, they shortchange their own abilities.

To police practitioners, community policing can yield a number of benefits, including (a) a sense of pride in their work, (b) a realization that patrol work can be more interesting than they thought, (c) a growth in their sense of efficiency, (d) a realization that citizens may welcome the opportunity to work with the police.[15]

Although community policing is not a "new" idea, but rather an old idea in a new package, it does differ remarkably from what is now viewed as the "traditional" model of policing. The basic theme is simple and straight-forward, but its total

[14]Robert Trojanowicz and Bonnie Bucqueroux, *Community Policing: A Contemporary Perspective* (Cincinnati: Anderson Publishing Co., 1990), p. 5.

[15]Mary Ann Wycoff, "The Benefits of Community Policing: Evidence and Conjecture," in Greene and Mastrofski, p. 111.

impact on the community can be significant. Basic differences between community-oriented policing and traditional forms of policing are illustrated in Table 6.2.

Table 6.2 Traditional vs. Community Policing: Questions and Answers

Question	*Traditional*	*Community Policing*
What is the Relationship of the police force to other public service departments?	Priorities often conflict.	The police are one department among many responsible for improving the quality of life.
What is the role of the police?	Focusing on solving crimes	A broader problem-solving approach.
How is police efficiency measured?	By detection and arrest rates.	By the absence of crime and disorder.
What, specifically, do police deal with?	Incidents.	Citizens' problems and concerns.
What determines the effectiveness of police?	Response times.	Public cooperation.
What is police professionalism?	Swift effective response to serious crime.	Keeping close to the community.
What is the essential nature of police accountability?	Highly centralized; governed by rules, regulations, and policy directives; accountable to the law.	Emphasis on local accountability to local needs.
How do the police regard prosecutions?	As an important goal.	As one tool among many.

Source: Malcolm K. Sparrow, "Implementing Community Policing," National Institute of Justice, Perspectives on Policing, (November 1988), pp. 8-9.

> For us, community policing refers to a philosophical position that holds that the goals of policing, the conditions it addresses, the services it delivers, the means used to deliver them, and the assessment of its adequacy, should be formulated and developed in recognition of the distinctive experience, mores, and special structures of local communities.[16]

Community-oriented policing strategies require not only different criteria upon which patrol-officer performance can be measured, but different strategies by which supervision can be exercised. Traditional methods of control and of monitoring the conduct of subordinate personnel are no longer relevant. In effect, this requires a greater degree of collaboration between the patrol officer and the supervisor in terms of identifying patrol goals and objectives, establishing priorities and developing problem-solving strategies. Thus, the relationship between the patrol officer and the

[16]*Ibid.*, p. 188.

sergeant becomes less formal and more collegial, and the emphasis shifts from controlling subordinate behavior to one of collaboration, cooperation, and mutual problem-solving.[17]

The "professional" model of policing emphasized the concept of decentralized control of the police. The lesson learned from the studies that questioned the value of the standards contained in the "professional" model was that the police should not invest so much of their resources in a limited number of practices which are based on an inaccurate understanding of the police role. Under the "professional" model, police administrators sought to make citizens feel safe from crime by ensuring rapid response to citizen calls for service.[18]

The return to foot patrol as a recognized means of providing basic police services during the 1970s marks an important and significant point in the evolution of policing philosophy. Once the dominant means of police patrol, foot patrol was discarded during the early part of the twentieth century as police administrators recognized the many advantages to be gained from the greater speed and visibility of motorized patrol. While foot patrols continued to be used in many of the large cities, particularly in the industrial Northeast, motorized patrol became the dominant mode of police patrol by the 1960s.

Foot patrol is an integral component of community policing. Indeed, foot patrol and community policing are often viewed as one and the same thing. This is not necessarily the case, however. Foot patrol is merely a tactic—a way of delivering police services. Foot patrol can be used primarily as a crime-prevention technique rather than as a fundamental part of a much broader community policing effort.

Community policing actually evolved out of two early experimental foot-patrol programs in Newark, New Jersey and Flint, Michigan. In Newark, the foot-patrol program was used in the late 1970s primarily as a crime reduction/prevention strategy. The foot-patrol program in Flint, which was implemented shortly after the one in Newark, was designed to accomplish more than a reduction in crime. Foot patrol officers were used "...as part of a strategy to involve officers directly in community problem-solving with the officers trained to do far more than to act as a viable deterrent to crime."[19]

Community policing, in its broadest sense, represents a significant departure from traditional styles of policing and should not be undertaken without careful planning and consideration of its consequences. Implementation of community policing in the true meaning of the term will have a dramatic impact on the organization and will require total commitment from the police chief executive as well as the full cooperation of management and supervisory personnel.[20]

[17]*Ibid.*, pp. 194-195

.[18]Herman Goldstein, *Problem-Oriented Policing* (Philadelphia, Pa.: Temple University Press, 1990), p. 19.

[19]Trojanowicz and Bucqueroux, *Community Policing*, P. 7.

[20]Malcolm K. Sparrow, "Implementing Community Policing," National Institute of Justice *Perspectives on Policing* (November 1988).

Community-oriented policing and problem-oriented policing run contrary to many traditional beliefs about the police role in society, about how the police conduct their business, about the way police departments are organized, and about the manner in which the performance of officers is evaluated. It is not surprising, then, to learn that this new policing philosophy threatens the status quo and is not always welcomed with open arms by police traditionalists, who are more comfortable with the style of policing they know best. Any attempt to convert a traditional police department to a community policing philosophy can expect to encounter significant resistance from all levels of the police organization, including top management.[21]

Community-oriented policing requires a change in the balance of power between the police and the community. Under the traditional policing model, the police alone determine what level of service is to be provided and the manner in which that service is to be delivered. Under community-oriented policing, these decisions are shared with community representatives. The notion that community groups should be involved in identifying problems and devising solutions to them may be difficult for traditional police officers, who are not accustomed to sharing their power, to accept.[22]

In addition, community policing requires that police departments be evaluated differently from the way they are under more traditional styles of policing. Rather than being evaluated on the number of crimes that are committed or the number of arrests and convictions they record, police departments may need to be evaluated on the level of citizen satisfaction with their services, citizen fear of crime in the community, or how sensitive they are to community concerns.

Similarly, community policing, with its emphasis on decentralizing the decision-making process and giving more authority to individual police officers to make decisions about what kind of services are to be provided to the public, may require that police officers be judged according to different standards of performance. Rather than being concerned with the number of arrests made or the number of self-initiated citizen contacts recorded, it may be necessary to examine more closely the quality of work. Officers in the future may come to be evaluated on the basis of their judgment and reasoning ability, problem-solving skills, human relations skills, and their sensitivity and responsiveness to community concerns and expectations.[23]

Despite the obvious advantages to be gained from adopting some form of community policing strategy, such a move should not be attempted without recognizing some of the difficulties involved. Since community policing involves a major shift in our traditional views about the role of the police in the community, how police priori-

[21]Timothy N. Oettmeier and Lere P. Brown, "Developing a Neighborhood-Oriented Policing Style," in Greene and Mastrofski, p. 130.

[22]Robert Trojanowicz and Bonnie Bucqueroux, *Community Policing: A Contemporary Perspective* (Cincinnati: Anderson Publishing Co., 1990), p. 12.

[23]David L. Carter, "Methods and Measures," in Robert Trojanowicz and Bonnie Bucqueroux, *Community Policing: A Contemporary Perspective* (Cincinnati: Anderson Publishing Co., 1990), pp. 177-178.

ties and objectives are established, how police organizations are structured and managed, and the manner in which their performance is to be evaluated, the adoption of community policing represents a major change in the delivery of police services. As such, a number of problems may be encountered.

Some have suggested that a possible negative effect of community-oriented policing is that the police run the risk of dividing the community between those who are willing to work with the police and those who are not. The police may thus find themselves caught between various community factions, causing greater alienation and antipolice sentiment.[24] Thus, any attempt to implement community policing should involve local community representatives in the planning stages so that the police cannot later be accused of implementing the program to favor or appease one particular segment of society to the detriment of another.

Since the duties of officers assigned to community policing or neighborhood policing may be dramatically different from those of officers assigned to traditional patrol, the criteria used to evaluate the performance of these officers must also be different. Criteria must be established, therefore, that will more accurately measure the **quality** of police services in terms of citizen satisfaction with the police and police commitment to community values and expectations.

> The new Community Policing Officer (CPO) serves as a generalist, an officer whose mission includes developing imaginative new ways to address the broad spectrum of community concerns embraced by the Community Policing philosophy.[25]

When the community-oriented policing program was instituted in New York City, one of the concerns was that requiring police officers to become more actively involved with community residents and merchants in neighborhoods might lead to a greater degree of familiarity that could, in turn, result in the opportunity for police corruption.[26]

It was this kind of familiarity that was cited as a root cause of police corruption during an investigation of the NYCPD in the early 1970s. Nevertheless, in initiating community-oriented policing, police officials decided that the potential benefits were worth the risk. To offset the possible risks involved, a number of control mechanisms were instituted including better screening of officers assigned to community-oriented policing and more intensive monitoring of their performance by supervisory and command officers.[27]

[24]Stephen D. Mastrofski, "Community Policing as Reform: A Cautionary Tale," in Jack R. Greene and Stephen D. Mastrofski, eds., *Community Policing: Rhetoric or Reality* (New York: Praeger Publishers, 1988), p. 57.

[25]Trojanowicz and Bucqueroux, *Community Policing,* p. 3.

.[26]Michael J. Farrell, "The Development of the Community Patrol Officer Program: Community-Oriented Policing in the New York City Police Department," in Greene and Mastrofski, *Community Policing,* p. 85.

[27]*Ibid.,* pp. 86-87.

Under community-oriented policing, decisions normally made much higher in the organization (that is, identifying police priorities, allocating resources) are made at the lowest levels of the organization by the neighborhood or community patrol officer, who is also charged with identifying problems and developing solutions to them. This can lead to supervisors and middle managers fearing that their authority in the organization is being undermined. Thus, community-oriented policing can place additional stress on supervisors and middle managers.

The traditional police organization, with its heavy reliance on a rigid chain of command, authority concentrated at the highest levels of the organization, conduct regulated by strict rules and regulations, and a lack of recognition of the decision-making responsibilities of those at the lowest level of the agency, is poorly equipped to accept or support the community-oriented policing philosophy.[28]

PROBLEM-ORIENTED POLICING

Police work is traditionally problem oriented, so the term *problem-oriented policing* would appear to be redundant. The police deal with problems of all kinds, from the serious to the minor. If there were no problems, then there would be no reason to have police. If this is so, then isn't all police work "problem-oriented?"

The term *problem-oriented policing*, as it is currently used, embraces a radically different style of policing than we have known in the past. In problem-oriented policing, the emphasis is on solving problems rather than simply dealing with them. It is certainly true that police work is and always has been problem-oriented. However, it has only been in recent times that the emphasis has been on solving the problem rather than simply responding to it.

The essence of problem-oriented policing is that more time and effort is devoted to identifying, analyzing, and eliminating problems that create the demand for police services. Often, the elimination of those problems calls for an approach involving more than typical law enforcement strategies. Too often, the police think about problem solving within the limited context of arrest and conviction. As many police departments using the problem-oriented strategy have demonstrated, alternative strategies may be much more effective in terms of long-term solutions.

Under the traditional style of policing, police officers are expected to respond to and deal with a variety of problems for which they have no solutions and for which they are poorly prepared and ill equipped. This tends to heighten their own level of frustration at their own ineffectiveness. Police officers, after all, tend to be result oriented and like to feel that their efforts are not without results.[29]

[28]David Weisburd, Jerome McElroy, and Patricia Hardyman, "Maintaining Control in Community-Oriented Policing," in Kenney, *Police and Policing,* p. 190.

.[29]George L. Kelling, "Police and Communities: The Quiet Revolution," National Institute of Justice, *Perspectives on Policing* (June 1988), p. 4.

In some respects, problem-oriented policing is neither new nor particularly revolutionary. In one way or another, police officers have always tried to deal with the underlying conditions that cause problems. In most cases, however, these efforts have been isolated, unsystematic, and sporadic. They have relied upon the creative instincts and ingenuity of individual officers rather than on the collective effort of the entire police organization.

Problem-oriented policing is not without its drawbacks, nor is it a strategy that can be incorporated into a police organization without making significant changes in the way the police agency is structured, the way police personnel are assigned, and the manner in which they carry out their duties. Frequent changes in job or shift assignments and shift rotation policies, for example, can impede the progress of problem-oriented policing strategies.

Problem-oriented policing may require a fundamental shift in how people in the police organization are treated and how their efforts to the goals and objectives of the organization are valued. Indeed, an entirely new way of thinking about the relationship between the organization and its human resources may be required.

> Changes in promotion and reward procedures, implementation of management-by-objectives, and explicit training in effective problem-solving techniques can both motivate officers to solve problems and show them that the administration is serious about its efforts.[30]

Problem-oriented policing recognizes the complexity of the police function in society and the futility of trying to evaluate police effectiveness simply on the basis of crime or arrest rates. As long as the police see themselves as being primarily crime fighters, their outlook on society and their view of their own mission in the community will be too narrowly defined. The value of problem-oriented policing is that it allows police officers to broaden their view of their role and their importance to the quality of life in the community.

Under the concept of problem-oriented policing, a number of methods might be involved in an effort to deal with a rash of burglaries in a low-income housing project. These might include, for example: (a) referring uncorrected conditions that are in violation of the law to building inspectors, zoning officials, or health authorities; (b) efforts to apprehend those responsible for the burglaries; and (c) working with school authorities regarding problems of truancy that might be related to burglaries, and with parks and recreation officials on problems relating to idle youth.[31]

The adoption of problem-oriented policing requires that police effectiveness be measured in different terms from those of the more traditional form of policing, with emphasis placed on results rather than on activities. Rather than relying on traditional measures of police effectiveness such as crime, arrest, and conviction rates, more attention needs to be given to assessing how well the police respond to community

[30]Eck and Spellman, in Kenney, *Police and Policing*, p. 109.

[31]Goldstein, *Problem-Oriented Policing*, pp. 44-45.

expectations and how well they deal with problems identified by the community. No longer can the police be the sole judges of their own performance.

Police departments are traditionally incident-driven. That is, they react to incidents rather than attempt to identify and solve the problems which create those incidents. As a result, they rely upon relatively little information concerning the incidents they handle and put forth little effort to identify resources that might be applied to resolving the underlying problems.

> "The presumed effectiveness of incident-driven policing rests on three assumptions: that marked patrol cars deter crooks; that rapid response catches them; and that when uniformed officers do not catch them, detectives will." Each of these assumptions has been proven false by empirical research, however.[32]

Some authorities nave described the traditional incident-driving police response mode as the dial-a-cop syndrome, in which the telephone system and call load dictate police priorities. "The dial-a-cop system is irrational because it prevents police from setting priorities and controlling crime more effectively."[33]

Experience has shown that, in most cities, a disproportionate number of "trouble" calls originate at a small number of locations. Police departments devote a great amount of resources toward handling problems at these "trouble" locations, but rarely solve the underlying problem. In Boston, 60 percent of the calls for service each year originated from 10 percent of the households calling the police.[34]

Problem-oriented policing is intended to serve a number of purposes including (a) tapping the accumulated knowledge and expertise of police officers, (b) enabling police officers to achieve a higher level of satisfaction in their jobs, and (c) allowing citizens to realize a higher rate of return on their investment in the police.[35] Through systematic analysis of calls to the police, the problems that give rise to them can be better addressed. In addition, police effectiveness in responding to citizen concerns can be improved greatly by addressing underlying problems.

Problem-oriented policing is not just another way of doing the same old thing, but represents a radical departure from traditional policing methods and management philosophies. At the heart of problem-oriented policing is the idea of decentralized decision making within the police organization and the notion of focusing on problems rather than symptoms.

[32]John E. Eck and William Spellman, "A Problem-Oriented Approach to Police Service Delivery," in Dennis Jay Kenney, *Police and Policing: Contemporary Issues* (New York: Praeger, 1989), pp. 99-100.

[33]Lawrence W. Sherman, "Repeat Calls for Service: Policing the 'Hot Spots'," in Kenney, *Police and Policing,* p. 151.

[34]George L. Kelling and Mark H. Moore, "The Evolving Strategy of Policing," in National Institute of Justice, *Perspectives on Policing* (November 1988), p. 10

[35]Goldstein, *Problem-Oriented Policing*, p. 29.

> First, problem solving is a central element of a problem-oriented approach, not a peripheral activity. Second, everyone in the department contributes to this mission, not just a few innovative officers or a special unit or function. And third, a problem-oriented approach places special emphasis on careful analysis of problems before officers develop solutions, and seeks to avoid instant answers that are unsupported by good information.[36]

Problem-oriented policing is intended to allow the community to define the problems to be addressed by the police. This approach recognizes the value of community residents and police working together to solve problems of mutual concern. "It is naive to assume only the police are in a position to determine neighborhood needs."[37] Problem-oriented policing allows police officers to work with those specific parts of the community that are in a position to assist in reducing or eliminating the problem.

The key element in problem-oriented policing is problem identification and analysis. Much reliance is placed upon the crime-analysis model in identifying problems and understanding them. The usual classification of events as felonies, misdemeanors, robberies, or thefts becomes less important than simply recognizing these events as problems to be solved. Information about these problems is collected from a wide range of sources and analyzed to better understand the scope and nature of the problems and to devise possible solutions to it.[38] Figure 6.1 portrays a basic model of the problem-solving system contained within the problem-oriented approach to policing.

The Problem-solving System

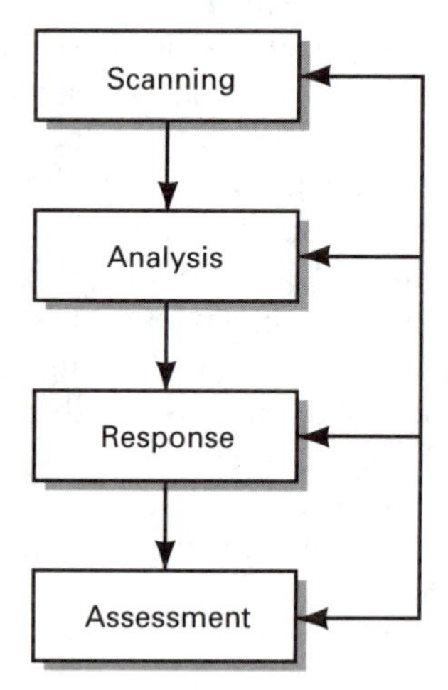

Figure 6.1 Source: William Spellman and John E. Eck, "Problem-Oriented Policing," in National Institute of Justice, *Research in Brief* (October 1986), p. 4.

[36]Eck and Spellman, in Kenney, *Police and Policing,* p. 96.

[37]*Ibid.,* p. 126.

[38]William Spellman and John E. Eck, "Problem-Oriented Policing," National Institute of Justice, *Research in Brief* (October 1986).

An important dimension of problem-oriented policing is determining what priority to assign to a particular problem so that limited resources might be put to the best use. In this regard, the following factors are important in deciding what priority to give to a problem: (a) the impact of the problem in the community, (b) the degree to which the problem adversely affects relationships between the police and the community, and (c) the interest of rank-and-file officers in the problem and their ability to address it.[39]

Because it radically alters the role of the patrol officer in defining police objectives and operational strategies, problem-oriented policing can be very threatening to supervisory personnel and command officers. Police officers on the street are in the best position to identify problems, but they rarely do so because it is not expected of them under traditional policing systems.[40] First-line supervisors who support problem-oriented policing encourage their officers to think of problems as the primary unit of work. In problem-oriented policing, officers should be rewarded for identifying, analyzing, and solving problems.[41]

COMMUNITY POLICING IN ACTION

There are literally hundreds of communities around the country that have adopted some form of community-oriented policing or problem-oriented policing.[42] Some of these are better known than others, and the results in some cities have been more remarkable than in others. The following summary provides an indication of the diversity of these programs:

Newport News, Virginia, was the first police department to attempt to implement problem-oriented policing agency-wide. The system was designed to adhere to five basic principles:[43]

1. Officers from all ranks and from all units should be able to use the system as part of their daily routine.
2. The system should encourage the use of a broad range of information including, but not limited to, conventional police data.
3. The system should encourage a broad range of solutions including, but not limited to the criminal justice process.

[39]Goldstein, *Problem-Oriented Policing*, p 77-78.

[40]Goldstein, *Problem-Oriented Policing*, p. 73.

[41]*Ibid.*, p. 1637.

[42]Robert Trojanowicz, and others, *Community Policing Programs: A Twenty-Four View* (East Lansing, Michigan: National Neighborhood Foot Patrol Center, Michigan State University, 1986).

[43]Eck and Spellman in Kenney, *Police and Policing*, p. 102.

4. The system should require no additional resources and no special units.
5. The system should be applicable to any large police agency.

In Newport News, a department task force designed a four-stage process for implementing problem-oriented policing: (a) scanning, (b) analyzing, (c) response, and (d) assessment.[44] The task force decided to test the system on two persistent problems: (a) burglaries from an apartment complex, and (b) thefts from vehicles parked adjacent to a factory.[45] Of all the experimental community-oriented policing programs, the Newport News program was the only one in which the officers addressing community problems were not separated from the rest of the organization.[46]

Houston, Texas: Neighborhood-oriented policing (NOP) is the Houston Police Department's version of community-oriented policing. Neighborhood-oriented policing is an interactive process between police officers assigned to specific beats and the citizens who work or reside in these beats to mutually develop ways to identify problems and concerns, and then to assess viable solutions by providing available resources from both the police department and the community to address the problems and/or the concerns.

To assist in the effort to ensure that police officers recognize the necessity of tailoring police services to meet community expectations, the Houston Police Department published a set of value statements that represent a set of beliefs that govern the delivery of police services in Houston. Included are the following:[47]

1. The Houston Police Department will involve the community in all policing activities which directly impact the quality of community life.

2. The Houston Police Department believes that it must structure service delivery in a way that will reinforce the strength of the city's neighborhoods.

3. The Houston Police Department believes that the public should have input into the development of policies which directly impact the quality of neighborhood life.

Baltimore County, Maryland: In 1982, the Baltimore County, Maryland, Police Department became one of the first police agencies in the United States to actually implement the problem-oriented concept. Three teams of 15 members each worked with community representatives to solve problems relating to fear of crime. The program was known as Citizen-Oriented Patrol Enforcement (COPE). An evaluation of

[44]John E. Eck and William Spellman, "Problem Solving: Problem-Oriented Policing in Newport News," in Dunham and Alpert, p. 431.

[45]*Ibid.,* p. 103.

[46]Goldstein, *Problem-Oriented Policing,* p. 63.

[47]Oettmeier and Brown, p. 122.

the COPE program showed reduced fear of crime by community residents as well as a reduction in crime in some neighborhoods.[48]

Reno, Nevada: There are many communities, large and small, with highly different needs and characteristics that have implemented some form of community policing program. The COP+ (Community-Oriented Policing-Plus) Program was implemented in the Reno, Nevada, Police Department in 1987 in response to what was perceived as declining public support for the police, an increasing population, and shrinking financial resources.[49]

San Diego, California: In San Diego, Problem-Oriented Policing (POP) is the practice of identifying and resolving neighborhood problems as the objective of the law enforcement agency. The focus is on solving community problems rather than using traditional enforcement to respond repeatedly to individual incidents. Police officers using a problem oriented approach routinely identify, analyze, and respond to the underlying conditions that prompt citizens to call for police assistance. By addressing the source of the problem rather than the symptoms (calls for service or citizen complaints), officers more effectively address neighborhood concerns. Some measures of success with POP include reduced calls for service, fewer complaints to policy makers, citizen satisfaction with police services, and officer satisfaction with solving community problems.

The process for systematically applying problem-oriented policing involves a simple four-step model. The four steps are: (1) identification of a problem, (2) analyzing or gathering of information to better understand it, (3) creating a custom-made response, and (4) assessing the effectiveness of the response. The model is not an inflexible process that officers must rigidly adhere to, but a guide to encourage officers to utilize their own problem-solving skills.

In 1988, the San Diego Police Department was one of five cities nationwide selected to participate in a demonstration project for problem-oriented policing.[50] The project was funded by the Bureau of Justice Assistance (BJA) and administered by the Police Executive Research Forum (PERF). The field test site was the Southeastern Command, one of seven area stations, where officers volunteered to use problem-oriented policing as part of their regular patrol duties. Officer enthusiasm and strong command support led to success in solving long-standing neighborhood problems. In July of 1989, it was decided to implement problem-oriented policing in all uniformed divisions.

Federal funding expired in September, 1989 but the department's commitment to expanding POP did not. PERF's San Diego field assistant for the BJA project was hired as the central coordinator to oversee the implementation effort. As the agency

[48]Eck and Spellman, in Kenney, *Police and Policing*, p. 98.

[49]Robert V. Bradshaw, Ken Peak, and Ronald W. Glensor, "Community Policing Enhances Reno's Image," *The Police Chief* (October 1990).

[50]The following program information was provided by the San Diego Police Department in 1991.

took ownership of POP, several structural changes were created to foster the development of this flexible and dynamic process.

Two committees were created to monitor the adoption of problem-oriented policing in the organization and to facilitate problem solving and resource utilization. The Implementation Advisory Group (IAG) is a committee of 13 comprised of two each from the ranks of commander, captain, lieutenant, sergeant, and officer. The central coordinator and a crime analysis supervisor are also on the committee. The IAG is tasked with ensuring the success of the implementation process by proposing changes in policies and procedures, conducting surveys and evaluations of the process, and promoting problem-oriented policing inside and outside of the organization.

The second committee is the Problem Analysis Advisory Committee (PAAC), a monthly forum providing guidance and support for department members using the POP process. Each month an agency is invited to attend the PAAC to give a half-hour presentation on issues of interest and relevance to patrol officers. The second half of the meeting involves individual officers or teams of officers who discuss their POP projects and engage the PAAC members in brainstorming ideas for creative analyses and responses. Attendance is multijurisdictional which promotes problem-solving at neighboring agencies.

Creating a coordination system for communicating and developing resources to solve beat problems was one of the first priorities of the expansion. The system that was created offers POP as an optional tool for solving beat problems, not a complicated process that increases the burdensome paperwork of officers and supervisors.

Training in problem-oriented policing continues to expand involving new groups of officers, other agencies, and community groups. Several officers who have accepted and used problem solving routinely to handle beat problems, have also assumed the role of trainers at their area divisions. As these officers eventually transfer into supervision and investigations, POP will gain a stronger foothold in the organization. The continuity and stability of the program depends greatly on the expertise and experience of the patrol officers and supervisors who use and support problem solving.

Elgin, Illinois: In Elgin, the Resident Officer Program is but one example of the unique and ambitious ways police administrators are bringing police officers into closer contact with the community. The Resident Officer Program was implemented in the Elgin, Illinois Police Department in 1991 in an attempt to combine the basic elements of community policing and problem-oriented policing in several specific neighborhoods of the city. The program began with three officers who volunteered for the program. The official mission statement for this program is as follows: "To provide proactive, community-oriented police services to a geographically defined neighborhood, said services to be provided by a sworn officer of the Elgin Police Department who lives in the neighborhood."[51]

[51]Elgin Police Department, Memorandum 91 D 37, dated 5-22-91.

The Elgin Resident Officer Program has a number of important goals, including the following:

1. To identify police, community, and quality-of-life problems in the targeted neighborhood, working directly with the residents of the neighborhood to determine the nature and magnitude of each problem.
2. To prioritize the problems as identified by the residents, planning courses of action to be implemented to effectively address same.
3. To implement the courses of action most likely to result in the solution to the identified problems.
4. To work directly with residents toward the solutions of identified problems, assessing the courses of action as work progresses and making alterations in the action plan as necessary to accomplish the solution to the identified problem.
5. To act as direct liaison with various city departments and social agencies to ensure that identified problems receive the appropriate attention.
6. To expand on the program, building on successes demonstrated in the early stages of the program.
7. To serve as the neighborhood's police officer by developing a close working relationship based on trust and understanding between the officer and residents of the neighborhood.
8. To become personally acquainted with as many neighborhood residents as is reasonably possible.
9. To coordinate the delivery of all city services in the targeted neighborhood for the benefit of residents and the success of the program.
10. To become active in all organizations in the targeted neighborhood that promote the improvement of the quality-of-life and quality structures therein.
11. To act as ombudsman for the neighborhood on issues of mutual concern.
12. To invest "sweat equity" in the neighborhood to improve same while at the same time demonstrating a dedication on the part of the police department to really improve the quality of the targeted area.
13. To maintain at all times a relationship with the residents of the neighborhood that gives reality to the historic tradition that the police are the public and the public are the police.
14. To strive to get citizens actively involved in the improvement of the targeted neighborhood.
15. To make decisions and seek solutions by working directly with residents for the improvement of the neighborhood.
16. To mold police services in the neighborhood so that the police become *a part* of the solution as opposed to being *apart from* the solution.

One of the first steps involved in the implementation of the Resident Officer Program is the identification of target neighborhoods. These must obviously be selected on the basis of some predetermined criteria. In Elgin, neighborhoods are selected on the basis of the following considerations:

1. The neighborhood is such that it could benefit from the presence and efforts of a resident officer.
2. The neighborhood is in some way defined, either geographically or by some other distinction, such as common language, ethnicity, or heritage.
3. Appropriate housing is available for the resident officer.
4. Quality-of-life issues are sufficient to warrant the presence of a resident officer.
5. Criminal activity is present and addressable by a resident officer.

The selection of officers who will be assigned to the Resident Officer Program is another important decision. Officers assigned to the program are selected according to the following criteria:

1. The ability to work successfully in a community-oriented policing atmosphere.
2. The desire to "make a difference" in a neighborhood through the identification and resolution of neighborhood problems.
3. The ability to work closely with neighborhood residents, social agencies, and other city departments for the common good of the neighborhood.
4. The desire to work toward the aforementioned program goals and objectives by voluntary association with the program.

In order to carefully monitor and evaluate the performance of the Resident Officer Program, minimum performance standards are set forth for resident officers. Within departmental rules and policies and applicable laws and ordinances, resident officers are charged with the following responsibilities.

1. To provide *basic* police services to the targeted neighborhood or area during identified hours.
2. To work proactively with neighborhood residents as is reasonably necessary to identify, address, and resolve crime and quality-of-life issues.
3. To respond to critical, key calls for service as off-duty availability allows.
4. To create effective liaison between residents and established local or neighborhood organizations (such as the Neighborhood Housing Services and/or the Housing Authority of Elgin) on matters of criminal activity and/or quality-of-life issues.

5. To visit every resident or apartment within the targeted area, delivering at the time of the visit a business card or some other form of identification.
6. To attend all school/parent and neighborhood meetings within the targeted area.
7. To develop and implement a foot-patrol program within the area.
8. To react to unforeseen circumstances and issues in a professional, forthright, caring manner.

Resident officers are responsible for their assigned neighborhoods and must remain flexible and be accessible to facilitate the success of the program. Once assigned, officers are expected to complete a minimum six-month tour of duty in their identified neighborhoods.[52]

Ashwaubenon, Wisconsin: Ashwaubenon, a town of 16,000 population located near Green Bay, is proof positive that size is no limitation on the implementation of community policing. Ashwaubenon's Community Assistance Program is an excellent example of how community policing can be implemented on a small scale and with limited resources.

Ten to twenty years ago, almost every police department had officers walking a beat. If you wanted to know what was happening on Main Street you would ask an officer. Police Officers were available to talk to members of the community and listen to complaints. With the increase in call volume, cutbacks in personnel, and mobilization of the patrol force, the officer has lost everyday contact with the citizens.

The Ashwaubenon Public Safety Department is one of the few "true" public safety departments in the United States in which all members of the department are totally cross-trained as police officers, firefighters, and emergency medical technicians or paramedics. The department has developed a program called the Community Assistance Program (CAP) designed to bring the officer and citizen back together to identify and address community concerns.[53]

The program assigns an officer to each ward, with the trustee of that ward as a partner. Together they make a team that allows citizen access to any resources necessary to solve a problem. It also provides a two-way avenue of communication between public safety department officers and the village board.

If a citizen has a problem that cannot be resolved by normal police procedure, a CAP officer can be assigned. This officer will make personal contact and will assist the citizen in taking steps toward resolving the problem. The problem may not necessarily be of a police nature. If this is the case, a referral will be made to the appropriate agency. A citizen can access the CAP program by contacting either the Ashwaubenon Public Safety Department or their village trustee.

[52]For a good overview of this innovative program, see "Their Neighborhoods Are Their Beats," *American City and County* (April 1992), pp. 32-34.

[53]The information contained in this section was provided by the Ashwaubenon Department of Public Safety.

SUMMARY

If we have learned nothing else after all the years of trial and error, experimentation, and research, it is that there are no quick fixes in law enforcement. The police are an integral element of society and are thus affected by all the influences that affect society itself. Any attempt to improve the level or quality of police services must take into consideration the environment in which the police operate.

It remains to be seen whether community-oriented policing or problem-oriented policing will have a significant long-term impact on the way police services are delivered or whether they will eventually be replaced by yet another "new" way of providing police services. What is certain, however, is that the techniques and strategies which comprise the basic principles of community-oriented policing and problem-oriented policing hold great promise and deserve to be given a chance.

REVIEW QUESTIONS

1. Why do you think this chapter is entitled "Back to the Basics"? Does this mean that American policing is starting all over?
2. Is it feasible, in light of the "new era" of policing, to eliminate the traditional, reactive method of policing altogether? Why or why not?
3. Discuss the major elements in the evolution of policing in terms of the political era, the reform era, and the community-policing era.
4. What is meant by the "professional model" of policing and what are its principal characteristics?
5. What is the basic difference between community-oriented policing and more traditional forms of policing?
6. What is the basic premise underlying the community-policing concept?
7. Explain why community policing requires a change in the "balance of power" between the police and the community.
8. Why might it not be a good idea for the police officer to become too familiar with the people on their beat? Do you consider this a problem? Why or why not?
9. What is the basic difference between community-oriented policing and problem-oriented policing?
10. What kind of changes, if any, might be required in the traditional police organization in order to implement either community-oriented policing or problem-oriented policing?
11. Explain how problems are identified and priorities established with problem-oriented policing.

12. What makes the "Resident Officer Program" of the Elgin Police Department unique?
13. What are the basic limitations, if any, of problem-oriented policing and community-oriented policing, and how might they be overcome?

REVIEW EXERCISES

The following exercises are designed to reinforce or amplify comprehension of the material contained in this chapter.

1. As a class project, research a large city in your area and pool your information concerning its crime problems, culture, political base, and other related factors. Compare the possible advantages and disadvantages of traditional policing versus either community-oriented policing or problem-oriented policing in this city.
2. Conduct independent research to identify one or more cities not mentioned in this chapter that have implemented either community-oriented policing or problem-oriented policing. Prepare a paper summarizing the reasons the program was implemented, obstacles to implementation, how the obstacles were overcome, and the results of the program.
3. Using the references at the end of this chapter, conduct a literature review of one or more community-oriented or problem-oriented policing projects. Prepare a brief paper outlining the impact on traditional principles of police organization and management that may result from the implementation of either of these programs.

REFERENCES

Barnett, Camille Cates, and Robert A. Bowers, "Community Policing: The New Model for the Way the Police Do Their Job," in *Public Management* (July 1990), pp. 2-6.

Bradshaw, Robert V., Ken Peak, and Ronald W. Glensor, "Community Policing Enhances Reno's Image," in *The Police Chief* (October 1990).

Brown, Lee P. "Community Policing: A Practical Guide for Police Officials," National Institute of Justice, *Perspective on Policing* (September 1989), p. 1.

Carter, David L., "Methods and Measures," in Robert Trojanowicz and Bonnie Bucqueroux, *Community Policing: A Contemporary Perspective* (Cincinnati: Anderson Publishing Co., 1990), pp. 177-178.

Dunham, Robert G., and Geoffrey P. Alpert, *Critical Issues in Policing: Contemporary Readings* Prospect Heights, Ill.: Waveland Press, 1989.

Eck, John E., and William Spellman, "A Problem-Oriented Approach to Police Service Delivery," in Kenney, pp. 95-111.

Eck, John E., and William Spellman, "Problem Solving: Problem-Oriented Policing in Newport News," in Dunham and Alpert, pp. 425-439.

Elgin Police Department, Memorandum 91 D 37, dated 5-22-91.

Farrell, Michael J., "The Development of the Community Patrol Officer Program: Community-Oriented Policing in the New York City Police Department," in Greene and Mastrofski, pp. 73-88.

Goldstein, Herman, *Problem-Oriented Policing*. Philadelphia, Pa.: Temple University Press, 1990.

Green, Jack R., and Stephen D. Mastrofski, eds., *Community Policing: Rhetoric or Reality*. New York: Praeger Publishers, 1988, pp. 47-67.

Kelling, George L., and Mark H. Moore, "The Evolving Strategy of Policing," in National Institute of Justice, *Perspectives on Policing* (November 1988).

Kelling, George L., "Police and Communities: the Quiet Revolution," National Institute of Justice, *Perspectives on Policing* (June 1988).

Kenney, Dennis Jay, ed., *Police and Policing: Contemporary Issues*. New York: Praeger, 1989.

Mastrofski, Stephen D., "Community Policing as Reform: A Cautionary Tale," in Jack R. Greene and Stephen D. Mastrofski, eds., *Community Policing: Rhetoric or Reality*. New York: Praeger Publishers, 1988, pp. 47-67.

Oettmeier, Timothy N., and Lere P. Brown, "Developing a Neighborhood-Oriented Policing Style," in Greene and Mastrofski, pp. 121-134.

Reuter, Heidi, "Not Everyone is Sold on Community Policing," *The Milwaukee Sentinel*, November 21, 1991, pp. 1A, 8A.

Sherman, Lawrence W., "Repeat Calls for Service: Policing the 'Hot Spots'," in Kenney, pp. 150-165.

Skolnick, Jerome H., *Justice Without Trial: Law Enforcement in Democratic Society*. New York: John Wiley and Sons, Inc., 1967.

Skolnick, Jerome H., and David H. Bayley, *The New Blue Line: Police Innovation in Six American Cities*. New York: Basic Books, 1985.

Sparrow, Malcolm K., Mark H. Moore, and David M. Kennedy, *Beyond 911: A New Era for Policing*. New York: Basic Books, 1990.

"Their Neighborhoods Are Their Beats," *American City and County* (April 1992), pp. 32 ff.

Trojanowicz, Robert, and Bonnie Bucqueroux, *Community Policing: A Contemporary Perspective*. Cincinnati: Anderson Publishing Co., 1990.

Trojanowicz, Robert, and others, *Community Policing Programs: A Twenty-Year View*. East Lansing, Michigan: National Neighborhood Foot Patrol Program, Michigan State University, 1985.

Weisburd, David, Jerome McElroy, and Patricia Hardyman, "Maintaining Control in Community-Oriented Policing," in Kenney, *Police and Policing*, pp. 188-200.

Wilson, James Q., and George L. Kelling, "Broken Windows," in Dunham and Alpert, pp. 369-381. (No. 8)

Wycoff, Mary Ann, "The Benefits of Community Policing: Evidence and Conjecture," in Greene and Mastrofski, pp. 103-120.

CHAPTER 7
POLICE PATROL METHODS

CHAPTER OUTLINE

LEARNING OBJECTIVES

After completing this chapter, the student should be able to

1. List the methods used in performing the police patrol function.
2. Explain what variables influence the selection of a method of patrol for a given area.
3. List the advantages and disadvantages of patrolling in an automobile compared to patrolling on foot.
4. Describe the specialized patrol methods used by the police in different situations to achieve different purposes.

According to one authoritative source, the word *patrol* is derived from the French word *patrouiler,* which means, roughly, "to travel on foot."[1] It is no coincidence that the foot-patrol officer continues to be the mainstay of the police patrol operation. At one time, before the development of the automobile, nearly all police departments patrolled on foot. In modern times, of course, foot patrol has been either replaced altogether or significantly supplanted through other patrol methods. As was pointed out in Chapter 6, however, the value of foot patrol has experienced a renewed appreciation in recent years.

In most modern police agencies, patrol is performed in a variety of different ways and through the use of many different methods. By far, automobile patrol is the most dominant form of police patrol. Automobile patrol may be supplemented by other patrol methods such as canine patrol, bicycle patrol, helicopter patrol, horse patrol, or marine patrol. Each of these patrol methods offers distinct advantages, but also has limitations. Developing a community patrol strategy demands creativity, flexibility, and ingenuity on the part of those managing and supervising the police patrol effort.

Various questions have been raised concerning the effectiveness of routine, random patrol in combating crime. Patrol officers have become frustrated by the lack of attention given to the patrol function by police administrators. They are not content to simply patrol their beats randomly without a specific purpose or mission.

More importantly, police officials are being forced to justify and defend their budgets and to compete with other municipal services for shrinking dollars. As a result, they are being forced to reassess the value and efficiency of all police operations, including the patrol effort. No longer can police patrol be taken for granted, as it has been for so many years. The patrol function typically accounts for the greatest portion of all police resources and is a logical place to begin the process of critical analysis.

[1]"Introduction," in Samuel G. Chapman, ed., *Police Patrol Readings*, 2nd. ed. (Springfield, Illinois: Charles C. Thomas, 1970).

It is important to point out that there is no ideal patrol method. As indicated previously, a variety of patrol methods are available to the police administrator. These can be tailored and refined to meet nearly any circumstance. In addition, there is still much to be learned about the effectiveness of different kinds of patrol methods. Thanks to research conducted during the 1970s and 1980s, we have learned something about what works and what does not, and what works better under certain circumstances. Even so, we still have much to learn about how various kinds of patrol methods can be applied to meet new and different conditions.

A number of variables will influence the type of patrol method chosen for a particular location. These variables include

1. **The size, location, and composition of the area being served.** Large cities with high density, extensive retail interests, or large areas devoted to manufacturing will require different kinds of patrol methods than will rural or suburban communities. Cities with large harbors or waterways, for example, pose a unique problem, so do communities with extensive undeveloped areas covering mountainous or heavily forested terrain.
2. **Type of crime and police problem experienced.** The kind of police problems and crimes experienced by different communities will affect the patrol method used. A high incidence of automobile thefts at a suburban shopping center will require a different patrol strategy than will an outbreak of sexual assaults in a residential area. Crime-analysis programs can help to identify offender characteristics and crime patterns that will, in turn, lead to the development of an effective patrol strategy.
3. **Community expectations.** How community leaders and public officials view the role of the police in the community and the type of services to be rendered will also affect what kind of patrol method is used. Some communities expect their departments to be primarily service oriented and feel that law enforcement and crime control are secondary objectives. Others place more emphasis on traffic enforcement and keeping the community safe from crime. These values will play a role in the selection of patrol methods.
4. **Financial, material, and human resources.** Patrol methods are also influenced by the level of resources available to support them. Patrol is the single largest operation within the police organization and accounts for most of its resources. A police organization with limited funding will be forced to rely upon rather simple patrol methods. Nevertheless, imagination and creativity can be used to make even the most simple patrol strategies effective.
5. **Relationships with other law enforcement agencies.** The existence of formal mutual-aid agreements or informal arrangements to provide and receive assistance from neighboring law enforcement agencies may affect the type of patrol method required. For example, it may not be necessary for two neighboring communities that share a common boundary on a waterway to have a marine patrol when one agency can provide this service for another. Similarly, a small-

er suburban police department may be able to call upon a larger city or county agency to provide helicopter patrol or some other specialized kind of patrol assistance. Sharing of patrol services is especially important in urban areas with a multitude of police jurisdictions.

For far too long, the police patrol effort has been taken for granted. Police administrators have paid lip service to the fact that the patrol force is the backbone of the police department, but have not acknowledged this in practice. In too many cases, the patrol force is viewed as the tailbone of the police organization rather than as its backbone. As a result, not a great deal of effort is devoted to planning, designing, supervising, monitoring, managing, or evaluating patrol activities.

In all too many cases, police patrol is conducted in a haphazard, unplanned, and unsystematic manner. Equal shift staffing is still used even when it is clear that calls for service and police activities occur at disproportionate intervals. Frequent shift rotation is used even when we know that it is counter-productive and adversely affects morale. Officers are assigned to shift according to seniority or individual preference rather than on the basis of need or the unique contributions each person can bring to the shift. Briefing sessions, if held at all, are a time for exchanging gossip or planning social events rather than seriously discussing patrol goals and objectives. Patrol supervisors are more concerned about keeping things quiet on the shift than they are about making sure the officers are doing their job. Police administrators often share this view.

Clearly, much more attention needs to be given by police administrators to the importance of police patrol and to the selection of patrol methods. This is just as true in small agencies, with limited resources, as it is in much larger cities with considerably more resources. Some issues police administrators should consider when planning or supervising police patrol efforts are: whether to use one-person or two-person patrol units; the feasibility of supplementing automobile patrol with some combination of foot, bicycle, canine, or equestrian patrol; and the circumstances under which decoy patrol or low-visibility patrol might be best employed.

This chapter examines some of the more common methods of police patrol. It attempts to point out the diversity of patrol methods that are available and the manner and circumstances in which they can be employed. The underlying theme is that patrol methods are limited primarily by the imagination and ingenuity of the police administrator.

AUTOMOBILE VERSUS FOOT PATROL

The invention of the automobile and its rising popularity in the early years of the twentieth century revolutionized the police patrol operation in many ways. The adoption of the automobile as a standard item of police equipment was one of the first attempts by police administrators to employ modern technology in carrying out the police mission. Use of the automobile for police purposes was also a recognition that it was no longer possible for a relatively small number of officers to provide the services expected of them without greater mobility. The rapid growth of the cities, and the proclivity of criminals to use the automobile to their own advantage made it necessary for the police to adapt the automobile as a dominant means of patrol.

Once introduced into the police service, the automobile quickly became the standard means of transportation for the patrol force. Although walking beats were continued in many large cities as a means of maintaining vital public contact, it soon became clear that the automobile was becoming the primary tool of the patrol officer. In some cities, it was not uncommon for rookie police officers to walk a patrol beat for a year or two before being allowed to "graduate" to automobile patrol. Riding in a patrol car was a benefit reserved only for those officers who had proven their worthiness.

By the 1970s, it was generally acknowledged by most authorities that walking beats were of limited practical value and were primarily used as a means of maintaining contact with the public and fostering favorable public relations. Walking beats were maintained primarily in downtown areas with heavy pedestrian traffic. They were used mostly during daylight hours, but some cities continued to use them around the clock, 365 days a year. It was generally conceded, however, that "real" police patrol was performed by the officer in the automobile, who could respond quickly to calls for service and who could patrol a large area in a quick, efficient, and highly visible manner.

As was pointed out in Chapter 6, however, the viability of foot patrol as legitimate and useful police function experienced a resurgence in the 1980s with the experiments in Flint, Michigan and Newark, New Jersey. Today, foot patrol has gained wide recognition as an integral component of any community-oriented policing program. More about foot patrol will be said later in this chapter.

Notwithstanding the recent success of foot patrol as an important tool in the police arsenal of crime-fighting and service-delivery methods, the automobile continues to be the dominant mode of transportation and is the primary tool of the police patrol officer. The automobile offers the patrol officer the ability to cover a large area, to respond to emergency calls quickly, and to employ a useful combination of stealth and mobility in the prevention of crime and the detection and apprehension of offenders.

It is also evident that the automobile is not the ultimate answer to providing effective police patrol. There are a number of disadvantages associated with automobile patrol, and these must be seriously evaluated when considering the most suitable method of patrol to employ in a given location. For example, one of the chief disadvantages of motorized patrol is that officers are not able to observe their surroundings when they are inside an automobile. Their ability to see, and more importantly, to hear what is going on around them is severely restricted in an automobile.

Traveling through residential streets, downtown alleys, or well-traveled boulevards at or slightly below the posted speed limit makes it almost impossible for the police officer in the automobile to detect conditions that might require police intervention. A door ajar, a broken window, a torn window screen, or a soft moan from an assault victim lying behind a tree may very well escape the attention of the officer in the automobile. An officer driving a patrol car must devote the greatest share of his or her mental faculties to the operation of the vehicle and to observing traffic conditions. Automobile patrol offers little opportunity for the detection of crime or criminals.

Another primary shortcoming of motorized patrol that has already been noted is that it severely restricts the ability of the patrol officer to come into contact with the people who reside, shop, travel through, or work in or around the beat. The automobile has had the effect of isolating the patrol officer from the public both physically and psychologically.

The patrol car is, in a very real sense, the officer's world. It is a mobile office as well as an equipment van which allows the officer to carry the tools of the trade from one place to another. The patrol car also affords the officer protection from severe weather. To a very great extent, the officer's safety and ability to perform depend upon the automobile. Many officers, therefore, take a great interest in seeing to it that their assigned automobile is kept clean and in proper operating condition.

The motor vehicle has become such an important part of the police patrol officer's arsenal of tools that it is difficult to imagine a police department operating without automobiles. Moreover, it is possible, through proper supervision, to remedy the primary disadvantages of the automobile as a means of police patrol. For example, officers can be taught more effective means of conducting motorized patrol so that they will not equate job performance with the number of miles they drive while on duty. Particularly in residential areas and on downtown streets during hours of darkness, slow, stealthy patrol is to be valued over rapid deployment over the entire beat.

Patrol officers need to be taught the value of parking the car for a few moments at various vantage points and simply listening and observing. In this way, they will be in a better position to identify sights and sounds that are out of the ordinary. They also need to be taught the value of parking and locking the car from time to time and simply patrolling for a while on foot. The officer may be surprised to learn how different the world appears outside the metal shield of the patrol car.

A number of police agencies have implemented park-and-walk programs that require patrol officers to park and lock their patrol cars at various intervals during the tour of duty and spend a designated period of time walking the beat. Since officers are usually equipped with portable radios, they are still available for emergency response and can contact the police dispatcher in the event of an emergency of their own. During the walking tour, officers may be assigned to perform specific functions such as conducting security checks of the area; visiting with area residents, shopkeepers, or passersby; or simply maintaining a high profile as a means of promoting the public's sense of security.

Officers need to develop techniques and strategies to overcome the inherent limitations of motorized patrol and, at the same time, take advantages of its best features. Rather than simply patrolling randomly from one end of the patrol beat to the other, officers need to concentrate on those areas of the beat that provide the greatest opportunity for criminal apprehension. This is not to say that the officer should entirely ignore some parts of the beat in favor of others, but rather that patrol efforts should be concentrated in those areas most likely to be targeted by criminals.

This "target-hardening" approach is simply a recognition that some areas, by their very nature, are more attractive to criminals than are others, and that these same areas should be given the greatest attention by the police. Automobile patrol can be very effective in allowing the officer to devote a disproportionate amount of time to selected areas without altogether overlooking other areas in the beat. If, for example, 45 percent of all criminal activity in a beat occurs within a two-or-three-block area, it would be advisable for the officer to devote at least half of his or her time to patrolling that area. The remainder of the officer's time can be spent in the remaining sections of the beat.

There is no question that the automobile is of vital importance to the patrol officer. Its contribution to the success of the patrol function cannot be overlooked. On the other hand, skilled police administrators will recognize that automobile patrol alone cannot provide the full range of patrol services needed in any community.

Alternative patrol methods must be considered if the patrol effort is to achieve its full potential. While the automobile will continue to be the mainstay of the patrol effort, its effectiveness will be enhanced by combining automobile patrol with other patrol methods.

SPECIALIZED PATROL METHODS

Specialized patrol, as used in this chapter, refers to any nontraditional type of police patrol method. That is, it would include any patrol method other than automobile patrol or foot patrol. A variety of specialized patrol methods are discussed in this chapter. These are viewed strictly as a means of supplementing traditional patrol methods.

In the purest sense, however, *specialized patrol* may be defined as patrol methods which are used under unique circumstances, or for specific purposes, or at particular locations to achieve desired objectives. In this case, *specialized patrol* is not viewed as simply a supplemental patrol method, but rather an entirely different way of patrol designed for purposes which might render routine patrol methods ineffective. For example:

> Specialized patrol is defined as the activities of officers who are relieved of the responsibility for handling routine calls for service in order to concentrate on specific crime problems. Its primary purposes are the deterrence of suppressible crime problems and the on-site apprehension of offenders.[2]

[2]Stephan Schack, Theodore H. Schell, and William G. Gay, *Improving Patrol Productivity, Volume II: Specialized Patrol* (Washington, D.C.: National Institute of Law Enforcement and Criminal Justice, 1977), p. 1.

Given this definition, one might question whether all police patrol efforts might be so classified. In fact, however, regular police patrol addresses a much broader area than that described in the preceding definition. Clearly, specialized patrol as defined in the preceding extract is directed almost exclusively toward the prevention of "suppressible" crimes and the apprehension of offenders, while general patrol has a much broader service orientation.

The kinds of specialized patrol method included in this definition are decoy patrol, surveillance of places and offenders, aggressive patrol tactics, and other measures intended to allow patrol efforts to be concentrated upon a specific type of offense, offender, or location. Although it is possible for these kinds of patrol strategies to be performed by members of the regular patrol force, they are most often employed by specialized patrol units, often called tactical units or special operations teams. More will be said about these specialized patrol methods later in this chapter.

Although specialized patrol can be a valuable supplement to the regular patrol force, it should not be viewed as a substitute for regular patrol efforts or occupy a position of greater importance than normal patrol activities. Every attempt should be made to coordinate specialized patrol efforts with normal patrol activities in order to achieve full harmony of effort.

It is also important to recognize that any kind of specialized patrol activities have certain inherent limitations.[3] For example, it is often difficult to coordinate the activities of specialized patrol units with those of regular patrol officers. In some cases, specialized patrol forces are under the supervision of an officer who is not in the chain of command of the regular patrol force. Thus, sharing of information and communication between units becomes extremely important.

Officers assigned to specialized patrol units sometimes are treated more favorably than those assigned to regular patrol in terms of shift assignment, car assignment, work schedule, pay, uniforms, and equipment. This tends to increase morale among specialized patrol officers, but to the detriment of morale among officers assigned to regular patrol. Once again, it is important to treat specialized patrol units as an adjunct to the regular patrol force rather than as a separate and distinct function within the department. Table 7.1 compares the relative advantages and disadvantages of specialized patrol.

There are a number of specialized forms of patrol that are available to the progressive patrol administrator. While some of these patrol methods are used only under specialized circumstances, they can be a valuable supplement to the regular patrol force. Some of these more specialized patrol methods will only apply to communities with unique characteristics, such as a waterfront or waterway, others can be employed by most police agencies in a variety of situations.

Bicycle Patrol

Many police departments have adopted bicycle patrol for the purpose of patrolling areas that are not easily accessible by automobile or that are too large to be patrolled

[3] *Ibid.*, pp. 10-13.

Table 7.1 Advantages and Disadvantages of Specialized Patrol

General Advantages	*General Disadvantages*
Clear placement of responsibility.	Problems of coordination and cooperation between specialized units and the general patrol force.
High levels of morale and *esprit de corps.*	Negative effects on the morale of general patrol officers.
Improved training and skill development.	Problems in maintaining unity of command.
High level of staff commitment to assigned responsibilities.	Reduction in general patrol coverage.
Positive public interest.	Empire building by specialized units.
	Interference with officer development.
	Negative public attention.

Source: Stephen Schack, Theodore H. Schell, and William Gay, *Improving Patrol Productivity, Volume II: Specialized Patrol*, (Washington, D.C.: National Institute of Law Enforcement and Criminal Justice, 1977), p. 14.

easily on foot. Bicycle patrols can be used for surveillance purposes, or to patrol parks, shopping centers, and other crowded areas where a uniformed officer would be too conspicuous. Bicycle patrol in such cases is usually targeted toward criminal detection and apprehension in crimes such as purse snatching, mugging, and sexual assault. Bicycles have the combined advantages of mobility, speed, and stealth. They are also easy to maneuver in crowded areas and cost very little to operate and maintain.

Bicycle patrols are often used by plainclothes patrol officers for conducting surveillance in high-crime areas where it is not possible to place a uniformed officer, either on foot or in an automobile. Officers assigned to such duties should be equipped with small, lightweight, portable radios so they can be in constant contact either with the police dispatcher or with other patrol units in the area. It is best, in such cases, to have officers assigned to bicycle patrol to work in tandem with regular patrol forces in the area. Although the bicycle is an excellent way of patrolling an area under special circumstances, back-up units from the regular patrol force need to be immediately available in the event it becomes necessary to chase an offender.

It is also important that officers assigned to bicycle patrol be rotated often. Although it has many advantages, bicycle patrol is also limited in scope and purpose, and officers should not be assigned to this kind of patrol duty for extended periods of time. While some officers may enjoy the relative freedom attached to bicycle patrol, others may find it too restrictive and boring. It is therefore important that bicycle patrol assignments be made, whenever possible, from the regular patrol force on a temporary basis.

Officers assigned to bicycle patrol must be physically fit and prepared for the physical exertion that may be required in the event of a long chase. Depending upon the kind of bicycle being used—-some are quite expensive and sophisticated—-it may be necessary to provide specialized training to officers assigned to bicycle patrol. Bicycle riding can be a real challenge for the person who is not accustomed to traveling on two wheels, and it should not be assumed that all persons are equally capable of performing this function with skill and ease.

In some communities, where the weather permits, bicycle patrols may be used all year and may be a permanent part of the patrol force. This would be true, for example, in California or Florida, where beach areas are particularly suited to bicycle patrol. This would also be true in cities in the southwestern United States where warm weather and sunny skies are prevalent. In communities with four seasons, however, bicycle patrol must necessarily be restricted to those few months of the year with sunshine and clear skies. Due to its obvious limitations, it is not recommended that bicycle patrols be used at night, except in areas where frequent automobile traffic will not pose a problem. In all cases, the safety of the officer must be the paramount consideration.

Just as patrol cars, weapons, and other types of police equipment should be checked periodically to ensure their proper working condition, it is also important that bicycles be taken in for maintenance on a regular basis. If there is a member of the department who is a bicycle enthusiast, it may be possible to assign that officer the additional duty of performing routine maintenance on the bicycles, either as an on-duty assignment or as a paid detail. Since most mechanics are not familiar with the delicate mechanisms of modern multiple-speed bicycles, however, it is possible that a local bicycle shop may need to be contacted to perform periodic preventive maintenance and adjustments on the bicycles.

Bicycle patrol offers many unique advantages and can be used as a valuable supplement to the regular patrol force. However, when thinking about implementing a bicycle-patrol program, the limitations as well as the advantages of bicycle patrol should be carefully considered. Careful planning should go into the objectives to be achieved, the methods and circumstances under which it will be used, and the means needed to coordinate bicycle patrol with the regular patrol force.

Special Terrain Patrol

While police patrol is generally considered an activity to be conducted on residential streets, busy highways, and busy downtown boulevards, police patrol operations are also conducted in areas where traditional forms of patrol are impractical. Some patrol operations, like those conducted in remote, mountainous, or heavily forested areas will require special-purpose vehicles to traverse the landscape. This is also true of those remote or rural areas in the northern parts of the United States which experience frequent and heavy snowfall. There, all-purpose vehicles with four-wheel drive and low-range transmission can be invaluable for conducting routine patrol operations as well as for emergency search and rescue.

Heavy and frequent snowfall poses a unique problem for police agencies in some parts of the United States. Lightweight, highly mobile snowmobiles have been found to be a particularly effective way of traversing areas where heavy snowfall impedes access by more conventional means. Officers can be trained to operate these vehicles safely with a minimum effort. A good program of preventive maintenance for these vehicles is critical, since an officer's life may depend upon reliable operation of the unit in a remote area. For this reason, it is also important that officers assigned to these vehicles be equipped with a reliable portable radio as well as a basic first-aid kit and other emergency equipment.

Military-type jeeps equipped with four-wheel drive have been used successfully by some police agencies for traversing sandy areas such as the desert and beach fronts. These vehicles are particularly well-suited for patrolling crowded beach areas as a means of enforcing local ordinances against drinking and the use of drugs as well as for emergency rescue purposes. These vehicles permit officers to respond quickly to emergency scenes which may be inaccessible to more conventional forms of transportation. Due to their size and construction, however, it is not recommended that these vehicles be operated routinely on city streets or for traditional patrol purposes. They should be viewed strictly as a specialized means of patrol.

Some limited use has been made of three-wheel all-terrain vehicles for patrolling areas where the unique characteristics of these vehicles are particularly useful. The City of Huntington Beach, California has eight three-wheel all-terrain vehicles that are used for patrolling beach areas and the adjacent parking lots. A ten-person team, headed by a sergeant, is assigned to this special function. Officers assigned to this unit wear special uniforms consisting of dark blue shorts and white short-sleeve shirts.[4]

Motorcycle Patrol

For many years, two-wheel motorcycles were used almost exclusively by police departments for the purpose of traffic enforcement, escort, and parade duty. The two-wheel motorcycle, supplemented by the automobile, has traditionally dominated the field of traffic law enforcement. This is largely due to the fact that the motorcycle offers several distinct advantages over the automobile, including maneuverability in heavily congested areas.

The use of the motorcycle has come under considerable criticism in recent years, however, and a number of police departments have sharply curtailed or abandoned their use altogether. This has been due in part to the relatively high cost of operating the motorcycle. A study conducted by the Metropolitan Police Department in Washington, D.C. in 1974, for example, revealed that per-mile costs for motorcycles were substantially higher than those for conventional automobiles. As a result of this study, the Metropolitan Police Department adopted slower, less-expensive, two-wheel and three-wheel motor scooters in lieu of standard police motorcycles. These smaller vehicles have been found to be just as maneuverable as their larger counterparts but less hazardous and costly to operate.

[4]Correspondence received from Lee Camp, Special Projects Officer, Huntington Beach Police Department, dated September 4, 1990.

Another factor leading to the decline in the use of the motorcycle has been the trend away from specialized traffic units in favor of returning the responsibility for overall traffic enforcement to the patrol force. This change has been made in recognition of the fact that in many cases, regular patrol units can perform accident investigation and traffic enforcement duties just as effectively as specialized units.

Another important disadvantage of the solo motorcycle is its limited utility under adverse weather conditions. Typical weather conditions in many parts of the United States make it difficult to use motorcycles more than a few months each year. During inclement weather, it may be necessary to assign the motorcycle officer to a regular patrol beat or to other duties, such as preparing monthly statistical reports, updating pin maps or similar work. As a result, the skills of the officer are not being properly utilized.

It is also true that motorcycles are dangerous. They add an element of increased risk to an occupation that has more than its share of risk to begin with. Even though officers assigned to motorcycle patrol must be specially trained in the use and operation of the motorcycle, and even though most motorcycle officers are very conscientious and safety oriented, they cannot control the behavior of other irresponsible drivers on the road. Most motorcycle officers are aware that they are very vulnerable to the actions of unthinking drivers. Most motorcycle accidents in which officers are injured or killed in the line of duty are the result of actions by other drivers. The tragic death or serious injury of a motorcycle officer may be all that is needed to bring about the end to a police department's motorcycle patrol program. This is unfortunate, since this program can, if properly used, be a tremendous asset to a police department's regular patrol activities.

Aircraft Patrol

Advanced technology and changes in society have substantially altered the nature and scope of police patrol operations in recent years. Police administrators can no longer afford to rely upon traditional means of performing the basic police functions, but must be alert for ways to adapt advancements in technology to new and changing police problems.

One of the most significant developments affecting the patrol function in the last two decades has been the adaptation of aircraft—both fixed-wing type and helicopters—for police use. Many law enforcement agencies have found aircraft patrol to be a valuable adjunct to regular patrol operations. Both fixed-wing aircraft and helicopters have been found to be very adaptable for police work, although they are used primarily by larger cities, county law enforcement agencies, state highway patrol, and state police agencies.

The Los Angeles Sheriff's Department was one of the first law enforcement agencies in the country to experiment widely with helicopter patrol, although the New York City Police Department is reported to have purchased its first helicopter in 1947. During the last two decades, a number of cities and counties in the United States, as well as many others abroad, have adopted both helicopters and fixed-wing aircraft for police use. State and county law enforcement agencies have found various forms of air patrol to be useful in patrolling large land areas, particularly those inaccessible to conventional groundcraft. Moreover, helicopters equipped with emergency medical personnel and equipment are ideally suited to evacuation and search-and-rescue duties.

Helicopters have disadvantages, however. The costs of purchase, operation, and maintenance of helicopters is beyond the resources of all but a few jurisdictions. In some instances, particularly in metropolitan areas, these costs may be diminished by cooperative programs involving several police agencies on a shared-cost basis. Since helicopters, as well as fixed-wing aircraft, have a considerable operating range, it is possible for a number of smaller police agencies to share the cost of one or more aircraft.

When considering the use of a helicopter for patrol operations, cost alone should not be the determining factor. There are a number of other points that must be considered. As the authors of one evaluation study of helicopter use in law enforcement pointed out:

> The case for or against the use of helicopters in police work is impractical in terms of pure economics. Increased officer security, better community satisfaction with police services, and such benefits as indirect improvement of ancillary services (such as crime prevention, public service, etc.) are not subject to objective measurement.[5]

Another disadvantage of the helicopter, as it is used in law enforcement, is that residents often complain about the noise created by the aircraft as they fly over residential areas, particularly during evening hours. This problem may be at least partially minimized in the future by the development of noise abatement devices by the helicopter industry.

It must also be recognized that many people are simply fearful or resentful of the helicopter as a police patrol method, owing to the greater powers of observation and surveillance that such craft provide. Even persons innocent of any wrongdoing may feel uncomfortable if they feel they are being "spied upon" by the police department from above. These fears can usually be put to rest by the police through carefully designed public information programs that stress the purpose and use of helicopter police patrol.

On balance, both helicopters and fixed-wing aircraft can be valuable assets in carrying out the police patrol mission under certain circumstances and if properly used. They offer the opportunity to patrol large land and sea areas and observe things that may be hidden from the view of officers on the ground. During crimes in progress, observers flying at several hundred feet above the ground can keep fleeing suspects under close observation and can direct the apprehension efforts of ground

[5]C. Robert Guthrie and Paul M. Whisenand, "The Use of Helicopters in Routine Police Patrol Operations: A Summary of Research Findings," in S.I. Cohn, ed., *Law Enforcement Science and Technology*, 2 (Chicago: IIT Research Institute, 1968), p. 554.

units. Aircraft patrols can also be used in spotting criminal activity not normally visible to ground-patrol units. While their operating costs put them beyond the reach of small jurisdictions, they can and do play an important function in the patrol effort.

The Ohio State Highway Patrol (OSP) has been one of the pioneers in the effort to adapt aircraft for general law enforcement purposes. The OSP purchased its first fixed-wing aircraft in 1947. It was used primarily for tracking fugitives, traffic control, disaster assistance, and errands of mercy. The first air-to-ground speed check occurred in 1952, using techniques that are, by modern standards, somewhat rudimentary. Much more advanced techniques relying upon coordination between patrol aircraft and ground support units are employed today. More recently, the OSP has developed another innovative program involving a cooperative effort between the OSP and local law enforcement agencies. The program is designed to assist local law enforcement agencies with their traffic safety problems and to promote interagency cooperation for the solution of regional traffic safety problems.[6]

The Los Angeles Police Department (LAPD) is another law enforcement agency that has a long history of using aircraft as a part of its patrol force. The department's use of aircraft dates back to 1957 when its helicopter unit logged 775 hours of operational time in support of ground unit operations. The LAPD has been especially successful in adapting helicopters to law enforcement purposes. By 1990, the department's Air Support Division included 15 turbine-powered helicopters, one military surplus helicopter and a six-place single-engine aircraft. The division has 77 sworn officers and civilian employees who log over 209,000 hours of flight time annually.[7]

Most police agencies obviously have neither the need nor the resources to justify utilizing aircraft for patrol purposes. On the other hand, practically any law enforcement agency can expect to have a need for aircraft patrol at one time or another. When such situations arise, most agencies seek assistance from a larger police agency or a state law enforcement organization. This is an ideal example of the kind of specialized patrol functions that can be provided on a limited, as-needed basis through interagency cooperation.

Canine Units

Canine units have been used for law enforcement purposes and by military and civil security agencies for years. Indeed, the use of canine units for security patrols dates back several centuries. According to Chapman's authoritative work on the subject, the South Orange, New Jersey Police Department and the New York City Police Department became the first two police agencies in this country to implement canine units in 1907.[8]

[6]Thomas W. Rice, "A Case for MACE," *The Police Chief* (July 1989), pp. 51-53.

[7]Barry A. Bowman, "Commemorating the 30th Anniversary of Airborne Law Enforcement in LA," *The Police Chief* (February 1989), pp. 17-18.

[8]Samuel G. Chapman, *Police Dogs in North America* (Springfield, Illinois: Charles C. Thomas, 1990), p. 15.

After World War II, the use of dogs in police work spread rapidly from England to the United States. By 1969, some 150 police agencies in the United States had adopted some form of canine patrol. Chapman estimates that there have been some 3,000 police canine programs in 3,500 cities, and that there have been as many as 7,000 dog-handler teams in those programs.[9] Moreover, canine-unit programs are not restricted in terms of the size of the city or the geographic area. Today, police dogs have continued to prove their usefulness in a variety of police-patrol applications and environments.

Canine units have many advantages over other forms of police patrol. They are often used in lieu of a second officer in a motorized patrol unit. This serves as a supplement to the regular patrol force and allows fewer officers to patrol the same or greater area. Properly trained dogs are virtually fearless and extremely loyal to their handlers. They also have a significant psychological effect on would-be trouble makers. Officers assigned to high-crime areas have little to fear when working with a well-trained police dog.

Dogs also have an acute sense of smell and can be used to track lost or wanted persons and to help in search-and-rescue missions as well. They can also be used to detect contraband. Dogs have been used quite successfully, for example, by federal customs agents and by state and federal drug agents in searching packages, automobiles, and persons for suspected narcotics. They have also been used to search aircraft and bus, rail, and air terminals for suspected explosives. Moreover, they have been found to be extremely adept in searching large warehouses and similar areas for thieves and burglars hiding within.

Local law enforcement agencies have used police dogs for crowd-control purposes. Their use in such situations has sometimes provoked public criticism, though, particularly when used in demonstrations involving great public emotion. It is well known that the use of police dogs against the civil-rights demonstrators in some southern cities during the 1950s not only brought criticism of the police, but resulted in great public sympathy for the demonstrators. As a result, the use of police dogs in such situations should be considered only with great care and recognition of the potential liability that might ensue if an innocent person were attacked or harmed by a dog.

While canine units represent a very useful addition to the police patrol force, they also have certain limitations. First, dogs are trained to work with only one trainer. If that officer is later transferred, promoted, or otherwise reassigned, both the officer and the dog must be replaced. In addition, dogs require special facilities, training, and handling. Officers assigned to canine units are often responsible for boarding, feeding, and training the dogs. This requires a very special commitment that not all officers can or are willing to make. In addition, cars assigned to canine patrol cannot be used for any other purpose. Regular in-service training for both the dogs and their handlers is required in addition to the officer's normal in-service training program.

On the positive side, a canine program is fairly inexpensive and simple to operate. Where canine units have been used, they have performed quite successfully.

[9] *Ibid.*, p. 27.

Dogs possess many natural abilities that are particularly well-suited to police patrol work.

The manner in which dog units are deployed will depend upon the needs of the individual police agency. Ideally, it is good to have one canine unit assigned to each shift so as to eliminate the need to call in a unit on an overtime basis. Usually, the canine unit will be assigned to a regular patrol beat and will be expected to answer calls for service like any other patrol unit. In some cases, however, the canine unit will serve as a backup unit or "floater" car, available to respond to particular situations in which the unique skills of the canine will be needed.

Before implementing a canine unit, it would be wise to contact other law enforcement agencies that have proven track records in the use of police dogs. Talk with the dog handlers and their supervisors to obtain their views about the program. Officers who express interest in being assigned as dog handlers should also be given first-hand information about the program and allowed to see how such operations are conducted. The selection of the dog or dogs to be used is just as important as the selection of the dog handler. Both dog and handler must be given extensive training before the unit can become fully operational. Careful planning and sound management are the key to any effective police canine program.

Marine Patrol

Marine, or water-patrol units, are a highly specialized form of police patrol and are used in those communities that have access to navigable waterways such as lakes, oceans, and rivers. Where their use is warranted, they represent an extremely valuable addition to the regular patrol force. Marine patrol units may be full-time or part-time operations, depending upon the particular situation. They may consist of one or more water-craft, ranging from small speedboats used for limited patrol and water-rescue operations, to much larger craft used for patrolling large areas of open water at great speeds.

Depending upon the requirements of the local area, police marine units may serve several distinct purposes. Their duties may include, for example, water safety and protection, search and rescue, antismuggling operations, general surveillance, and general enforcement of local or state marine regulations. In cities with large waterfront operations or extensive marinas, marine-patrol units play an important role in guarding against theft, sabotage, smuggling, and other illegal operations. In coastal areas where water recreation and tourism are major attractions, marine units may patrol waterways to enforce boating regulations and to protect life and property from the many hazards associated with recreational boating.

Search and rescue and fire detection are among the most important functions of marine patrol units. They are very useful in patrolling inland waterways that may be otherwise inaccessible. In many instances, police marine units are called upon to work closely with the United States Coast Guard; United States Customs Service; state conservation agencies; and other state and federal agencies in the enforcement of boating regulations, the prevention of narcotics trafficking, detection of smuggling, and other serious crimes which may be beyond the scope or capability of traditional patrol units.

Marine units can be very expensive to operate, especially those that are used only on a part-time basis. Depending upon their size and sophistication, boats are expensive to purchase and to maintain whether they are used all year or only during summer months. Officers assigned to marine units must possess special skills in oar handling, water navigation, and water-rescue operations as well as extensive knowledge of boating regulations and state and federal laws pertaining to the unique nature of their duties. In small police agencies, or those needing marine patrol only on a limited basis, qualified personnel may be hard to find; this should not be a major problem in larger police agencies.

Due to the costs involved in purchasing, equipping, staffing, and maintaining marine patrol units, small police agencies would be well-advised to find a way to share the costs of owning and operating a marine unit with one or more neighboring police agencies. Indeed, this kind of specialized patrol function is most efficiently managed through a combination of interagency cooperation, joint ownership, and contract services.

One-Officer Versus Two-Officer Patrol Units

Whether to use one-officer patrol units or two-officer patrol units is a question that has been debated among police authorities for many years. Despite a few empirical studies that have been conducted to assess the relative effectiveness, advantages, and disadvantages of one-officer versus two-officer units, the issue has never been satisfactorily resolved. It is clear that there are many good arguments on both sides of the question. In addition, it can be a very emotional issue in some quarters.

It is certainly true that there are many instances during a "normal" patrol shift when more than a single officer should be assigned to a call. Multiple officers are necessary, for example, at a family quarrel, a disturbance, a robbery in progress, a man-with-a-gun call, or a silent burglar alarm. There are dozens of other typical police calls where more than one officer will normally be dispatched. The issue is not whether more than one officer is required to handle the call, but rather whether the officers so dispatched should be assigned to one-officer or two-officer units.

There are clearly some areas in large cities where the kind of activity routinely encountered by patrol units almost certainly demands a two-officer unit. High-crime areas which are afflicted with a high incidence of violent crimes, weapons violations, and drug trafficking, and where respect for the police is not a prevalent attitude are locations where the police need to be constantly on guard and alert. A single-officer police unit in such an area would be largely ineffective.

All other things being equal, officer productivity and operational efficiency can be increased through the use of one-officer patrol units. There are many reasons for this, as will be discussed. Perhaps the real question is not whether to use one form or the other, but rather under what circumstances either is most appropriate. In most instances, the single-officer patrol unit is the rule rather than the exception. Police administrators, facing budget cutbacks, have had to find ways to make limited resources go much further than they used to; the harsh reality is that two-person patrol units are simply a luxury they can no longer afford

The use of two-officer patrol units should generally be restricted to those areas, shifts, and types of activities that are most likely to pose a threat to the safety of the officer. As a rule, two-officer units are used most frequently during the evening hours when hazardous calls occur with greater frequency. In addition, there are some sections of large cities where a patrol officer working alone, either on foot or in a patrol car, might be the target of an assault or other type of attack. It is in such situations that two-officer patrol units are clearly justified.

Those who support the single-officer concept stress the fact that such units are more cost effective since the same number of officers can patrol twice the area. Those who favor two-officer units, on the other hand, insist that two officers working as a team create an added element of safety and that this should be the paramount consideration. Although this is certainly true in some cases, it is also true that, in many cases, officers working alone are much more cautious when approaching dangerous situations since they know that they have only their own wits and skills to depend upon. Officers working together may prove to be a very good team, but their powers of observation are sometimes diminished because they spend time socializing with each other.

Although the issue of one-person versus two-person patrol units has been hotly debated, very little empirical research has been conducted to determine the relative merits of either form of police patrol. Indeed, there has been very little attention paid to this subject during the last decade. In one study, conducted by the University of Oklahoma Research Institute in 1973-1974, it was discovered that the number of officers present at the scene of a police-citizen confrontation did not necessarily reduce the probability of an officer's being assaulted. The same study also revealed that nearly two-thirds of the officers suffering on-duty assaults were assigned to two-officer units and that less than 15 percent of the reported assaults occurred with only one officer present. Researchers concluded that the evidence did not support the belief that officer safety is enhanced by the use of two-officer patrol units.[10]

Another study conducted in the San Diego Police Department under the auspices of the Police Foundation was designed to specifically evaluate the overall efficiency and effectiveness of the single-officer patrol unit. The San Diego study, featuring rigorously controlled conditions, compared single-officer patrol units and two-officer patrol units on several different measures. Among the findings of the study were the following:[11]

[10]Samuel G. Chapman and others, *Perspectives on Police Assaults in the South Central United States,* I (Norman, Oklahoma: The University of Oklahoma, 1974), p. 145.

[11]John E. Boydstun and others, *Patrol Staffing in San Diego: One or Two-Officer Units* (Washington, D.C.: Police Foundation, 1977).

1. While overall performance for calls for service and officer-initiated activities were judged to be equal, more traffic citations were issued by two-officer units.

2. Overall efficiency of one-officer units was judged to be higher than that of two-officer units, even though the former required more backup support.

3. While citizens resisted arrest less frequently when dealing with one-officer units, overall safety was judged to be equal between both types of units. Assaults on officers, officer injuries, and vehicle accidents occurred at the same rate for both single-officer and two officer units.

The authors of the San Diego study, while supporting the overall superiority of the single-officer patrol unit, pointed out that the results of the study could only be applied to the San Diego Police Department and that similar studies should be conducted in other cities before the findings could be deemed to have general applicability. Nevertheless, the results of the San Diego study lend strong support to the use of single-officer patrol units in most cases.

The relative advantages of one-person patrol cars in terms of police-public relations were revealed in another study by Decker, in which he found that:

> ...citizens were found to be more likely to be injured in hostile police-citizen encounters than the police, especially when two officers were present. Officers in two-officer units were more likely to arrest the citizen in that incident which precipitated the complaint and the arrest of that citizen was more likely to be for the more serious charge of assaulting an officer. The results suggest that lone officers resolve more disputes without resorting to formal outcomes.[12]

Because the issue is an emotional one, there is sometimes a tendency to make judgments based on rhetoric rather than on hard evidence. The tragic killing of a police officer responding to a family disturbance or to a robbery in progress, for example, may be cited as "evidence" of the need for more officers and the "beefing up" of patrol strength. Proponents of two-officer patrols often seize upon these situations to argue that single-officer patrol units are not safe, even though this is simply not the case. A careful evaluation of each police killing or serious assault on a police officer on a case-by-case basis will reveal that officer carelessness, improper procedures, inadequate training, defective equipment, or poor supervision are just as much to blame as the number of officers in the car.

Police unions frequently support the use of two-person patrol cars as a safety issue. Once again, emotion often speaks louder than fact or reason. It is difficult to argue against officer safety, and it is certainly not a good thing to do politically. As a result, some police unions have managed to have two-officer patrol units included in the union contract. As most police administrators know, once this kind of language gets into a contract, it is very difficult to have it removed. The best thing, obviously, is to keep this issue out of contract negotiations.

[12]Scott H. Decker, "The Impact of Patrol Staffing on Police-Citizen Injuries and Dispositions," *Journal of Criminal Justice*, 10 (1982), p. 375.

While there is clearly a role for two-officer patrol units, they should only be employed in those situations where the circumstances clearly warrant their use. At the same time, the following factors should be considered when employing single-officer units:

Training. Obviously, the tactics employed by single-officer patrol units are different from those employed by two-officer patrols. Training programs should be designed with this difference in mind. Special training should be provided in patrol tactics such as vehicle and pedestrian stops, field-interview techniques, arrest procedures, and other operations which may pose a hazard to an officer working alone. The emphasis should be clearly on protecting the safety of the officer while not detracting from individual productivity. Proper training can help reduce the risk of injury to the officer, while at the same time make the officer more prudent and less inclined to take unnecessary or unreasonable risks. In addition, officers should be properly trained in the transportation and handling of prisoners, since this is an activity which often leads to assaults against officers.

Policies and Procedures. Policies and procedures should be revised and updated, if necessary, to reflect the unique situations in which single-officer or two-officer units are to be employed. In particular, tactical situations which typically involve multiple-officer response should be addressed to define how either single-officer or two-officer units, or a combination of both, are to be employed. Responding to a silent alarm at a bank during business hours, for example, may require a different tactic if single-officer units are used rather than two-officer units. The same thing can be said of searching a large warehouse or office building at night where a suspected break-in has occurred.

Deployment. Since a greater number of officers can be deployed when single-officer units are used, beat plans should be designed to reflect this fact. Beat plans should be designed to accommodate a varying number of patrol units. Otherwise, it is often the case that officers are "doubled up" when there happen to be a greater number of officers working than there are available patrol beats or patrol vehicles. This does not reflect good planning or a judicious use of available resources.

Call Information. While it is always important that sufficient information concerning a call to which an officer is assigned is obtained as early as possible, this is especially important when single-officer patrol units are being used. The proper training and supervision of police dispatchers is vital to the patrol function. Police dispatchers should be trained in police deployment and response strategies so that they can mentally ride along with the officer responding to the call. By imagining himself or herself alongside the officer, the dispatcher can better appreciate the kind of information concerning the call that the officer is likely to need. It is equally important that officers assigned to single-officer patrol units keep the dispatcher fully informed as to their location and activity. In effect, the dispatcher becomes the officer's partner, since the dispatcher may be the difference between the life and death of the officer.

Field Supervision. It is important that field supervisors check periodically to ensure that proper field procedures are being observed by all officers. While sloppy work habits should not be tolerated under any conditions, it is especially important that officers working alone be alert and comply with established policies and procedures. The manner in which officers conduct themselves is often a reflection of the attitude and competence of the field supervisor. Supervisors who don't care about their officers will find that attitude reflected in the work of their subordinates.

Field supervisors should not be reluctant to observe officers from time to time and to critique them on how they handle particular calls. Officers need to know that their supervisors care enough to keep an eye on their performance. This will lead to a more conscientious attitude by the officers and improved field performance.

Take-Home Patrol Cars

A rather unique approach to police patrol operations that has been tried with varying degrees of success in a limited number of police agencies is the take-home car. Under this plan, patrol officers are assigned patrol cars on a permanent basis and are allowed to use them off duty as well as on duty, with certain exceptions. The purpose of this plan is to increase police visibility in the community and to provide a greater response capability in emergency situations. An added objective of the take-home car plan is the reduction in overall maintenance costs due to the greater care exercised over the car by the individual officer.

The take-home car plan was initiated by the Indianapolis Police Department in August 1969, although the same basic concept had been utilized by a number of state police agencies for several years prior to that time. Under the Indianapolis Police Fleet Plan, as it is called, marked police cars are assigned to individual patrol officers on a permanent basis. These officers are allowed to use the cars for personal as well as police business. While the city bears all costs of operating the program, certain exceptions are placed on the use of the cars. For example, officers are not allowed to operate the cars outside the county limits without special permission, they are required to respond to any emergency within their immediate area while in their cars, and they must keep the police radio on at all times while the cars are being driven. In addition, officers are expected to keep the cars in good operating condition and appearance at all times.

The Indianapolis Police Fleet Plan was intended to meet a number of important objectives, including:

- Prevention of crime.
- Increase in rates of crime clearances.

- Prevention of automobile accidents and personal injuries resulting from such accidents.
- Creation of a more favorable public image of the police.
- Improved police morale.
- Development of greater feelings of community safety and security.

Unfortunately, the Indianapolis program was not designed with the intention of scientifically testing its merits or determining the extent to which its intended objectives were achieved. Even though the program was evaluated several months after it was implemented, the lack of controlled conditions and the paucity of reliable performance date by which program objectives could be measured substantially limited the scope and utility of the evaluation results. While there were a few indications that the program objectives were partially achieved, the evaluation failed to produce any definitive conclusions.[13]

A more definitive evaluation of the take-home car program was conducted in 1977 by the Lexington-Fayette (Kentucky) Urban County Police Division. Among the factors considered were overall operating costs, replacement-unit costs, number of off-duty hours expended providing police service, and overall performance in emergency situations. While the Lexington-Fayette County program evaluation suggested that over a period of several years the take-home car plan would be more cost effective than traditional methods, it was also discovered that the Department realized an estimated savings of $42,438 in one year by off-duty officers responding to various types of police incidents. The Lexington-Fayette County take-home car plan was judged to be quite successful by the agency.[14]

While the take-home car plan might not be appropriate for all agencies, and while start-up costs of the program may be quite high, there are several definite advantages of the plan: heightened police visibility, increased arrests, and improved vehicle maintenance. Although such programs have been in existence for several years, the take-home car plan has yet to be adopted on a widespread basis. Any police agency seriously considering the adoption of the take-home car plan should first conduct carefully controlled experiments on a limited basis to determine its suitability in meeting predetermined goals and objectives. Among the factors that should be considered in any evaluation of the take-home car plan are (a) overall operating costs, (b) citizen attitudes, (c) police performance, (d) employee morale, and (e) cost savings realized as a result of off-duty officers responding to police incidents. Once the expected costs and benefits of the take-home car plan have been determined, then a decision can be made as to whether to adopt this patrol method.

[13]Donald M. Fisk, *The Indianapolis Police Fleet Plan* (Washington, D.C.: The Urban Institute, 1970); Charles G. Vanderbosch, "24-Hour Patrol Power: The Indianapolis Program," *The Police Chief,* (September 1970), p. 34.

[14]A Comparative Analysis of the Lexington-Fayette Urban County Police Division's Home Fleet Program Versus the All Pool Plan (Lexington, Ky.: Lexington-Fayette Urban County Division of Police, 1977).

THE REBIRTH OF FOOT PATROL

As was pointed out previously in this chapter, foot patrol has regained acceptance in recent years as a viable patrol procedure. As discussed in Chapter 6, foot patrol is an integral element of many of the modern community-oriented policing programs.

In the 1970s, a number of experimental programs that focused on foot patrol as a means of reducing crime and fostering improved police-community relations emerged. One such program was conducted in the Newark, New Jersey Police Department as part of a statewide effort to reduce crime in the urban cities.

Five years after the Newark Foot Patrol Program was initiated, an evaluation was conducted by the Police Foundation. This evaluation revealed that foot patrol did not succeed in reducing crime, but that it did result in greater feelings of security by residents in the neighborhoods served by foot patrol officers. Foot patrol was also found to be related to more positive attitudes toward the police by residents of the affected areas. Moreover, it was also found that officers assigned to foot-patrol posts had higher morale, greater job satisfaction, and a more positive attitude toward citizens in their neighborhoods than did officers assigned to patrol cars.[15]

The police car represents both a physical and a psychological barrier between the patrol officer and the citizen. As Wilson and Kelling have observed, officers in patrol cars act differently toward people than do officers on foot.[16] Officers in patrol cars tend to treat people in a more formal manner than do officers on foot. The visibility and accessibility of the foot patrol officer, on the other hand, enables him or her to interact with people in a much more positive and friendly manner.

The foot patrol officer is usually able to develop a much closer relationship with the people who reside, shop, or work in the beat than is the officer assigned to a patrol car. Foot-patrol officers can identify with the people on the beat since they experience direct, interpersonal contact with them. The various roles of an officer—information provider, assistance giver, law enforcer, apprehender of lawbreakers, counselor, and friend—are facilitated by this direct, interpersonal contact. These functions can be performed, but with much less effectiveness, by an officer assigned to a patrol car.

Foot-patrol officers have another advantage over motorized patrol officers in that they

[15]James Q. Wilson and George L. Kelling, "Broken Windows," in Robert G. Dunham and Geoffrey P. Alpert, *Critical Issues in Policing: Contemporary Readings* (Prospect Heights, Ill.: Waveland Press, 1989), p. 369. See also, George L. Kelling, "Police and Communities: The Quiet Revolution," *National Institute of Justice, Perspectives on Police.*

[16]*Ibid.*, p. 375.

> ...are in a better position to "manage" their beats; to understand what constitutes threatening or inappropriate behavior; to observe it, and to correct it. Foot patrol officers are likely to pay more attention to derelicts, petty thieves, disorderly persons, vagrants, pan handlers, noisy juveniles, and street people who, although not committing serious crimes, cause concern and fear among many citizens.[17]

The highly mobile and urbanized nature of our society makes it impossible to even consider relying entirely upon foot patrol as the primary means of providing police services. However, the changing nature of police work and the new philosophies being advanced today about what law enforcement is and how it should be performed may advance foot patrol even further as an integral component of the police service delivery system. Clearly, there are many advantages associated with foot patrol that cannot be realized through more conventional patrol methods.

> Foot patrol will always be a support service for motor patrol. It will never replace motor patrol for obvious reasons—the speed of motorized response in serious situations as well as the automobile's ability to cover more area are indispensable to contemporary policing.[18]

Any attempt to implement a foot patrol program on anything but a very limited basis should be carefully planned in advance. The duties and responsibilities of foot-patrol officers compared with those of other patrol elements, the selection and training of officers assigned to foot-patrol duties, the manner in which they will be supervised, the method by which their performance will be evaluated, and the impact of foot patrol on other patrol functions and organizational activities need to be carefully considered.[19]

These are all very important issues and the manner in which they are addressed may very well determine whether the foot-patrol program will succeed or not.

SUMMARY

The methods of police patrol available to law enforcement agencies are many and diverse and are essentially limited only to the imagination and wisdom of the police executive and patrol commander. Traditional methods of police patrol can no

[17]Robert Sheehan and Gary W. Cordner, *Introduction to Police Administration,* 2nd ed. (Cincinnati, Ohio: Anderson Publishing Co., 1989), p. 380.

[18]Dennis M. Payne and Robert C. Trojanowicz *Performance Profiles of Foot Versus Motor Officers* (East Lansing, Michigan: National Neighborhood Foot Patrol Center, School of Criminal Justice, Michigan State University, 1985), p. 24.

[19]Robert C. Trojanowicz and Paul R. Smythe, *A Manual for the Establishment and Operation of a Foot Patrol Program* (East Lansing, Michigan: The National Neighborhood Foot Patrol Center, Michigan State University, 1984).

longer be accepted without question. Flexibility and suitability in meeting patrol goals and objectives should be the key considerations when deciding upon which patrol methods to employ.

A fluid, flexible, and situational approach to the delivery of police services will be necessary to accommodate the ever-changing environment of policing, the increasing demands and expectations of the public, and the need for increasing accountability by police administrators for the actions of their organizations. Traditional, conventional, or inflexible approaches are no longer suitable. The selection, planning, and design of police patrol methods and programs needs to be guided by an understanding of the changing nature of the police mission and role in society.

This chapter has identified a few of the options that are available, but there are others that can be equally successful. Imagination and creativity are the only limiting factors in designing and implementing a successful police patrol program.

REVIEW QUESTIONS

1. Why is there no single "best" way of patrolling?
2. What are some of the factors that should be taken into consideration when selecting police patrol methods?
3. Discuss the advantages and disadvantages of automobile patrol and foot patrol.
4. What, in your opinion, is the single greatest disadvantage of automobile patrol?
5. What are some of the reasons that a patrol supervisor might elect to use a bicycle?
6. Why have some cities chosen to abandon their motorcycle patrol units? How might these problems be overcome?
7. What unique features do canine patrol units offer and how can they best be applied to general patrol purposes?
8. What are some of the special considerations that should be made in forming a police canine unit?
9. Describe some of the situations or circumstances that may indicate the need for two-officer patrol units.
10. What kind of justification can a police administrator make to a city council considering the start-up costs of a take-home car program for the police department?
11. What are some of the arguments that can be made in favor of substituting walking beats for motorized patrol beats?
12. What are some of the characteristics of officers who would be most successful on a walking beat?

REVIEW EXERCISES

The following exercises are designed to reinforce or amplify comprehension of the material contained in this chapter.

1. Conduct a literature review of one of the specialized patrol methods described in the chapter. Prepare a brief paper outlining its principal advantages and disadvantages as well as the circumstances and conditions under which it might be successfully employed.
2. Conduct a survey of several police agencies in your area to determine how many of the specialized patrol methods described in the chapter are used by these agencies. Obtain information concerning the manner in which they are used and how successfully they are viewed by the agency. Present your results to the class in open discussion.
3. In classroom discussion, debate the relative advantages and disadvantages of two-officer versus one-officer patrol. Assign members of the class to represent the patrol officer, the union, and management in order to promote opposing views on this subject.

REFERENCES

Bowman, Barry A., "Commemorating the 30th Anniversary of Airborne Law Enforcement in LA," *The Police Chief* (February 1989), pp. 17-18.

Boydstun, John E., and others, *Patrol Staffing in San Diego: One or Two-Officer Units.* Washington, D.C.: Police Foundation, 1977.

Chapman, Samuel G., ed., *Police Patrol Readings* (2nd ed.), Springfield, Illinois: Charles C. Thomas, 1970.

Chapman, Samuel G., *Police Dogs in North America.* Springfield, Illinois: Charles C. Thomas, 1990.

Chapman, Samuel G., and others, *Perspectives on Police Assaults in the South Central United States,* I. Norman, Oklahoma: The University of Oklahoma, 1974.

Decker, Scott H., "The Impact of Patrol Staffing on Police-Citizen Injuries and Dispositions," *Journal of Criminal Justice.* 10 (1982), pp. 375-382.

Fisk, Donald M., *The Indianapolis Police Fleet Plan.* Washington, D.C.: The Urban Institute, 1970. Charles G. Vanderbosch, "24-Hour Patrol Power: The Indianapolis Program," *The Police Chief,* (September 1970), p. 34.

Guthrie, C. Robert, and Paul M. Whisenand, "The Use of Helicopters in Routine Police Patrol Operations: A Summary of Research Findings," in S.I. Cohn, ed., *Law Enforcement Science and Technology,* 2. Chicago: IIT Research Institute, 1968, p. 554.

Kelling, George L., "Police and Communities: The Quiet Revolution," National Institute of Justice, *Perspectives on Police* (June, 1988).

Payne, Dennis M., and Robert C. Trojanowicz, *Performance Profiles of Foot Versus Motor Officers*. East Lansing, Michigan: National Neighborhood Foot Patrol Center, School of Criminal Justice, Michigan State University, 1985.

Rice, Thomas W., "A Case for MACE," *The Police Chief* (July, 1989), pp. 51-53.

Schack, Stephen, Theodore H. Schell, and William G. Gay, *Improving Patrol Productivity, Volume II: Specialized Patrol*. Washington,D.C.: National Institute of Law Enforcement and Criminal Justice, 1977.

Sheehan, Robert and Gary W. Cordner, *Introduction to Police Administration* (2nd ed.). Cincinnati, Ohio: Anderson Publishing Co., 1989.

Trojanowicz, Robert C., and Paul R. Smythe, *A Manual for the Establishment and Operation of a Foot Patrol Program*. East Lansing, Michigan: The National Neighborhood Foot Patrol Center, School of Criminal Justice, Michigan State University, 1984.

Wilson, James Q., and George L. Kelling, "Broken Windows," in Robert G. Dunham and Geoffrey P. Alpert, *Critical Issues in Policing: Contemporary Readings*. Prospect Heights, Ill.: Wavelandp Press, 1989.

CHAPTER 8
PATROL FORCE ORGANIZATION AND MANAGEMENT

CHAPTER OUTLINE

LEARNING OBJECTIVES

After completing this chapter, the student should be able to:

1. Understand the importance of principles of organization in managing the patrol force.
2. Know how the informal organization impacts the formal police organization.
3. Discuss the principles of organization as they apply to the patrol force.
4. Understand the importance of strategic planning in the management of the patrol force.
5. Learn how conflict may be properly managed and utilized in the police organization.
6. Understand the importance and nature of internal communications in the police organization.

To achieve maximum effectiveness, police patrol operations need to be properly organized and managed. Patrol administrators today are required to know more than ever before about contemporary principles of organization and management. As intelligence, creativity, and imagination replace age and seniority as the primary means by which patrol commanders and administrators are selected, more emphasis is placed on the efficient manager of patrol operations. Without the application of sound principles of organization, duties and responsibilities may not be clearly defined; duplication of effort may occur; vital resources may not be utilized in the most efficient manner; and essential tasks may not be properly coordinated.

For too long, the police patrol function was taken for granted. It was generally assumed that managing the patrol force consisted of little more than assigning officers to shifts and beats, making sure that there were enough personnel working to cover the minimum number of patrol beats, and ensuring that beat officers were properly equipped to do the job expected of them. Roll-call training sessions were rarely used for training, but rather to brief officers on past occurrences that might require their attention. Very little attention has been given to providing direction to patrol officers or to properly managing the patrol effort.

As the years have passed, we have learned a great deal about what works and what does not work in police operations. We know, for example, that faster response times by patrol units do not necessarily ensure successful apprehension of an offender, nor does it ensure greater citizen satisfaction with the police. We know that increasing the size of the patrol force, by itself, does not necessarily mean a reduction in crime. We know that directed, deterrent patrol strategies are more effective in reducing crime than routine, random patrol. Clearly, there is much more we need to know about what works and what does not in terms of police patrol, but we do have a growing body of knowledge that needs to be translated into effective management of the patrol force.

This chapter examines a number of important issues associated with the effective organization and management of the patrol force. It is the intent of this chapter to emphasize the importance of applying the basic principles of good organization and management to the operation of the patrol force.

PRINCIPLES OF ORGANIZATION

Due to the nature of the work they perform, the conditions under which they operate, and the manner in which they are expected to perform their duties, most police departments are organized along semimilitary lines. They adhere very closely to a rigid chain of command, specific assignment of duties and responsibilities, and emphasis on accountability commensurate with authority. Although individual discretion is inherent in the nature of the police function, standard operating procedures are used to prescribe how routine as well as extraordinary situations are to be handled.

There have been a number of attempts over the years to develop nontraditional organizational models for the police. These have had varying degrees of success, but have not had a major impact upon how most police departments are organized. There is, however, probably no single best way to organize a police department. The best organizational structure in any police department is the one that works best for that particular agency. One source suggests that the difficulty in developing an effective organizational model for the police is related to the lack of a commonly accepted police mission.[1]

Although there have been a few sporadic attempts to break away from the traditional principles of organization that have characterized the police service for years and years, these have been largely unsuccessful. There have been attempts, for example, to replace military rank designations such as sergeant, lieutenant, and captain with more businesslike titles such as manager, agent, director, and other titles which more closely reflect job duties and responsibilities. In more recent years, there have been a few successful attempts to replace traditional civil-service promotional procedures with innovative techniques, and to create "exempt" positions which are filled on the basis of merit, qualification, and performance instead of on the basis of time in grade, seniority, and knowledge-based tests.

As the nature of police work changes, it is likely that the basic organizational principles that have worked so well for so long will need to be either modified or altered altogether. It is difficult to predict what the future holds for law enforcement. We do know, however, that there will be many changes and that those things we accept as common practice today will be rejected as outmoded or old fashioned in the years ahead. So too, the principles of organization upon which we have relied for so many years may no longer be functional. Nevertheless, they remain today as important guidelines for the effective management and organization of the modern police agency.

[1]Robert H. Longworthy,"Organizational Structure," in Gary W. Cordner and Donna C. Hale, *What Works in Policing? Operations and Administration Examined* (Cincinnati, Ohio: Anderson Publishing Co., 1992), pp. 96-97.

Organizing is the process of the grouping of people, things, functions, activities, or processes according to some logical or systematic plan or procedure so that work is carried out in the most effective and efficient manner. In other words, the process of organizing helps to ensure that work is carried out in the most effective manner by achieving maximum benefit of available resources. There are a number of principles of organization which help to see to it that the goals of organizing are achieved. These basic principles have met the test of time and have been developed over the years by the application of theory to real-word situations. Still, there is no guarantee that they will all apply equally in all situations. That is, the police administrator needs to be able to modify any of these principles as may be necessary to better satisfy the requirements of the situation at hand.

Simplicity

An organization plan should be simple enough to be clearly understood by all concerned, yet detailed enough to provide clear lines of authority and responsibility. Overly-complex structures thwart the free flow of communication, confuse organizational relationships, hamper unity of operations, and impede the proper coordination of operations.

One of the simplest and most effective ways to delineate duties and responsibilities is to prepare an organization chart that clearly depicts organizational relationships and lines of authority.

> An organization chart is a diagram of a system. It shows how all subsystems are expected to relate formally within an organization, and it assigns each subsystem a specific task to perform. It improves understanding in that it shows at a glance how information is formally communicated within an organization and who reports to whom within an organization. It prescribes relationships and thereby facilitates communications.[2]

A properly drawn organizational chart can show, in a simple and straightforward manner, the organization arrangement of different functions by title as well as reporting relationships. Figure 8.1 shows the organization of a medium-sized police agency. This kind or organization chart clearly portrays the lines of authority among elements and organizational grouping of the various divisions within the agency.

On a smaller scale, the organization chart might simply show the arrangement of specific persons within the organization according to shift or job assignment, as shown in Figure 8.2. This kind of chart provides much more detail than that shown in Figure 8.1 and would obviously be possible only in the smallest agencies, although it could be used within smaller divisions of a larger agency.

The larger an organization grows, the more complex its organizational structure becomes. There is no way to avoid this, unfortunately. In order to make the organization as uncomplicated as possible, as well as to promote internal coordination of

[2]Robert Sheehan and Gary C. Cordner, *Introduction to Police Administration,* 2nd ed. (Cincinnati, Ohio: Anderson Publishing Co., 1989), p. 31.

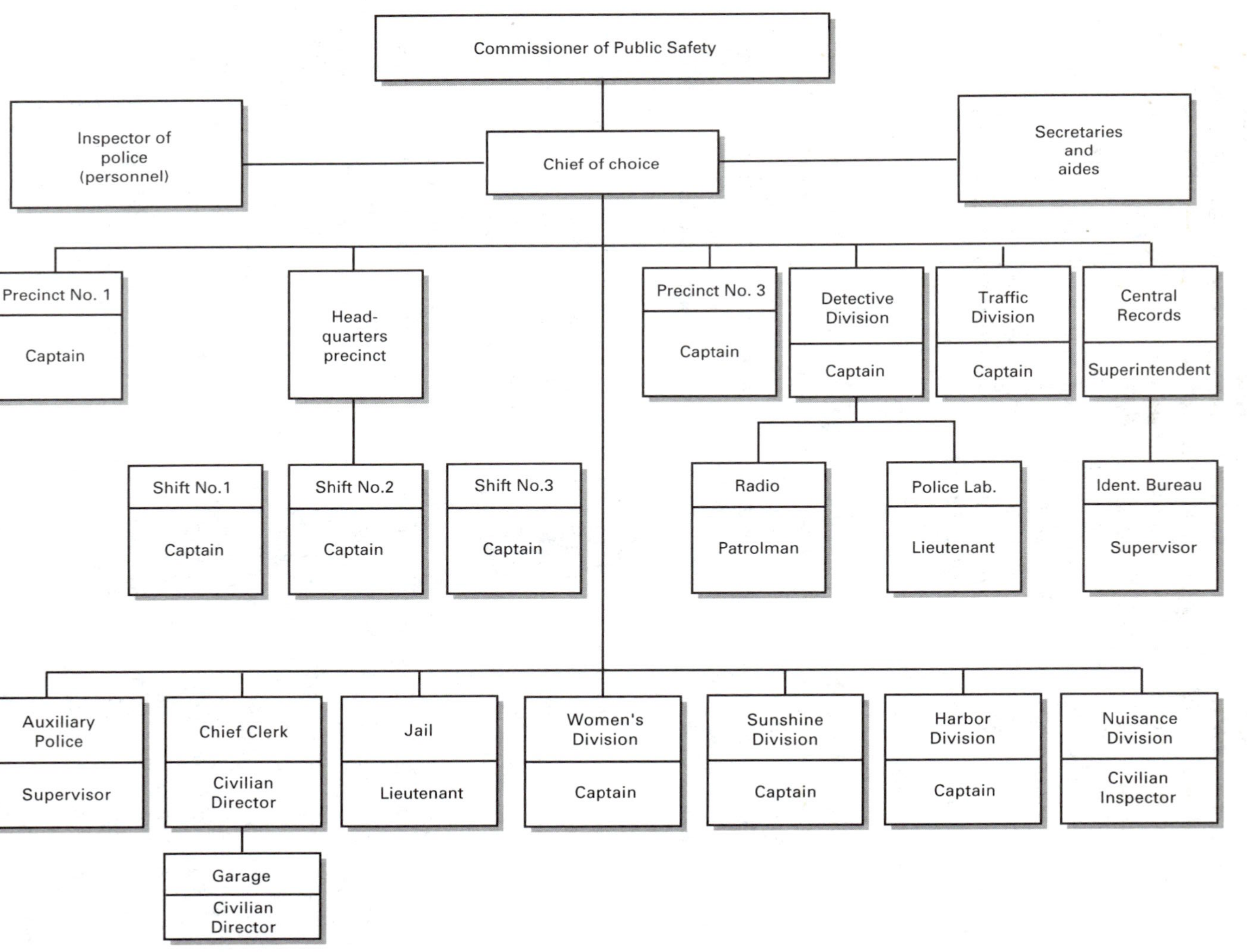

Figure 8-1 Functional Organization Chart

HOMETOWN POLICE DEPARTMENT

Organization Chart

Martin Bourke, Chief of Police Marilou Fussy, Adm. Asst.

Operations Division Commander J. House	Investigations Division Commander F. Mulaney	Support Services Commander R. Joy
Sgt. Nancy Ducey Sgt. Harry Hillary Sgt. Steve Corsey Sgt. Larry Longtree Off. John Townley Off. Fred Bluenose Off. James Jones	Sgt. John Haley Det. David Green Det. Ray Cox Det. Sylvia Pressly Off. David Jenkins CPO Cindy Stevens Inv. Sally Witsky	Marilyn Turney Dennis Anderson Shirley Snafu Marcia Gilmore Cindy Sweet Nancy Dewhurst Terry Binder

Figure 8-2 Detailed Organization Chart

operations, tasks should be grouped according to their similarity of function, purpose, or method. This helps to provide logic and consistency in the conduct of operations and guards against overlapping or conflicting command responsibilities. Grouping of tasks also helps to ensure that persons assigned to specific tasks will be familiar with the manner in which they are to be performed.

Organizing by Function

Grouping similar tasks, job assignments, and functions together and placing them under a single supervisor or command officer is a basic principle of organization. Organizing by function helps to promote efficiency and eliminate duplication of effort. It also promotes logic and clarity in the organizational structure. In the police organization, tasks are normally grouped according to their function, process, method, or clientele. They may also be organized according to geographic area (police districts or precincts), and by time of day (patrol shifts).

In the police organization, responsibility for carrying out the basic police function is normally assigned to a single organizational element, usually called the patrol bureau, field operations division, or some similar designation. Located within this same organizational element may be other support units or specialized functions such as the traffic bureau, the parking-enforcement unit, the tactical-operations team, and so on. These specialized, or support, units will normally be under the command of the same person who commands the patrol force simply because this helps to ensure better coordination of effort and communication among those units having a like or similar mission.

A common exception to the principle of organizing by function is the need to find a place in the organization for "special people." "Special people" are those individuals who, for one reason or another, do not "fit" into the logical place in the organization. This commonly occurs, for example, when a member of the department becomes disabled and may not qualify for a police pension. In order to protect the individual, a job suited to his or her abilities must be found within the organization.

In one such situation, a police lieutenant who commanded the Professional Standards Division of a police department, was no longer, due to surgery, able to perform regular police duties. The lieutenant was made responsible for overseeing the city's alarm ordinance, which included sending out and collecting fines for excessive alarms. Since this function did not occupy 100 percent of the lieutenant's time, he was also assigned to coordinate and supervise the department's fleet-maintenance operations. Because the department's training officer worked for this same lieutenant, the unit was called the "Professional Standards Division." It would not be difficult to think of a dozen or more titles that would have been more descriptive of this unit's duties.

Making room for "special people" is not an uncommon practice in a police organization, and is often done in other municipal departments as well. Clearly, it is a positive aspect of the organization, since it shows that it takes care of those who have served it for so long, However, there ought to be a way of dealing with these extraordinary situations in a more logical and forthright manner than that described in the previous example.

In keeping with the principle of organizing by function, it is important to observe the basic distinction that exists between line and staff functions. **Line functions** such as patrol, traffic enforcement, or criminal investigation are those that are designed to meet the basic police mission. **Staff functions,** on the other hand, are those that exist to support line functions, either directly or indirectly. For example, the planning and research unit is clearly a staff function, but its mission is primarily to support the line operations in either a direct or indirect manner.

Depending upon the size of the organization and its unique characteristics, staff functions may be grouped under a single organizational element that is separate and apart from line operations, or may be placed in a subordinate position within the operations division. For example, a crime-analysis unit may be organizationally located in the Support Services Division, as shown in Figure 8.3, or placed within the Operations Division itself, as shown in Figure 8.4. The advantage in placing the staff function within the same organizational division that it is intended to support is that it is more responsive to those persons who are the principal beneficiaries of its services. The advantage of placing a staff unit in a separate organizational element is that it may also serve other parts of the organization.

Chain of Command

Information must be transmitted through the organizational structure in a systematic manner in order to ensure that all concerned personnel will be properly informed. Moreover, the chain of command permits each person in the hierarchy of authority to take appropriate action at the proper level before passing the matter upward or down-

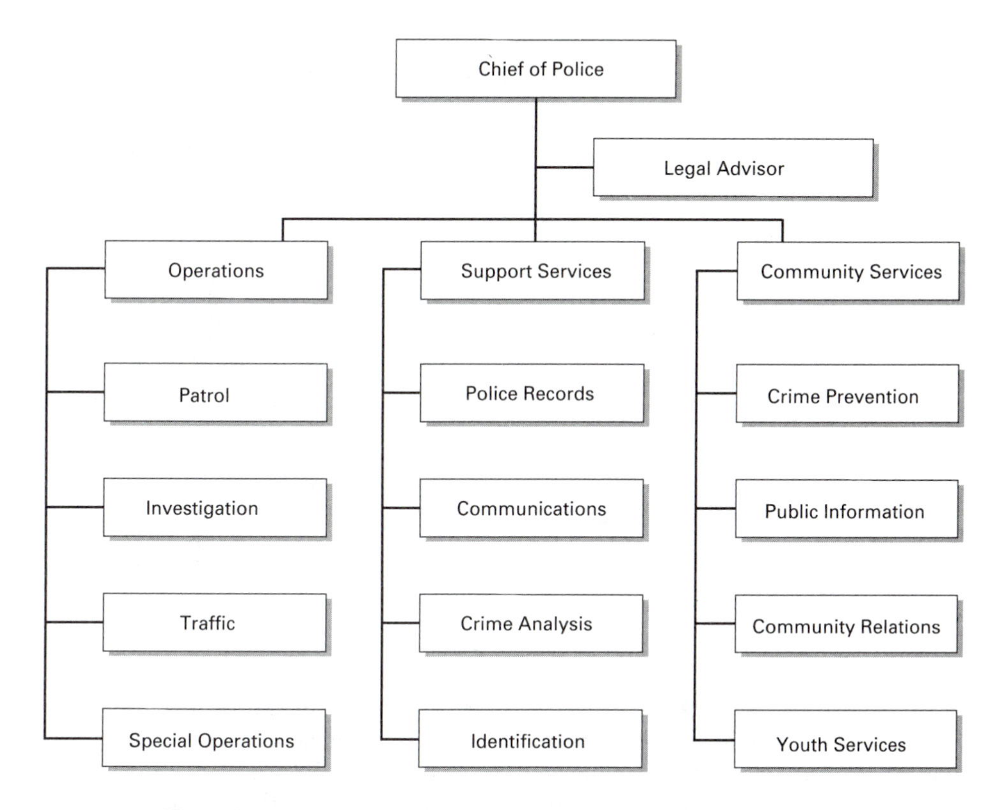

Figure 8-3 Organization of Crime Analysis Under Support Services Division

ward in the line of command. Violations of the chain of command create internal disharmony; they create confusion in the minds of subordinates as to whom they are responsible. Frequent violation of the chain of command will also undermine the authority of responsible supervisors.

The principle of chain of command helps the supervisor establish and maintain necessary control over the actions of subordinates. All orders and directives from higher authority should be communicated through the supervisor to his or her subordinates. This helps to keep the supervisor in the "information stream" and also helps to reinforce the role of the supervisor as decision-maker.

To be fully effective, the chain of command must be observed and enforced by all members of the department, including command and management personnel. Unfortunately, they are the ones who are most likely to violate the chain of command, although they usually do not do so willfully or maliciously. Often, they simply forget why the chain of command is important and how their actions can affect the welfare of the organization. In some cases, it is much easier and quicker to go directly to someone rather than to go through that person's supervisor. The adverse consequences of these kinds of actions may not be felt until much later, long after the damage has been done.

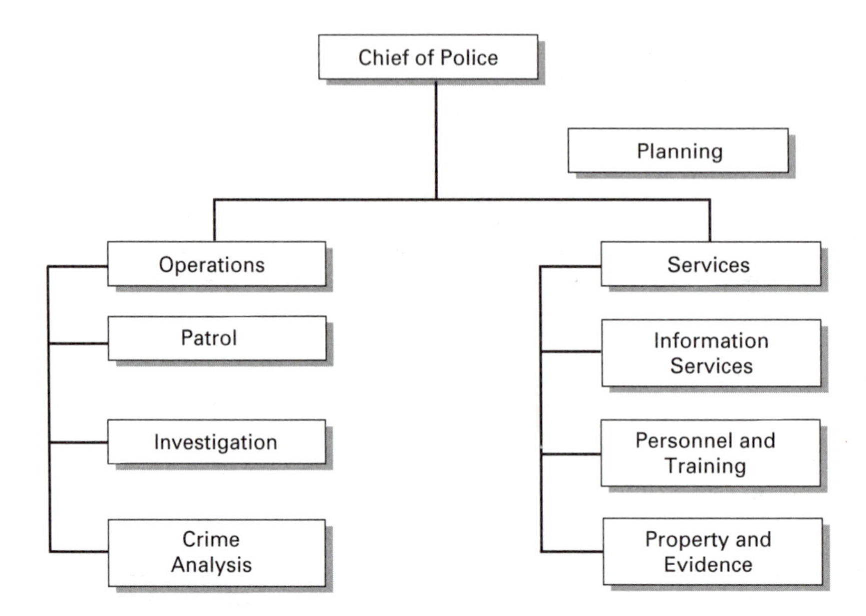

Figure 8-4 Organization of Crime Analysis Under Operations Division.

Chain of command is especially important in a police organization because accountability is critical. The principle of chain of command holds that each successive person in the chain of command, from the first-level supervisor to the chief of police, must be allowed the opportunity to deal with those incidents for which he or she is responsible. A person cannot be held accountable if the chain of command is violated by allowing persons either below or above that person to handle the situation.

A good example of how adversely this kind of violation will affect the functioning of the organization is as follows:

Lieutenant John Marsh is the shift commander on the day shift. He has been given orders by the patrol commander to use directed patrol strategies to deter a growing incidence of purse snatching occurring near and around the bus stop at 4th and Main. Lieutenant Marsh directs two officers to conduct high-visibility patrol in the area between 12 PM to 2 PM, when most of the incidents have occurred. After three days, the number of incidents has been reduced by one-third, and three arrests have resulted in over a dozen cases being cleared.

Lt. Marsh decides to leave the two-person unit in the area for another week. However, without the lieutenant's knowledge or consent, the patrol commander directs that they be reassigned to traffic-enforcement duties in another sector. Lieutenant Marsh learns of this decision after returning to work from his two days off. A week later, the purse snatchings return to the same level as before the directed patrol efforts were initiated.

In this example, the patrol commander violated the chain of command by countermanding the orders of the responsible supervisor who had been given the responsibility to handle a particular situation. Thus, the shift commander can no longer be held accountable for the purse-snatching problems at 4th and Main, because the patrol commander has decided that traffic enforcement elsewhere is a higher priority. Not only does this kind of action undermine individual accountability, it also undermines the authority of the shift commander. If this kind of action were to continue, it would be very easy for Lieutenant Marsh to simply defer all ideas and responsibility to the patrol commander and to remove himself from the decision-making process. It is through unthinking actions such as these that police organizations often self-destruct.

Span of Control

The ability of one person to supervise the affairs of subordinates is limited by such factors as the level of difficulty of the work being performed, whether supervision is direct and continual or irregular and indirect, and the degree of judgment and initiative exercised by the employee. A span of control that is too wide tends to weaken the control exercised by the supervisor. Conversely, a narrow span of control does not provide for optimum use of available supervisory personnel.

The principle of span of control is based on the assumption that there is a limit to the number of individuals that one person can effectively supervise. The optimum span of control in any organization depends upon a number of things, including: (a) the type and complexity of work being performed; (b) the job skills, and training, and experience of those performing the work; (c) the degree of specialization involved in the work being performed; and (d) the knowledge, skill, and experience of the supervisor.[3]

The span of control in a police patrol division should be neither too broad nor too narrow. A broad span of control is undesirable because it inhibits the ability of the supervisor to effectively direct, monitor, and control the work of subordinates. A span of control that is too narrow, on the other hand, does not provide for the most efficient utilization of available resources. Generally speaking, a span of control greater than eight would probably be considered excessive in a patrol force. Given the typical duties performed by patrol officers, the volume and nature of calls for service, as well as the other duties being performed by the supervisor, it is highly unlikely that a single supervisor could effectively supervise more than eight patrol officers.

In most small and medium-sized police departments, the field supervisor—usually a corporal or sergeant—will also be called upon to perform a variety of administrative duties including report review, scheduling, dispatching, handling citizen complaints, and investigating officer misconduct. Consequently, the first-line patrol supervisor will not have a great deal of time to spend observing and evaluating the operations of subordinates.

[3]V. A. Leonard and Harry W. More, *Police Organization and Management,* 7th ed. (Mineola, New York: The Foundation Press, Inc., 1987), p. 155.

One of the problems that may be encountered when attempting to narrow the span of control is that of increasing levels of supervision, or "layering" the organization. In other words, one must sometimes choose between a "flat" organization with minimum levels of supervision and an increased span of control, and a shorter span of control with increased distance between the top and bottom of the organizational pyramid. The relationship between levels of supervision and span of control is shown in Figure 8.5.

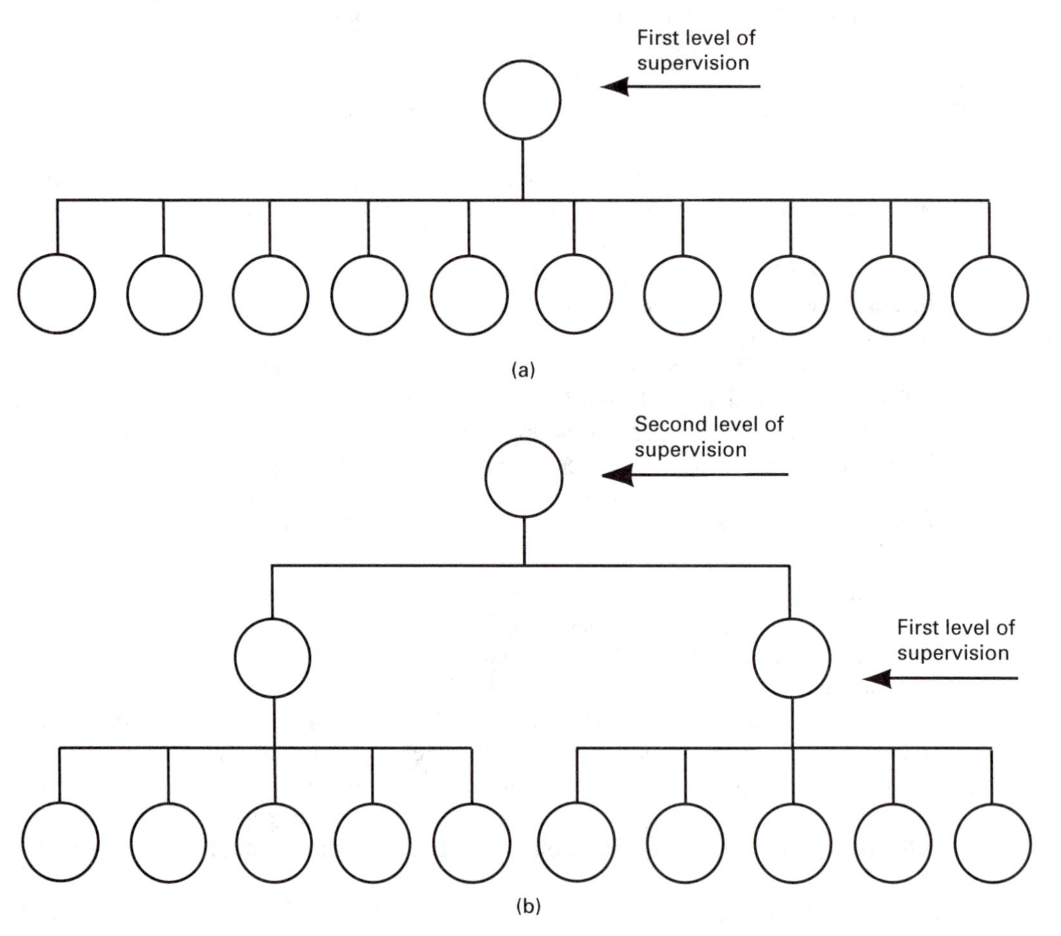

Figure 8.5 Examples of Span of Control

Unity of Command

In any situation, there must be only one person in command. To put it quite simply, there can be but one captain if the ship is to remain on its course, Moreover, each person should be responsible to only one person in the organization. Each person in the organization must know clearly to whom he or she is responsible. Department policies should specify clearly who is in command of any given situation.

The principle of unity of command is based on the belief that an individual should be responsible to one and only one person at any given time and in any given situation. This seems simple enough, but in practice it is not always easy to achieve. The principle of unity of command recognizes that a person cannot function effectively if he or she is expected to receive and follow orders from more than one person at a time. To do otherwise opens up the opportunity for internal conflict, confusion, and lack of cohesion and coordination of effort. It is therefore important in the police organization that lines of authority and responsibility be clearly delineated so that everyone knows to whom they are responsible at all times.

There may be instances when the principle of unity of command may be temporarily relaxed or modified, however. This might occur, for example, when members of two or more operating units (for example, patrol and traffic) operate together in a tactical situation under the command of a single supervisor. Under such circumstances, patrol officers might be placed under the command of a supervisor of another organizational unit for the purposes of the situation at hand. Once the situation has been resolved and individual operating units return to their normal duties, the principle of unity of command would once more apply.

The primary purpose of the principle of unity of command is to eliminate the possibility of conflicting and contradictory orders that would interfere with the smooth and effective functioning of the unit. A secondary purpose is to ensure that supervisors and command officers do not overstep their authority by attempting to issue orders to persons who do not report to them. This also helps to eliminate any doubt in the mind of the patrol officer as to who is in charge in any situation and whose orders are to be followed.

In applying the principle of unity of command, it is important to understand the difference between functional supervision and administrative supervision. **Functional supervision** is exercised by the person who is formally assigned to supervise or command a unit or person to the organization chart or established directive. **Administrative supervision,** however, may be exercised by the person who is in charge of a particular function at a particular time and place. For example, in a small police department, the senior dispatcher may be responsible for supervising the dispatching function and all subordinate dispatchers are responsible to the senior dispatcher. However, at any time of the day when the senior dispatcher is not on duty, the duty dispatcher reports to and is under the supervision of the shift supervisor, who is not otherwise in the dispatcher's chain of command.

This kind of exception to the rule does not weaken the principle of unity of command, but rather is necessary in view of the unique conditions under which the police must work. It usually does not create a problem, since supervisors and subordinates understand that there can only be one person in charge in a given situation in order for the organization to operate effectively.

THE INFORMAL ORGANIZATION

The formal structure of the organization is that presented in the organization chart, as discussed previously in this chapter. The organization chart represents how the organization functions in a formal sense by delineating lines of authority, responsibility, and communication flow. The organization chart does not, however, show how things work informally within the organization. In other words, the organization chart shows how things are supposed to work, but not how they actually do work on an everyday basis.

The informal organization is just as important (if not more so) as the formal organization, since it reflects the department as it **actually** exists. In a sense, the informal organization represents the human side of the organization. At times, there may be a big difference between how the organization is supposed to function according to the organization chart and how it actually operates according to the informal organization. When this happens, conflict is likely to occur. Managers, for example, may become agitated and frustrated when communications are passed through the informal organization more efficiently than they are through the formal system of communication.

The intelligent police manager, however, will view the informal organization not as a threat, but as a fact. The informal organization can actually play a very vital role in the functioning of the organization, and its existence should not be seen as an adverse consequence or as a failure of management. Informal organization operates in any formal organization and can, if treated properly, contribute to effective management.

Once of the greatest dangers posed by the informal organization can also be its greatest asset. That is its ability to distort or filter communication within the organization. There are a number of ways in which information can be distorted or impeded within an organization, and the informal organization can either be part of the problem or part of the solution. The manager who insists that all information be communicated through official channels may very well be contributing to the problem by not recognizing the value of the informal organization in aiding the transmission of information.

The successful police manager should recognize rather than ignore the existence of the informal organization and the importance it plays in the lives, feelings, and attitudes of the people in the organization. The successful police manager will attempt to utilize the informal organization by harnessing its strengths, such as information flow, and channeling them toward the goals of the organization. Above all else, the police administrator must recognize that the informal organization is a very real part of the formal organization and that it can work either to the advantage or disadvantage of the organization's goals.

MANAGEMENT FUNCTIONS

Management is the process of getting the most out of the organization: using resources in the most efficient manner to achieve the desired goals of the organization. Management can be both an art and a science, depending upon the skills, capabilities and experience of the manager. Some managers seem to be very adept at what they do and appear to have a natural ability to manage. Other managers are more technically adept. What they may lack in natural ability, they more than make up for in technical skill.

Management is a complex and interdisciplinary function which requires a number of skills. Very few people possess all these skills naturally and must acquire them through formal training and experience on the job. Today, more than ever before, good police managers gain their positions through years of hard work, long hours in the classroom, and faithful attention to the lessons they learn as they work their way up through the ranks of the organization.

The Human Equation

Organizations are not inanimate objects, but are living, breathing things because they consist of and represent people. Although we tend to speak of them in objective and impersonal terms, organizations are just as human, just as caring, just as sensitive, and just as prone to being hurt by improper treatment as are the people they represent. Organizations need tender loving care, and suffer if neglected. The care and nurturing of organizations is one of the most important tasks of management.

When we speak of caring for organizations, we are really talking about caring for the people that comprise them. A successful manager must possess good "people skills." People are the most important resource of any organization. Moreover, people represent that one resource that can be cultivated, molded, adapted, motivated, and made to respond to new and changing conditions. No other resource available to the manager has this unique quality.

Dealing effectively with people is not an easy task. People can be insensitive, moody, temperamental, rude, uncaring, selfish, greedy, snobbish, wicked, dishonest, and may suffer from every other vice known to mankind. An employee who exhibits any one of these imperfections can have an adverse effect on other people working in the organization. On the other hand, people have many good qualities—and we prefer to think that these dominate their behavior—which make them a pleasure to work with most of the time. It is the imperfections in behavior and personality that pose the most significant challenge to the manager.

Although there are many skills that contribute to successful management, none is more important than the ability to deal effectively with people and their problems. Unfortunately, this seems to be one of the most difficult tasks to master. The more technical skills such as budgeting, planning, and establishing internal controls seem to be easier for some people to develop. A manager with weak technical skills may survive, the manager with poor people skills rarely will.

> Many of our most critical problems are not in the world of things but in the world of people. Our greatest failure as human beings has been our inability to secure cooperation and understanding with others.[4]

The task becomes more complicated as the number of people one has to deal with increases. As the number of people for whom a manager is responsible increases, his or her ability to give each person individual attention becomes strained. There is only so much time that can be devoted to "people problems." Other problems within the organization demand as much and sometimes more time. Unfortunately, dealing with "people problems" is one task that cannot be easily delegated by the manager. A manager's success in motivating employees toward the goals of the organization depends upon remaining in touch with the people who work there.

Ironically, people skills sometimes are more important in smaller organizations than they are in larger ones. This is due to the fact that in small organizations, the manager is able to remain in contact with a large percentage of the employees on a fairly regular basis. Because the manager sees these people from time to time, he or she is available to meet with them to discuss problems and to resolve any differences between them. As the organization grows in size, the manager is able to maintain less contact with the people at the lower levels of the organization, and his or her contact is limited mostly to those in the upper echelons. At this level, people skills become less important than technical skills.

Team Building

Managers need to be team builders. They need to recognize that they are limited in their own abilities, in the amount of time they have to devote to the demands placed on them, and in their knowledge of specific responsibilities for which their organizations are responsible. They must treat the organization as a team rather than as a collection of individuals. In any team, each person has a designated area of responsibility, and the team is only as strong as its weakest member.

The manager is the captain of the team. It is his or her responsibility to provide leadership and to direct the actions of the team. The success of the team depends very much on the ability of the manager to motivate the members of the team. This motivation will come in part by members of the team sharing common goals and objectives. It is not necessary that team members look alike, think alike, or even have the same motivation. What is important is that they share common goals and objectives for the organization, and that the goals of the organization complement their own personal goals.

[4]Paul Hersey and Kenneth H. Blanchard, *Management of Organizational Behavior: Utilizing Human Resources,* 5th ed. (Englewood Cliffs, N.J.: Prentice Hall, 1988), p. 3.

The word *team* is not synonymous with "organization." Teams are very special kinds of organizations because they work closely together in a spirit of coordination, cooperation, and teamwork. Members of a team enjoy a special kind of relationship that is much closer than people who simply work in the same organization. Team members dedicate themselves to a common purpose and assist each other when they need to. They devote their energies to the good of the team, and thereby to the other members of the team.

The effective police manager knows how to develop a team and how to encourage and motivate teamwork among members of the team. Through personal example, the manager motivates members of the organization to commit themselves and their energies to the goals of the organization and to the welfare of each member of the organization. The effective manager must be able to instill in each member of the organization a sense of purpose and pride in the organization. This spirit of teamwork, of common commitment to the goals of the organization, and pride in accomplishment, will set that organization apart from all others.

Vision

The successful manager is a person with the vision to see how the organization relates to the world around it, to project the role of the organization into the future, and to anticipate the problems to be encountered as the organization progresses along its established path. The successful manager does not have a crystal ball to predict the future, but rather has the ability to sense what is to come and to know what the organization must do to prepare for the future.[5]

People with vision are proactive rather than reactive. They are capable of anticipating the need for doing something and getting it done before it is too late. People with vision rarely are caught unawares or unprepared. People with vision are sensitive to the environment in which they work, including the organization and its people. They have a knack for knowing what needs to be done and when it needs to be done; they rarely need to be reminded to do something. Vision is a unique gift that not all people possess. It is one of those qualities that distinguishes the highly successful manager from the average manager.

Leadership

Leadership is the ability to motivate others toward the achievement of organizational goals. This definition implies that employees work toward the achievement of organizational goals because they want to, not because they are forced to. Thus, the ability to get others to do what has to be done through the application of force or through some other means other than their own desire is not leadership. This is an important distinction, because in the police organization, leaders occupy positions of power and authority. They do not have to be motivators to get people to do some-

[5]See, for example, Witham, p. 3.

thing—they just tell them to do it. People in the organization will normally follow orders because they know that the penalty for doing otherwise is usually some form of punishment.

Our definition of leadership, however, suggests a strong linkage between motivation and behavior. An effective leader understands how to motivate people, that is, how to get them to do something because they want to, not because they have to. Just as people differ greatly in their ability to do something, they also differ greatly in their desire to do it.

> Motives are the "why's" of behavior. They arouse and maintain activity and determine the general direction of the behavior of an individual. In essence, motives or needs are the mainsprings of action.[6]

The importance of effective leadership in the police organization cannot be emphasized too strongly. Strong leadership is what makes the difference, more often than not, between a highly professional and productive organization and a mediocre or poorly performing one. The leader sets the tone for the entire police organization and establishes a "quality of life" in the organization that sets it apart from others. Poor leadership, on the other hand, can have disastrous results. One source, for example, has blamed poor leadership as being the single most important factor in "...the irrational development of American police organizations..."[7]

The same point is made by Leonard and More:

> Leadership is the most important factor in the success or failure of police operations. Invariably, in observing a successful police organization one finds a strong executive who has been the driving force in elevating the level of performance. Conversely, where mediocrity or failure characterizes the work of a police organization, it generally can be traced to incompetence in management.[8]

There have been volumes written about leadership: what makes a good leader, the essential qualities of leadership, styles of leadership, skills of leaders, and so on. What seems to come out of the literature on the subject is that there is no such thing as a single set of traits that identifies success as a leader, nor is there any common pattern to how leaders become successful.[9] Although leaders may share common skills and qualities, it appears that the organizational context as well as the individual situation, as much as anything else, determines success by the leader. We shall refer to this phenomena as "situational leadership."

[6]Hersey and Blanchard, *Management of Organizational Behavior*. p. 19.

[7]Donald C. Witham, "Transformational Police Leadership," in James J. Fyfe, ed., *Police Practice in the 90's: Key Management Issues* (Washington, D.C.: International City Management Association, 1989), p. 4.

[8]V. A. Leonard and Harry W. More, *Police Organization and Management,* 7th ed. (Mineola, New York: The Foundation Press, Inc., 1987), p. 229.

[9]See, for example, Hersey and Blanchard, *Management of Organizational Behavior,* p. 88.

By our definition, **situational leadership** means simply that the successful leader knows how to adapt his or her particular skills and abilities to the organizational environment and to the particular situation at hand.[10] For example, we know that there are many ways to motivate people, ranging from their desire to achieve to their desire to avoid punishment. We also know that the ability to motivate people depends upon their own emotional level and physical condition at the time. A person who is on an "emotional high" (feeling satisfied, content, and wanting very little) poses a much different motivational challenge from one who is destitute, suffering from emotional trauma, or completely lacking in self confidence. What motivates each of these people to work toward the goals of the organization, as articulated by the leader, will obviously be quite different in each case.

The skill of a leader lies in recognizing the differences in people, knowing what motivates them and under what circumstances, and applying the right motivational tactics to each person in a way that will make him or her want to do the job. This is the essence of situational leadership. Thus, the skills that mark a successful leader are those of sensitivity, insight, patience, and self-confidence.

Accountability

The person to whom authority has been delegated must be held accountable for its proper use. Supervisors have the authority to direct and monitor the actions taken by their subordinates and to correct improper performance. They must also be held accountable for the actions of their subordinates, and their failure to take corrective action when improper performance is discovered will reflect on their own supervisory ability.

Delegation of Authority

There are limitations on the abilities of persons charged with administering the affairs of an organization. Authority must sometimes be delegated to subordinate personnel. Delegation helps to increase the commitment of subordinate personnel to the goals of the organization and fulfills their aspirations of achieving higher levels of responsibility. The effective manager is one who recognizes his or her own limitations and who can rely upon subordinates to get the job done.

Failure to delegate is harmful both to the organization and to the members of the organization. Failure to delegate creates a bottleneck at the top of the organization because decisions cannot be made in a timely manner in the absence of the manager. When only the manager can make decisions, no one else has much of anything to do. Delays caused by lack of decisions when the manager is absent create resentment and frustration among middle-managers who are supposed to have something to do with running the organization. When they realize that they have no real decision-making

[10]In this regard see V. A. Leonard and Harry W. More, *Police Organization and Management,* 7th ed. (Mineola, New York: The Foundation Press, Inc.,1987),p. 81.

authority, they first become resentful, then resigned, and finally complacent. They lose interest in their jobs and in the organization as a whole. This attitude is carried on down to the people under them and before long the entire organization suffers from a lack of interest and commitment.

Managers who fail to delegate usually lack confidence in their own abilities and are therefore afraid to allow subordinates to have any more power or authority than is absolutely necessary for them to do their job. Managers who lack confidence in themselves are reluctant to allow any one person or group of persons to know more than they know or to be able to do anything that they cannot do themselves. They fear that, given the right opportunity, someone may try to take their jobs away from them.

Confident and competent managers, on the other hand, seize the opportunity to delegate as a means of preparing their subordinates to rise higher in the organization and to accept increasingly responsible positions. They know that they look good only when their subordinates look good and that they are made stronger having well-prepared and capable people working under them.

Decision Making

Police administrators make decisions of all kinds and of varying degrees of importance every day. The successful manager is usually the one who makes the best decisions. That is, the outcomes turn out to be favorable to the organization. A poor decision, particularly one that has significant consequences, will tend to make the manager appear ineffective. A good decision makes the manager look good. In many cases, however, the decision is little more than a calculated gamble. The manager makes the decision based upon what he or she thinks or hopes will happen.

Whether a decision is good or bad really depends upon what happens as a result of the decision. When the desired outcome is produced by the decision, then the decision is viewed as a good one. When negative consequences result from the decision, the decision is seen as a poor one. Often, however, the final outcome may be beyond the control of the manager and the decision, when viewed from the context within which it was made, was probably a reasonably good one.

The mark of a good manager is not just to be able to make a good decision, but to be able to make the best possible decision under the prevailing circumstances. A poor manager is not necessarily a poor decision maker, but a person who cannot make decisions when the situation requires them to be made. Inability to make decisions indicates lack of self-confidence. A manager has to be able to accept the consequences of the decisions that he or she makes even when those decisions turn out to be poor ones. It is usually easier to defend making a poor decision than no decision at all.

The key to good decision making is the ability to objectively and rationally evaluate alternative courses of action and to anticipate the outcome that will be produced by a particular action. Too often, managers make decisions on the basis of emotion and personal bias rather than on an objective evaluation of the facts. They act hastily in reaction to the pressures of the moment rather than think the situation through logically. Or, they make decisions on the basis of incomplete information, in which case the outcome usually is not the one that was anticipated by the decision maker.

Police organizations vest most of the formal authority in those persons who occupy the top of the organizational hierarchy, while police officers on the street make many more critical decisions each hour that have greater consequences for the organization. Decisions made by police officers on the street have a very direct and immediate impact on the citizens. Decisions regarding arrests, for example, as well as those involving the application of deadly force, are made by the police officer rather than by the police administrator.

Although we have paid lip service to the idea of "participative management" for many years, most formal decisions continue to be made by those in upper management. Rank-and-file members of the organization are often not consulted, even when those decisions have a very direct impact on them. Involving police officers and other lower-echelon members of the police organization in the decision-making process is good for the individual as well as for the organization. "The organization gains because they obtain the use of a resident expert and the individual gains in that his sense of professionalism and stress-coping ability are enhanced."[11]

Police policy making, particularly concerning operational matters, is much more effective when input is sought from the lower-ranking members of the organization. "When street-level officers are made part of the decision-making and implementation process, their commitment to the change can become internalized. Without the shared input and involvement, officers may view the change as unwarranted and unnecessary and abide by it only when reminded and required."[12]

Internal Controls

Internal control mechanisms are necessary to ensure that the police organization operates in accordance with established operating policies and procedures. Proper training and effective supervision are two of the fundamental ways to ensure that proper operating procedures are being observed. However, violations and lapses will still occur from time to time. These often do not create a problem and are easily corrected when they are observed. Other defects, however, can be more serious and do not lend themselves to easy correction.

Many police agencies make it possible to assign patrol officers to specific geographic areas of responsibility, as will be discussed in the following chapter. In most cases, officers are allowed to leave their assigned area only to back up a car in an adjacent area, or to handle an assigned car or for another authorized purpose. They are not allowed to routinely wander from their assigned area whenever the mood strikes them.

[11]William H. Kroes, *Society's Victim—The Policeman: An Analysis of Job Stress in Policing* (Springfield, Ill.: Charles C. Thomas, 1976), p. 105.

[12]Geoffrey P. Alpert and Lorie A. Fridell, *Police Vehicles and Firearms: Instruments of Deadly Force* (Prospect Heights, Ill.: Waveland Press, 1992), p. 2.

Although "beat discipline" may not be enforced rigidly in all cases, laxity on the part of a supervisor to reinforce the rules of the beat-assignment policy can result in a total breakdown of that policy. Over time, this can have serious effects, such as when an officer is found to be some distance from the assigned patrol beat and unable to handle an emergency call. Internal controls are in place to make sure that this kind of deviation from established policy does not go unchecked.

Planning

The effective police manager must be a good planner and must be able to instill proper planning techniques in the police organization. Police operations are too important to be left to chance. Too much depends upon police performance to allow police activities to be conducted in a haphazard, unsystematic manner. While sound planning is ultimately the responsibility of the police executive, it is equally important that proper planning be conducted at all levels of the police organization.

While planning can be, and often is, a very complex and complicated process, there is nothing very mysterious about it. In very simple terms, planning is the process of determining where the organization is going and how it can get there. Police administrators engage in some form of planning each day, whether deciding on a shift rotation or the type of weapons to purchase for the department. Both decisions will affect the organization in the long term, and both decisions can be thought of as part of the planning process. The annual budget, the work schedule, and the vehicle-maintenance policy of the department are all examples of plans.

Planning can be viewed as the process of looking ahead to determine what actions need to be taken for the organization to accomplish its specified goals. This definition assumes that the organization itself has identified goals. Without goals, the planning process lacks meaning and substance.

Plans consist of a series of decisions. Each time a decision is made, others are presented. For patrol purposes, planning may consist of determining how to attack a specific crime problem, which in turn may lead to decisions about resources. How many? What kind? How should they be deployed? These are the kinds of decisions that are made by the patrol manager on a regular basis, even though they may not be recognized as being part of the planning process.

Planning in a police agency need not be a highly formalized process, nor should it be viewed as primarily the responsibility of the agency administrator or the command staff. Every supervisor and shift commander should be responsible for planning the affairs of their units and should be trained in the techniques of good planning. Training should emphasize decision making, identifying and evaluating alternatives, and predicting the probable outcome of those actions.

Police administrators have been engaged in planning of one form or another for years. The preparation of a work schedule is an example of planning, as is the development of the annual budget. Unfortunately, planning is often not recognized as a management responsibility, and planning tasks are sometimes performed without full recognition of their importance.

Sound police organization and procedures depend upon good planning. Frequently, the emergency nature of police work and the constant attention that must be given to day-to-day operations do not leave enough time for effective planning. Much planning is done daily in all police departments, but, primarily, it is to serve an immediate need.[13]

Planning should be the cornerstone of the decision-making process in a police agency. For every problem encountered, for every new program to be implemented, for every policy or procedure to be developed, the police administrator may be faced with several alternative courses of action. Planning is the means by which the best course of action can be chosen.

Planning in a police agency need not be a highly formalized process. Indeed, planning in the most basic sense should be carried out routinely in a variety of situations. Moreover, planning should not be the sole responsibility of a single person or unit within a police agency, but should be viewed as a basic responsibility of all management and supervisory personnel. In larger departments, of course, responsibility for specialized planning purposes will usually devolve upon a planning unit or a designated individual. Still, this should not relieve all persons in authority from the responsibility for planning within their respective spheres of authority.

Although the planning process may range from quite simple techniques to much more sophisticated ones, planning usually consists of several distinct elements, as follows:

Problem Identification. Planning usually begins with the identification of a specific problem to be solved, such as the need to provide expanded police protection during a large demonstration or similar gathering. The exact nature and scope of the problem must be clearly understood.

Determining Objectives. The agency must have the objectives of the planning process in mind when analyzing the problem. In the example cited previously, the objectives would probably include: (a) ensuring maximum public protection and safety, (b) providing for the speedy and safe passage of vehicular and pedestrian traffic through or around the demonstration area, and (c) guarding against outbreaks of violence or lawlessness. Once the objectives of the planning process have been identified, planning may proceed in an orderly fashion.

Establishing the Facts. The third step in the planning process entails the gathering of all relevant information concerning the problem under consideration. Information concerning who, what, when, where, and how must be obtained. Using the same example previously cited, information may be needed concerning (a) the number of persons expected to attend the demonstration, (b) the nature of the groups involved and their purpose for demonstrating, (c) intelligence information about previous demonstrations involving the same groups (Did violence occur, and if so, what

[13]President's Commission on Law Enforcement and Administration of Justice, Task Force Report: The Police (Washington, D.C.: U.S. Government Printing Office, 1967), p. 78.

kind and how extensive?), and (d) probable resources needed to contain the demonstration within prescribed limits.

Developing Alternate Courses of Action. Most good plans involve more than one course of action. Although a primary course may be chosen, it is important to provide suitable alternatives in the event that the primary course proves unworkable for some reason.

Implementing the Plan. Plan implementation should involve representatives of all groups or units likely to be affected by the plan. Coordination of activities is essential. Involved units or groups must be provided with advance information concerning what is to be done, how it is to be done, scheduling of activities, and expected results. Ample opportunity should be provided for reaction to the plan by affected participants, and amendments to the plan if necessary.

Evaluating Results. Very few things go exactly as planned. Minor and sometimes major changes in the plan may be necessary as the plan unfolds. For this reason, it is important that the plan and its results be carefully evaluated on an ongoing basis. Specific provisions should be made to obtain feedback from plan participants during and after the planned action in order to improve operations in the future. Evaluation may be a highly formalized process, or a rather simple one, but it is important that feedback concerning the suitability of the planning process and its outcome be obtained.

The Mission Statement. It seems rather obvious to say that all organizations need a mission or purpose to guide them, and it might seem equally obvious that the mission of the police is clear. However, as has been pointed out in previous chapters of this book, all police departments are not alike, nor do they all have the same goals and objectives, nor are their priorities or operational strategies the same. Therefore, their mission is not necessarily the same. A **mission statement** is a statement of purpose. It outlines the major task of the organization and defines organizational values. It gives the public, as well as the members of the organization, a sense of what the organization stands for.

Ironically, very few police departments have defined mission statements. The reason for this may be that the police take their mission for granted, which may also mean that they have never really spent a great deal of time thinking about what they are supposed to be doing or what they represent. This is unfortunate, since anything taken for granted almost always becomes less than what we expect it to be. Organizations that take themselves or their services for granted often find that they have serious problems that had gone too long unnoticed.

Strategic Planning. To be effective, an organization—any organization—must operate according to a defined plan of action. Without a strategic plan, the organization is like a ship without a rudder. It has no clear destination or any defined purpose.

The organizational plan should include a mission statement which outlines the short-term and long-term goals and objectives of the organization.

Very few police departments have a set of clearly-defined goals and objectives. While the goals of a police department (to enforce laws, apprehend criminals, and protect the public) may appear obvious, they are vague and need to be much more clearly delineated. In addition, there needs to be a set of rather specific objectives by which organizational performance can be monitored and evaluated. For example, one objective might be to reduce the number of automobile thefts by 5 percent in one year, or to increase the number of arrests for driving under the influence by 3 percent during the next 12 months. Virtually every operating unit of the department should have its own set of objectives which can be incorporated into the overall mission and goal of the organization.

Contingency Planning. Contingency plans are used by police departments to prepare for a variety of emergency situations which they hope will never occur. These include, for example, both natural and man-made disasters (airplane crashes, chemical spills, tornadoes) as well as criminally-motivated incidents (bank robberies, hostage situations). Contingency plans for these and other types of critical incidents need to be carefully developed and periodically reviewed and modified. Moreover, key personnel need to be acquainted with the provisions of such plans and their roles in them in order that they may be prepared if and when the plans are activated.

Many police agencies operate from contingency plans which have not been revised or updated. Due to the rapid developments in the technology of disaster preparedness, hazardous materials, mass disorder, natural disasters, and other kinds of emergencies of unusual proportions, it is important that contingency plans be regularly revised and updated. In metropolitan areas, this should generally be done on a regional basis to ensure necessary coordination and cooperation with other municipal departments and allied public safety agencies,

Internal Communications

Open communication is essential to a healthy and well-functioning organization. Information must be communicated vertically and horizontally throughout the organization in order that all members be informed of decisions that affect them. In addition, members should be encouraged to communicate their thoughts and ideas concerning management policies and operating procedures to those in a position to act on them. The police organization should be an "open system" in which information flows freely. Communication must be a two-way, interactive process involving all members of the organization.

The principle of communicating seems so simple, yet it turns out to be one of the most difficult management principals to accomplish successfully. Communicating is more than simply exchanging information, or speaking clearly and distinctly. Communicating also means listening and understanding what someone says to you, as well as making them understand what you mean. When we say that someone is an effective communicator, we don't necessarily mean that the person speaks clearly and distinctly, but that he or she is able to convey information to others effectively. A good communicator is also a good listener—a person who gives full attention when someone else is speaking.

There are many impediments to effective communication in a police organization. First, since police work is a 24-hour-a-day activity, people in the organization are assigned to various shifts and may not see each other often. They are also assigned to various details which place organizational barriers between them. In a highly structured organization, vertical communication between levels of supervision and management is sometimes impeded due to personality conflicts, overwork, lack of concern, or a dozen other things.

Good communication in a police organization—or in any organization for that matter—does not just happen but must be carefully cultivated and nurtured. If left to chance, communication in a police organization will not be effective. In other words, communication is something that must be planned, monitored, supervised, and managed. More than one otherwise-able police manager has fallen from grace due to what was later described as "poor communication."[14]

In a police organization, information can be communicated in a variety of ways, both formal and informal. For example, the most common means of providing information to officers concerning past events and other matters of importance to them is during roll call prior to the beginning of a shift. Staff meetings involving supervisors and command officers are used to discuss new management policies, problems affecting various units in the department, and suggestions for changes in the manner in which the department operates.

On a more formal basis, a good written-directive system is the key to keeping members of the organization informed of new policies and procedures as well as more immediate types of information. A written-directive system will normally include a series of **general orders** (dealing with long-standing policies and procedures), **special orders** (dealing with more immediate concerns or short-term events), **personnel orders** (dealing with assignments, promotions, disciplinary actions, terminations, and other personnel matters), and **memoranda** (information of a more general nature).

Written communications are necessary to properly direct the activities of any organization. A modern police department is no exception. **Rules and regulations** govern the responsibilities, conduct, and appearance of employees. They must be written in a manner to properly guide employees in their daily activities. Rules and regulations are firm guidelines to ensure that the conduct of employees is within acceptable standards as established by the department.

Policies and procedures direct employees in the performance of assigned duties. In order to assure consistency throughout the department, they should be written to describe the "whys" and "hows" of each particular operation.

[14]For a discussion on effective communications in a police agency, see Edward A. Thibault, Lawrence N. Lynch, and R. Bruce McBride, *Proactive Police Management,* 2nd ed. (Englewood Cliffs, N.J.: Prentice Hall, 1990), pp. 118-124.

Policy may be defined as a broad statement of purpose or intent. In a police department, policy is the means whereby the operating principles of the organization are defined. Policies attempt to define the "why" of a particular action or state of being. For example, a policy statement might indicate to members the philosophy of the department regarding traffic violators, drunk drivers, or juvenile offenders.

Procedures, on the other hand, describe how a particular function is to be performed. Procedures indicate to police officers what is expected of them in particular situations. They often provide step-by-step guidelines regulating specific types of actions. For example, procedures might describe how a prisoner is to be booked or the manner in which evidence or property is to be marked and stored.

Often, policies and procedures are issued together in a single document. The first part of the document might describe the department's policy on a particular matter, while the remaining portion of the document may describe how particular situations are to be handled. In the case of a policy and procedure regarding police pursuits, for example, the first section should indicate what the department's policy is concerning such pursuits (that is, areas of concern, intended purposes). The remainder of the statement would include detailed instructions concerning the manner in which police pursuits are to be conducted (how many vehicles permitted, role of support vehicles, use of emergency equipment).

Managing Conflict

Conflict is inherent in the nature of organizations, simply because organizations consist of people who have emotions, drives, needs and desires. It is virtually impossible for any single organization to satisfy the needs and aspirations of all of its members. In fact, if their needs were totally fulfilled, they would lack motivation and ambition. Because human beings have different needs and ways of satisfying them, they create conflict within the organization. Managing conflict in an organization is therefore another way of saying "managing people." As Leonard and More have observed, conflict is a sign of a healthy organization and should not be kept hidden, but should be brought out in the open and dealt with properly. Conflict can actually be a positive force and can be channeled into productive purposes within the organization.[15]

The task of the manager is to recognize sources of conflict in the organization, and to learn how to minimize the negative consequences of conflict. It is also important to recognize that not all conflict is harmful to the organization. To the contrary, an organization with no conflict is probably one in which not much of anything happens. A healthy, dynamic, and productive organization is bound to produce a reasonable share of conflict, some of which may contribute to the goals of the organization and some of which may be harmful to the organization. The successful manager knows how to distinguish between the two.

[15]V. A. Leonard and Harry W. More, *Police Organization and Management,* 7th ed. (Mineola, New York: The Foundation Press, Inc., 1987), p. 53.

The typical police organization has many sources of potential conflict that, if not recognized and dealt with properly, can pose a real problem for the police manager. A lack of clearly defined rules, regulations, policies, and procedures, for example, leaves police officers uncertain about what is expected of them by management. If there are no standards of performance, officers become anxious in the job.

Conflict is also created when there are perceptions of double standards, that is, when some people appear to be treated differently, depending on who they are or who they know in the organization. These kinds of perceptions can undermine morale and create a great deal of interpersonal conflict which can lead, in turn, to poor job performance.

Miscommunication and lack of complete communication is another source of conflict in a police organization. Regardless of how good the information system may be, it will occasionally result in members of the organization receiving incorrect, incomplete, or misleading information. People often apply their own translations to information that is incorrectly communicated to them. One person interprets the same information differently from the way someone else does. The result is often conflicting actions which result in interpersonal conflict.

Since the potential for organizational conflict is almost unlimited, the manager should strive not to eliminate conflict but to recognize those kinds of conflict that are most harmful to the organization and develop methods for either eliminating them or minimizing their impact on the organization. Learning how to deal with people and with their hopes, fears, anxieties, disappointments, desires, and frustrations is an important step in learning how to deal with conflict in the organization.

Managing Change

We live today in a dynamic environment in which change is occurring rapidly. Chances are, if a worker, supervisor, or manager is doing something—anything—today the same way he or she did it ten years ago, there are better ways it can be done. Managers need to be constantly on the alert for signs that their methods of performing tasks are becoming outmoded and need to be replaced. Moreover, they need to be alert to indications that the basic nature of the job is changing. This may require a fundamental change in the way managers think about, approach, and perform their jobs.[16]

The need to be alert for changes in the work place and in the nature of the job being performed is nowhere more important than in the police service. Managing change is closely associated with managing conflict, with the exception that, unlike conflict, change is not inevitable in an organizational setting. Some organizations are quite stable and have operated in very much the same fashion over a long period of time. While some degree of stability is necessary and desirable in an organization, the ability to change and adapt to new conditions and pressures is even more impor-

[16]See Woolridge and Wester

tant. This is particularly true in a police organization, where changing demands of society, changes in the law, and changing public attitudes toward the police and the police function make it imperative that the police organization be adaptable to change. It is also a fact that the attitudes of the people who are employed in the police organization are changing, and this change also creates the need for greater flexibility in management styles.

The police administrator must not only manage the forces of change within the police organization, but play a pivotal role in the change process. Indeed, the police administrator should act as the change agent in enabling the organization to meet new demands and public expectations. As Goldstein has observed:

> While not the exclusive initiator of change, the police administrator remains a central figure in the process of change, for it is upon him and through him that both the internal and the external forces of change exert their pressure.[17]

Police administrators should carefully consider the purpose of any proposed change to ensure that the results will be substantive rather than cosmetic. For example, a police agency wishing to ease tensions between itself and minority community members must do more than create a public relations program designed merely to inform the members of the community of all the positive things the department does on their behalf. Instead, very real efforts should be made to develop genuine understanding and mutual support between minority representatives and members of the police organization.

Change should not be itself an objective, but should rather be the medium through which the organization is made more responsive to new conditions and better able to achieve emerging goals and objectives. Change needs to be directed toward perceived needs in the organization, whether they be concerned with programs, policies, procedures, or the organization itself. Change should not be undertaken lightly or in haste, but should be based on a solid foundation prepared by careful and thoughtful planning and analysis.

The police administrator's role should be not only to **initiate** change but to **manage** the change process. By this it is meant that the forces of change are carefully controlled and that the inevitable conflict that results from the change is minimized. Conflict is usually a byproduct of change, and it is important that changes are effected in such a way as to produce the least amount of conflict possible. It is also important that the change process be carefully planned and directed toward identifiable goals and objectives. This is the essence of managing change.

[17]Herman Goldstein, *Policing a Free Society* (Cambridge, Mass.: Ballinger Publishing Co., 1977), p. 309.

SUMMARY

Managing is the process of getting the most out of available resources in the accomplishment of organizational goals. Organization is the process of creating a structure and environment in which things get done in the most efficient manner possible. Thus, organization and management go hand in hand and are integral components in the process of achieving organizational goals.

While there are many components of management, none is more important than the managing of people, because people are what makes the organization work and are the organization's most important resource. The effective manager is the one who sees his or her most important responsibility as the effective management of people. By "management of people" we do not simply mean the process of getting people to do what needs to be done, but the caring for, looking after, understanding, and caring about the needs and sentiments of the people in the organization.

REVIEW QUESTIONS

1. Discuss the purposes of organization and the problems that arise from improper organization.
2. Why is it important to have an organization chart? What purposes does it serve?
3. Explain what is meant by the term *organization by function.* How does this apply to a police department?
4. Explain the difference between staff functions and line functions. Describe how these functions apply to the police organization.
5. Why is span of control important in a police department? What kinds of things in a police agency will dictate optimum span of control?
6. Explain the principle of unity of command. Give examples of how problems can develop if this principle is violated in a police agency.
7. What is the informal organization and what impact does it have on the formal organization? What kinds of problems can be created by the informal organization and how can they be solved?
8. What is your definition of leadership? What are some of the leadership qualities you think are important in a police agency? Why are these important?
9. What are the elements of the planning process? Give examples of how the planning process is or might be used in a police agency.
10. Explain why a mission statement is important in a police agency.
11. What are some of the problems that might be encountered in maintaining good internal communications in a police agency? How can these problems be resolved?

12. Describe the difference between managing conflict and managing change. Give examples of how these processes might apply in a police organization.

REVIEW EXERCISES

The following exercises are designed to reinforce or amplify comprehension of the material contained in this chapter.

1. Select one of the principles of organization described in the chapter that most interests you. Conduct library research and a literature review to obtain more detailed information on this principle. Prepare a paper summarizing the results of your work.
2. Prepare an organization chart for a hypothetical police organization, showing all major organizational components including lines of command and control. Identify the various functions shown as either line or staff functions.
3. Conduct independent research on the informal organization. Prepare a paper discussing the informal organization and how it may impede as well as enhance organizational management.
4. In small groups or in open classroom discussion, deliberate the importance of delegation, why some managers may be reluctant to delegate, and how this may harm the organization.

REFERENCES

Alpert, Geoffrey P., and Lorie A. Fridell, *Police Vehicles and Firearms: Instruments of Deadly Force*. Prospect Heights, Ill.: Waveland Press, 1992.

Goldstein, Herman, *Policing a Free Society*. Cambridge, Mass.: Ballinger Publishing Co., 1977, p. 309.

Hersey, Paul, and Kenneth H. Blanchard, *Management of Organizational Behavior: Utilizing Human Resources* (5th ed.), Englewood Cliffs, N.J.: Prentice Hall, 1988.

Kroes, William H., *Society's Victim—The Policeman: An Analysis of Job Stress in Policing*. Springfield, Ill.: Charles C. Thomas, 1976.

Leonard, V. A., and Harry W. More, *Police Organization and Management* (7th ed.), Mineola, New York: The Foundation Press, Inc., 1987.

Longworthy, Robert H, "Organizational Structure," in Gary W. Cordner and Donna C. Hale, *What Works in Policing? Operations and Administration Examined*. Cincinnati, Ohio: Anderson Publishing Co., 1992, pp. 87-105.

President's Commission on Law Enforcement and Administration of Justice, Task Force Report: The Police. Washington, D.C.: U.S. Government Printing Office, 1967, p. 78.

Sheehan, Robert, and Gary C. Cordner, *Introduction to Police Administration* (2nd ed.), Cincinnati, Ohio: Anderson Publishing Co., 1989.

Thibault, Edward A., Lawrence N. Lynch, and R. Bruce McBride, *Proactive Police Management* (2nd ed.), Englewood Cliffs, N.J.: Prentice Hall, 1990.

Witham, Donald C., "Transformational Police Leadership," in James J. Fyfe, ed., *Police Practice in the 90's: Key Management Issues*. Washington, D.C.: International City Management Association, 1989.

Woolridge, Blue, and Jennifer Wester, "The Turbulent Environment of Public Personnel Administration: Responding to the Challenge of the Changing Workplace of the Twenty-First Century," *Public Personnel Management,* 20 (September 1991), pp. 207-224.

CHAPTER 9
PATROL FORCE STAFFING AND DEPLOYMENT

CHAPTER OUTLINE

LEARNING OBJECTIVES

After completing this chapter, the student should be able to

1. Explain the process involved in determining patrol-force staffing requirements.
2. Describe the techniques and methods used in developing patrol force schedules and deployment plans.
3. Describe the methods and processes involved in conducting workload analysis for a police patrol force.
4. Explain the importance of developing and maintaining up-to-date beat plans for patrol-force deployment.
5. Describe the different kinds of shift schedules that may be used by a police patrol force to accommodate different purposes.

The patrol force is typically the largest single element in the police department and is the primary point of contact between the police agency and the public. All other police functions are a result of or are designed to support the efforts of the patrol force. To achieve maximum effectiveness, the patrol force must be properly staffed and deployed in a manner that will ensure the most efficient utilization of resources. Unfortunately, there are no reliable standards or precise formulas by which to determine optimum staffing levels for the patrol force. Each police agency is different and has different needs that will affect patrol staffing requirements. This chapter addresses the issues of patrol-force staffing, deployment, and utilization in an effort to provide some general guidelines that can be adopted and altered for use by local police agencies in a variety of environments.

Most of a patrol officer's time is uncommitted. That is, patrol officers spend relatively little time handling calls for service, investigating traffic accidents, arresting drunken persons, chasing criminals, or investigating crimes. Most of their time is spent on random patrol. Thus, there is a great opportunity for the utilization of innovative and creative patrol strategies to make better use of their time.

PATROL-FORCE STAFFING

The optimum staffing level for a police patrol force will depend upon a number of factors including population, size of area served, particular crime problems, the level of service demanded by the community, contractual requirements with the bargaining unit (which may affect personnel deployment and scheduling), and, of course, local financial constraints. Moreover, staffing levels will be influenced very directly by the efficiency of deployment practices.

As indicated previously, there are no exact standards that can be used to determine patrol staffing levels. From time to time, one may come across references to the

FBI Standards, referring to the police staffing ratios published in the FBI's annual *Uniform Crime Report*. These are not standards, however, nor are they even recommended guidelines. They are merely national averages, which means that actual practices vary considerably from these figures.

In general, cities on opposite ends of the population range report higher-than-average police staffing to population ratios, ranging from around 2.5 officers per 1,000 population to over 4.0 officers per 1,000 population. Small communities usually exceed the average simply because they need to maintain a minimum staffing level just to have 24-hour-a-day police protection. Large cities, on the other hand, tend to exceed the national averages simply due to the unique crime problems faced in large cities and the need to develop a myriad of special units to deal with them.

For cities having more than 10,000 people and less than 100,000 people, somewhere between 1.5 officers and 2.5 officers per 1,000 population is the norm, although there are many exceptions to this so-called standard. There are many communities with populations between 25,000 and 50,000 that get along very nicely with less than 1.5 officers per 1,000 population. Likewise, there are many communities in the same population range that have more than 2.5 officers per 1,000 population and that provide a very high level of police service with a reasonable degree of efficiency. In other words, the number of officers per population is not a reasonable measure of either police efficiency, effectiveness, or productivity.

Community characteristics such as location and economics will have a much greater bearing on police-staffing levels than population alone. Cities with large industrial areas interspersed among residential and commercial parts of the city will require a different level of police protection than bedroom communities with very little industry and only limited commercial activity. Areas with recreational and cultural features such as lakes, beaches, parks, golf courses, and historic attractions will require a level of police staffing that is different from communities without such attractions. These variables have little to do with population. Nevertheless, some police administrators still point to population as being the standard by which their departments should be staffed. Nothing could be further from the truth.

Workload Analysis

The only logical and acceptable means of determining how many persons should be assigned to patrol duty is through a thorough and systematic analysis of the duties being performed by the patrol force. Workload analysis is an extremely important function that should be performed periodically in all police agencies having at least 25 uniformed patrol officers. Smaller police agencies need to perform workload analysis too, but on a less frequent basis, simply to determine whether changes in workload may indicate the need for either additional personnel or a reallocation of existing personnel.

The larger the police agency, the more complicated the process of workload analysis becomes. In large police departments, workload analysis will normally be performed on a regular basis by the research and development unit, crime-analysis section, or planning unit. In police agencies where workload is measured in terms of calls for service per day, workload analysis will require a mainframe computer and a fairly sophisticated management information system designed to record, retrieve, and analyze calls for police service, arrests, traffic accidents, police-initiated incidents, and other police activities.

In smaller police departments, workload analysis can be accomplished with much less difficulty and with much smaller computer systems. Although it is possible to conduct workload analysis manually, computers have become an integral component of most police agencies, and it is difficult to imagine a police agency being able to function effectively without at least a basic computer system.

Purposes of Workload Analysis

Workload analysis can and should be performed at all levels of the police department and for all major functions for the following reasons: (a) to determine the need for additional resources or the reallocation of existing resources, (b) to assess individual and group performance and productivity, and (c) to detect trends in workload that might indicate changing activity levels and conditions.

Types of Workload Indicators

The process of workload analysis begins with the identification of suitable workload indicators which can be just about anything that can be used to measure workload. These indicators will vary from one department to the next and from one division of the same department to another. No two agencies will want to measure workload in the same way because the kind of work they do is often quite different. There are, however, some common indicators that will apply more or less equally to all police agencies. These include arrests (misdemeanor, felony, traffic); traffic accidents (injury, noninjury, fatal); calls for police service (broken down by type and severity); citizen contacts (traffic stops, field interrogations, suspicious persons); open doors investigated; silent alarms; and a variety of other activities.

Collecting Workload Data

Once workload indicators have been identified, it is necessary to develop a reliable and accurate way of collecting the necessary information. Again, computerized activity reporting makes this relatively easy. Some police agencies continue to use daily work logs to record various activities performed by the officers. However, the reliability of these is often questionable, since they are not often audited to determine their accuracy. A more reliable way of capturing workload data is by obtaining it from the original dispatch card prepared by the dispatcher. In police departments with computer-assisted dispatching, this information is automatically recorded and can be later retrieved for analysis.

Determining Staffing Needs

Once the appropriate workload indicators have been identified and a reliable and accurate way of capturing the necessary data has been established, it is then necessary to develop a formula by which activity data can be translated into staffing needs. In other words, it is necessary to determine how much of what kind of activity can be performed by how many people. Here again, there are no precise formulas that can be employed to determine average workload per officer. Once a system is established, it should be periodically reviewed and, if necessary, adjusted, to make it more workable.

By systematically examining both the number and type of patrol activities performed, as well as the length of time required to complete them, it is possible to develop standard workload measure which can then be included in a formula to determine staffing requirements. For purposes of this discussion, the standard workload measure will be called the **effective level of effort** or ELE. It should be remembered that the ELE is not an actual figure, but is rather an average figure, and that the actual level of effort for various kinds of activities may be greater or less than the ELE.

Once the ELE for various categories of activities has been established, it is then necessary to establish past activity trends and project future activity levels. This information, along with the ELE, can then be used to develop basic patrol staffing requirements.

It should be obvious that the ELE will vary from one type of patrol incident to another. In addition, it is not unusual for the ELE to vary somewhat from one shift to another, from one day to another, or from one area to another for the same type of activity. These variations need to be taken into consideration when calculating staffing needs. The preliminary investigation of a burglary of a large appliance store, for example, may very well require a number of hours to complete, whereas the theft of a bicycle from a garage may require only a few minutes to investigate.

Variations in Workload Indicators

In order to account for variations in the ELE for different kinds of activity, the workload analysis should take into consideration the following: (a) time of day, (b) day of week, (c) location, (d) type of incident, and (e) number of officers or police units required to handle the incident.

The first step in workload analysis is to obtain necessary information on a representative sample of patrol activities. The size of the sample will depend upon the total number of activities, but should cover a period of not less than one year to account for any deviations in activity levels that may occur on a seasonal basis. For a large data base of several thousand incidents, a 10 percent sample size will usually be satisfactory. The greater the sample size, the greater will be the accuracy of the analysis, but there is a point at which increasing the size of the sample will not yield a proportionate increase in the accuracy of the data. On the other hand, the smaller the original data base, the larger the sample size should be to ensure that it is representative of the entire population of incidents.

Sampling Methods

Either systematic or random sampling may be used to ensure the representativeness of the sample. To obtain a systematic sample, it is necessary to determine what percentage of the total data base will be included in the sample and then to select, in a systematic and uniform manner, the number of incidents corresponding to that percentage. For example, if the total data base consisted of 12,268 incidents and a 25 percent sample was desired, every fourth incident in the data base would be selected, beginning with the first and ending with the last in the order in which they occurred (that is, chronologically). This would yield a sample of 3,067 incidents. Systematic sampling is the best way to ensure that the sample is a representative sample of the entire data base.

Random sampling, if properly conducted, can also yield a representative sample of the entire data base, but is more time consuming and requires additional effort if done manually. Under random sampling, each incident in the data base is assigned an identification number Then, using a table of random numbers (found in most elementary statistics texts), each incident is selected according to the identification number corresponding to the number in the table of random numbers until the desired sample size is achieved. Since all incidents have exactly the same probability of being included in the sample, the sample is theoretically representative of the entire data base.

After the sample has been drawn, it is then necessary to code each incident according to the type of incident, day of week, hour of occurrence, and total time involved in the incident. If the analysis is being conducted manually, simple scoring sheets can be issued to record the data by category and in any format desired. For example, it may be desirable to know the distribution of incidents by type, number of shifts, hour of the day, or day of the week. As incidents are collected for the sample, they can be recorded on the tally sheets in the manner desired

If computerized analysis is being used, the computer program will need to be set up ahead of time to provide the kind of breakdown needed to complete the analysis. All too often, computer programs are written to satisfy one set of requirements, but not another. Thus, a software program that offers greater flexibility to the user is to be desired when attempting to conduct any kind of workload analysis.

When recording the amount of time devoted to each incident, it is also necessary to take into consideration those incidents involving multiple responses. The total committed time for each unit involved in the incident, exclusive of supervisory personnel, should be included in the analysis. For example, a domestic dispute will almost always require more than a single officer, as will a public disturbance, a bank alarm, and a myriad of other kinds of police incidents. These kinds of incidents need to be weighed more heavily than those which require only a single officer to handle. The handling of a fatal traffic accident, for example, may require one officer one hour to conduct the investigation, two additional officers 30 minutes each to handle traffic control and highway cleanup, and another officer 30 minutes to take statements from witnesses. As a result, that one incident could require in excess of 2.5 hours to conclude, not including any later follow-up action that may be required.

In identifying workload indicators, it is desirable to distinguish between incidents referred to the police for action and those initiated by the police themselves. An obvious example of the latter kind of incident would be a traffic stop or a field interrogation. If we are interested solely in examining workload, then it would not be appropriate to count officer-initiated activities as part of the workload indicators, since these incidents will vary with the number of officers on duty.

Once the sample has been collected and the ELE for each type of activity has been determined, it is then possible to compute the number of patrol hours required to handle the workload. A simplified illustration of this computation is shown in Table 9.1, which shows that a total of 80,000 patrol incidents of all types are projected for a one-year period. It is also estimated that approximately 45,350 patrol hours will be required to satisfactorily handle the 80,000 incidents.

Table 9.1 Simplified Workload Analysis

Incident Type	*Number In Sample*	*Actual Number*	*Average Time*	*Total Hours*
Traffic Arrests	900	9,000	1.1 Hours	9,900 Hours
Misdemeanor Arrests	695	6,950	1.0 Hours	6,950 Hours
Felony Arrests	475	4,750	1.4 Hours	6,790 Hours
Calls for Service	1,385	13,850	.5 Hours	6,925 Hours
House Checks	1,480	14,800	.2 Hours	2,960 Hours
Traffic Stops	1,750	17,500	.3 Hours	5,250 Hours
Miscellaneous	1,315	13,150	.5 Hours	6,575 Hours
Totals	8,000	80,000	------------	45,350 Hours

Note: Total Incidents are Based on a 10 percent sample.

After the estimated number of patrol hours required to handle the projected workload is known, it is then necessary to allow additional time for such functions as directed deterrent patrol, traffic enforcement, school patrol, and general administrative duties, to include training, vehicle maintenance, meal breaks, personal service, court, and the like. The allocation of patrol time for directed patrol activities, calls for service, and administrative duties will have a significant impact on staffing requirements. Local preferences should dictate just how busy the patrol force should be and how much time should be left over for attending to directed patrol functions and administrative duties.

Some authorities have suggested that about one-third of a patrol officer's total available time should be devoted to calls for service and the remaining two thirds of the time should be more or less evenly distributed between directed deterrent-patrol duties and administrative activities. Given what we now know about the effectiveness of "routine" patrol, it appears that any time allowed for preventive patrol should be directed toward specific crime prevention or deterrent activities rather than routine, random patrol.

Distribution of Patrol Time

In any allocation of patrol time, the greatest consideration should be given to the time needed to handle calls for service. Experience has shown that somewhere between 25 and 35 percent of time devoted to calls for service represents an acceptable workload. Anything in excess of 35 percent of available time devoted to calls for service tends to become excessive, while less than 24 percent of an officer's time devoted to calls for service indicates an excessive amount of available time. For the purposes of illustration, the following time allocations will be used in determining patrol staffing requirements:

Calls for Service	35 percent
Directed Patrol	35 percent
Administrative Duties	30 percent

Since we know that total calls for service will require 45,350 hours to complete, and that we want this figure to represent no more than 35 percent of the total time devoted to the patrol force, it is possible to calculate the total time that should be allocated to the patrol effort by dividing 35 percent (.35) into 45,350. The result is:

$$\frac{45,350}{.35} = 129,570$$

Thus, the total time that should be devoted to the patrol effort is 129,570 hours. It is now possible to show how this time should be allocated among the three primary categories: calls for service, preventive patrol, and administrative functions. According to the percentages indicated previously, this distribution would be as follows:

Calls for Service	35 percent	45,350 hours
Directed Patrol	35 percent	45,350 hours
Administrative duties	30 percent	38,870 hours
Total Activity	100 percent	129,570 hours

According to the figures, a total of 129,570 hours will be required to carry out patrol operations over a one-year period, assuming that the level of activity remains constant from previous years or that the projections made on the basis of the incident sampling are correct. If a police agency is experiencing a steadily increasing workload as a result of growth or other factors, this should be taken into account. In this case, it will be necessary to take this increase into account when formulating staffing requirements.

Since we now know how many patrol hours will be required to perform basic patrol duties, it is possible to calculate the number of personnel actually needed to staff the patrol function. To do this, it is necessary to divide the total number of hours of patrol time required by the total number of hours one person can work in one year, assuming no time off for sickness or holidays. Assuming that one person worked eight hours a day, 365 days a year, he or she would work a total of 2,920 hours a year.

$$365 \times 8 = 2,920$$

By dividing the total number of hours required for patrol purposes by 2,920, it is then possible to determine the number of positions required to staff the patrol force. This calculation is as follows:

$$\frac{129{,}570}{2{,}920} = 44.4$$

The Availability Factor

We know now that it will require 44.4 **positions** to provide an acceptable level of patrol service based upon the data shown in the workload analysis. The next step in the process of calculating staff requirements is determining the availability factor, which tells us how many people are required to staff one position 365 days a year, 24 hours a day. The **availability factor** takes into consideration regular days off, vacations, holidays, sick leave, court time, injury leave, and the various other factors that detract from full availability of personnel.

The availability factor can usually be computed quite easily, using past personnel performance records. In larger departments, this information might already be available from either the city personnel office or the finance office. If not, the information will have to be calculated using data drawn from personnel files. Calculating the availability factor begins with subtracting from 2,920 the average number of hours that will be lost each year per person due to all known or anticipated reasons. The remaining figure is then divided into 2,920 to produce the availability factor. A brief example will help to illustrate this calculation.

Maximum Number of Working Hours Per Year (365 X 8)		2,920 hours
Less time off for:		
Regular days off (52 X 16)	832 hours	
Holidays (12 X 8)	96 hours	
Vacation	120 hours	
Training	60 hours	
Sick Time	48 hours	
Court Time	60 hours	
Miscellaneous	86 hours	
Total Unavailable Time	1,302 hours	

		2,920 hours
Divided by		
2,920 less 1,302	=	1,618 hours
	=	1.8

The computation indicates that it will require 1.8 persons to fill one position 365 days a year, eight hours a day. In order for this position to be filled around the clock, the availability factor must be multiplied by 3, which means that it will require 5.4 persons to fill one position all year around the clock.

Once the availability factor has been calculated, it must be multiplied by the number of positions required (44.4) to determine the actual staffing level of the patrol force:

$$44.4 \times 1.8 = 79.9$$

This tells us, then, that we will need 80 officers to staff the patrol function, using the workload data presented in Table 9.1. We can double-check this figure by simply dividing the total number of required patrol hours (129,570) by the actual number of hours that each patrol officer will be expected to work in one year (1618):

$$\frac{129{,}570}{1618} = 80.1$$

The availability factor is important for three principal reasons. First, as has just been shown, it is a very simple and effective way of calculating staffing requirements, assuming that the necessary personnel data is available. Second, it is an important management tool and can be used to gauge whether there are personnel problems in the agency. Obviously, the higher the availability factor, the fewer number of hours are being worked by available personnel and the greater number of people will be required to get the job done.

Third, the formula used to calculate the availability factor can be used to estimate the impact of changes in collective bargaining agreements that may affect staffing levels. For example, if additional holidays, personal days, or vacation days are granted through negotiations, the formula used to calculate the availability factor can be used to estimate what increases in staffing may be required to offset them.

PATROL-FORCE DEPLOYMENT

The previous section focused on the methods used to calculate the number of officers needed to provide an acceptable level of patrol service. Local conditions and needs will vary, but the basic method of workload analysis should provide a good framework for determining staffing needs. Once the needed staffing level has been established (and presumably attained), the next task for the patrol administrator is to develop a deployment strategy that will ensure maximum effectiveness and productivity as well as optimum utilization of personnel resources.

The goal of proper patrol-force deployment is to distribute the patrol force in such a way as to (a) provide equitable distribution of workload, (b) provide for an acceptable level of response to calls for service, (c) account for the expected variations in activity that occur by hour of the day and by day of the week, and (d) provide a work schedule that will reasonably accommodate the needs of the department as well as those of the officers. While it may be impossible to devise a patrol-force deployment strategy that will satisfy all of these goals perfectly, this should be the ultimate objective.

Like all other functions of management, patrol-force deployment cannot simply be taken for granted, nor should it be held captive by the traditions of the past. Simply because something has always been done a certain way is no excuse to maintain the status quo. Indeed, if the present method of deployment has remained unchanged for five years or more, there is a very good chance that it needs to be revised or updated. Proper patrol-force deployment includes developing a logical beat plan and shift schedule that will adjust for variations in activity levels and provide reasonable periods of relief for affected personnel.

Shift Scheduling

Shift scheduling is not usually one of the patrol administrator's favorite pastimes simply because "you can't please all of the people all the time." Due to the 24-hour-a-day, 365-day-a-year nature of police work, people must be scheduled to work nights, weekends, and holidays regardless of their personal preferences or the demands of their families. There are never enough "good" shifts to go around, and it is virtually impossible to find a way to fill all the shifts that need to be filled without having someone grumble and complain.

Shift scheduling is one of the more unpleasant tasks faced by the patrol administrator, but it is also one of the most important. While there may be no perfect solutions, there are techniques that can be employed to minimize the disruption of personal lives and provide for an equitable distribution of personnel according to workload.

One major factor that may need to be taken into consideration in terms of shift scheduling is the collective bargaining agreement. In many cities, police unions have managed to insert provisions into the contract that place limits on the scheduling prerogatives of the department. Seniority and permanent shifts, for example, may dictate which people and how many people will be working on a particular shift and may also prohibit any consideration of shift rotation. Such provisions should be avoided if at all possible, simply because they hamstring the police administrator and eliminate the shift-scheduling process as a viable management tool.

One indication of the effectiveness of a patrol-shift schedule is how well it matches variations in workload by day of the week and by hour of the day. This can easily be determined by dividing the number of available patrol hours into the number of hours actually devoted to calls for service or other workload indicator. The result should be the percentage of time actually devoted to calls for service. Typically, this figure will vary somewhat from one day of the week to the next, and from one shift or hour of the day to another. The greater the degree of variance, however, the greater the imbalance between duty schedule and workload. The goal should be to provide the least possible variance in those figures.

It is generally acknowledged that effective scheduling in a police agency can achieve a number of purposes, including[1]

1. reduced sick leave
2. more efficient use of personnel and equipment
3. reduced overtime
4. improved service levels
5. improved morale

The scheduling practices of police agencies vary widely and there is probably no such thing as a typical or standard method of scheduling police personnel. However, one survey of police scheduling practices revealed that the usual practice among those police agencies surveyed included the following elements:[2]

1. The work schedule consists of three 8-hour shifts and averages 40 hours a week.
2. The daily work pattern consists of five days on and two days off, with officers having the same days off each week.
3. Shift assignments change monthly
4. Officers average 11 holidays and 10 days of vacation after one year of service.

It is also true that there is no single best method of scheduling patrol personnel. There are too many local variables that may impact upon scheduling including local contract provisions, as just mentioned. What works well in one locale might not work at all in another. There are, however, an almost endless number of possibilities which exist that will serve almost any purpose under almost any condition. The only limiting factor is the imagination and ingenuity of the person responsible for designing the work schedule. There are departments that work 8-hour, 10-hour, and 12-hour schedules. There are departments that have fixed-days-off schedules and others that have rotating-days-off schedules. There are those that rotate shifts every 28 days and those that do not rotate at all. The possibilities are endless.

Police agencies also vary in the length of the duty period chosen for scheduling purposes. While the 8-hour day seems most common, many police agencies have experimented with 10-hour and 12-hour days. Generally speaking, departments should be cautious of the adverse effects that can result from extended shifts.

> Weariness and fatigue can take their toll at the end of a long shift. An overload of stressful situations can dull the edges of an officer's alertness or the simple willingness to get involved and do the job.[3]

[1]William W. Stenzel and R. Michael Buren, *Police Work Scheduling: Management Issues and Practices* (Washington, D.C.: U.S. Department of Justice, 1983), p. xvii

[2]*Ibid.*, p. 1.

[3]William W. Stenzel and R. Michael Buren, *Police Work Scheduling: Management Issues and Practices* (Washington, D.C.: U.S. Department of Justice, 1983), p. 23.

The 5-day-on, 2-day-off work schedule continues to enjoy great popularity among police departments. In the survey conducted by Stenzel, the 5-day-on, 2-day-off work schedule was reported as being used by the patrol divisions in 42.4 percent of the departments surveyed.[4]

Fixed-day-off work schedules are also popular among police departments offer several advantages including[5]

1. They are viewed with favor by officers who find it easier to schedule off-duty activities.
2. They make it easier to achieve proportional staffing levels, which is usually a goal of any scheduling technique.
3. They are relatively easy to design, implement, and administer.
4. They are generally compatible with work schedules used in other municipal departments and in private industry.

The 10-hour shift became very popular among police agencies during the late 1970s and early 1980s. It appears to be used most frequently in connection with the four-day duty cycle. It provides a number of advantages, including the availability of additional personnel during overlap periods; a common day that can be used to schedule training, court appearances, or days off, thus reducing overtime; improved employee morale stemming from working four days followed by three days off. On the other hand, the 4-10 shift plan has been found to have some rather distinct shortcomings:[6]

1. More officers are required to fill shifts under the 4-10 plan than under more traditional scheduling plans. As a result, there is a false economy achieved by having additional personnel available during overlap periods.
2. In some cases, overlap periods provide for more personnel than are really needed or that can be effectively used. For example, going from six officers on duty to 12 on duty for a period of two hours may not provide the most efficient utilization or distribution of available personnel.
3. When the 10-hour shift schedule is used in conjunction with a fixed day off cycle, as is the case in the 4-10 plan, it is difficult to provide for proportionate staffing based upon workload.

[4]Stenzel and Buren, *Police Work Scheduling,* p. 43. See pages 149-152 for examples of different shift schedules using fixed and rotating days off with fixed and variable staffing patterns.

[5]*Ibid.,* p. 62.

[6]*The Four-Ten Plan—California Law Enforcement,* California Commission on Peace Officer Standards and Training, 1981, as cited in Stenzel, pp. 72-74.

In addition, one detailed evaluation of the four-ten plan conducted by the California Commission of Peace Officer Standards and Training found that:[7]

1. While overtime, sick leave, and court time initially decreased under the 4-10 plan, overtime returned to or surpassed the levels experienced before the adoption of the plan.
2. Problems were experienced in terms of unity of command and supervision.
3. It became difficult to provide training in ten-hour blocks of time.
4. The 4-10 plan became difficult to maintain without increasing the number of personnel.

Relatively few police agencies have experience with 12-hour shifts and there is little data to evaluate to indicate whether such schedules have any practical value in law enforcement. However, it would seem that, while 12-hour shifts might appeal to some officers simply because they would provide extended day-off periods, over a period of time fatigue could become a real problem.

Shift Rotation

Monthly or periodic shift rotation affects the body's natural rhythm and takes a toll on an officer's ability to perform at optimum capacity.[8]

It is not uncommon for police departments to rotate shifts every 28 days, and some rotate as often as every week. The more frequent the rotation, the more significant are the consequences to the individual. It usually takes a human being several days to adjust the eating, working, and sleeping cycle to a new schedule.

It should be remembered that the human body is governed by a sort of biological clock that controls and regulates daily cycles of sleep, hormone release, alertness, and short-term memory. This biological clock normally synchronizes itself to the earth's day-night cycle. Its ability to adapt itself to variations in this basic time cycle is limited. Attempts to radically alter this basic cycle can have counter-productive effects on the human body over a period of time. While some individuals may react to frequent changes in work schedule without negative effects, others may suffer serious consequences. Thus, any work schedule featuring regular rotation of shifts should take into consideration the impact these changes may have upon the individual.[9]

[7]*Ibid.*

[8]Gail A. Goolkasian, Ronald W. Geddes, and William DeJong, "Coping With Police Stress," in Robert G. Dunham and Geoffrey P. Alpert, *Critical Issues in Policing: Contemporary Readings* (Prospect Heights, Ill.: Waveland Press, 1989), pp. 498-507. See also William H. Kroes, *Society's Victim—The Policeman: An Analysis of Job Stress in Policing* (Springfield, Ill.: Charles C. Thomas, 1976), pp. 29-33.

[9]See Timothy Monk, "Advantages and Disadvantages of Rapidly Rotating Shift Schedules: A Circadian Viewpoint," *Human Factors* (October 1986), pp. 553-557.

One-third of the departments surveyed by Stenzel do not use rotating shifts. Of those that do, more than 20 percent rotate either weekly or biweekly.[10]

> Among departments that use rotating schedules, 19% reported they had problems with rotation. Frequently mentioned were disruption of outside education and employment commitments, problems with home life, and fatigue from short off-duty periods at changeover from one shift to another.[11]

Police departments may very well discover that rotating shift assignments make it difficult to proportionately distribute the patrol force according to workload. In addition, there are other distinct advantages of fixed shifts:[12]

1. Efficient use of available personnel.
2. Fewer adverse physiological problems.
3. Provide shift preference as a reward to senior officers.
4. Most officers seem to prefer fixed shifts.
5. Court scheduling is made easier.
6. More effective supervision and evaluation of personnel.

Proportional Scheduling

It has already been pointed out that one of the measures by which the effectiveness of patrol-shift scheduling methods can be evaluated is the degree to which the shift schedule provides for a distribution of available personnel that is proportionate to the distribution of the workload, measured either by calls for service or by some other means. The essence of proportionate scheduling is that the largest number of people will be on duty when and where the highest frequency of activity occurs. Proportionate scheduling, then, helps to provide for the most efficient utilization of available personnel resources and equalized distribution of workload. As the National Advisory Commission on Criminal Justice Standards and Goals noted:

> Effective deployment of patrol personnel must begin with distribution of personnel on a proportionate need basis. Through careful study of crime occurrences, calls for service, and other selected factors, available manpower can be systematically distributed geographically and chronologically according to the relative need for police service throughout the community. This process in itself will not guarantee the best use of

[10]Stenzel and Buren, *Police Work Scheduling*, p. 2.

[11]*Ibid.*, pp. 10-11.

[12]*Ibid.*, p. 12.

patrol personnel, but it is the foundation for developing the most effective deployment strategy.[13]

Attempting to maintain proportional scheduling is not easy, and it is one of the reasons that scheduling sometimes tends to be a headache for some police administrators. It is so much easier to simply divide the patrol force up into three equal shifts and assign personnel to those shifts on the basis of seniority, or by random order or in some other manner that provides some degree of fairness. Proportional scheduling, on the other hand, requires the patrol administrator to make decisions about how people are to be assigned: to what shift, with what days off, and with what frequency of rotation, if any.

Proportionate scheduling is as much an art as it is a science, and requires a great deal of experimentation with various combinations of schedules and deployment methods. As indicated elsewhere in this chapter, it is important that the collective bargaining agreement provide as much latitude as possible in this regard, since flexibility is an important element in designing a proportionate shift schedule.

To design a proportionate scheduling plan, it is necessary to return to the previous discussion of workload analysis, since we need to know not just what the actual workload is, but the manner in which it is distributed. For the purpose of this discussion, we use the same figures presented earlier in this chapter and assume that the times required to complete the various calls are the same from one shift to the next and from one day of the week to another.[14]

Table 9.2 shows the same data that was presented in Table 9-1. In this case,

Table 9-2 Police Workload Distribution

Incident Type	*Night Shift 12AM-8AM*	*Day Shift 8AM-4PM*	*Afternoon Shift 4PM-12AM*	*Total*
Traffic Arrests	2,470	3,455	3,975	9,900
Misdemeanor Arrests	750	3,500	2,700	6,950
Felony Arrests	3,075	1,400	2,315	6,790
Calls for Service	1,025	3,020	2,850	6,925
House Checks	450	825	1,715	2,960
Traffic Stops	1,820	1,340	2,090	5,250
Miscellaneous	1,875	2,080	2,620	6,575
Totals	11,465	15,620	18,265	43,350
Percent of Total	25.3	34.4	40.3	100.0

[13]National Advisory Commission on Criminal Justice Standards and Goals, *Police* (Washington, D.C.: U.S. Government Printing Office, 1979), p. 201.

[14]In reality, this will probably not be the case and detailed analysis may be required to determine shift-by-shift and day-by-day variations

though, the data is presented to show the actual distribution of the time devoted to various patrol activities on each shift. To simplify the process of illustration, we have chosen to use three standard eight-hour shifts, although the same process can be used for any other breakdown that is required.

According to the figures shown in Table 9.2, we can see that the night shift accounts for about 25 percent of the total activity, the day shift handles almost 35 percent of the total activity, and the afternoon shift is responsible for about 40 percent of the total activity. Although these figures are hypothetical, they are very representative of the distribution of patrol activity found in many police departments.

The idea of proportional scheduling simply means that, if 40 percent of the workload occurs on the afternoon shift, then 40 percent of the patrol strength (not including supervisory personnel) should be assigned to that shift. Although simple in theory, putting the idea into practice is not always easy. Workload figures are easy to break down into percentages. People, however, are another matter. Finding a work schedule that will provide the desired level of service by hour of the day and by day of the week is not an easy task, nor is it even always possible.

Since we now know, based upon the calculations described earlier in this chapter, that it will require 80 patrol officers to provide an acceptable level of service, we are now in a position to decide how those personnel ought to be distributed according to shift. By simply multiplying the number of patrol officers (80) by the percentage of workload for each shift, the following breakdown can be obtained:

	Night Shift 12AM - 8AM	***Day Shift 8AM - 4PM***	***Afternoon Shift 4PM - 12AM***
Percent of Workload	25.3	34.4	40.3
Number of Officers	20	28	32
Percent of Total	25.0	35.0	40

By assigning 20 officers to the night shift, 28 officers to the day shift, and 32 officers to the afternoon shift, we can achieve a distribution of the work force that is very close to the distribution of the workload. It should be obvious from this discussion that achieving proportionate distribution of personnel by shift becomes easier as the number of personnel in the pool increases. Achieving proportionate scheduling for a patrol force of less than 20 people, on the other hand, becomes much more difficult, simply because it becomes more difficult to divide people into exact percentages.

Determining the number of officers needed on each shift is a relatively easy task, as has just been demonstrated. Finding a work schedule that will meet the varying needs of the shift on a day-to-day basis, on the other hand, is quite another matter. As has been discussed previously, there are a variety of work schedules that are available to the police administrator. There seems to be a growing tendency toward the four-day work week, using either eight, nine, or ten-hour work days and varying day-off cycles. The shorter work week, of course, is a big attraction to the rank and

Table 9-3 Distribution of Patrol Activity

	Sun	Mon	Tue	Wed	Thu	Fri	Sat
Calls for Service	1,125	1,310	1,700	1,800	1,905	1,960	1,675
Percent of Total	9.8	11.4	14.8	15.7	16.6	17.1	14.6

Note: Figures for Calls for Service Represent Hours

file, as is the rotating day-off cycle, since this ensures that no one will have an exclusive right to weekends off while others have to have days off during the week for an indefinite period.

The five-day-on, two-day-off work cycle continues to provide the greatest versatility in terms of scheduling, since it allows the patrol administrator to match up days off with periods of the least amount of activity. Any duty schedule that does not provide fixed days off makes it difficult to match up staffing levels with activity lev-

Table 9-4 Duty Schedule for 20-Person Shift

Off. No.	Sun	Mon	Tue	Wed	Thu	Fri	Sat
1.	0	X	X	X	X	X	0
2.	0	X	X	X	X	X	0
3.	0	X	X	X	X	X	0
4.	0	X	X	X	X	X	0
5.	0	X	X	X	X	X	0
6.	0	0	X	X	X	X	X
7.	0	0	X	X	X	X	X
8.	0	0	X	X	X	X	X
9.	0	0	X	X	X	X	X
10.	0	0	X	X	X	X	X
11.	X	0	0	X	X	X	X
12.	X	0	0	X	X	X	X
13.	X	0	0	X	X	X	X
14.	X	0	0	X	X	X	X
15.	X	X	0	0	X	X	X
16.	X	X	0	0	X	X	X
17.	X	X	X	0	0	X	X
18.	X	X	X	X	0	0	X
19.	X	X	X	X	0	0	X
20.	X	X	X	X	X	0	0
On duty	10	11	14	17	17	17	14
Off duty	10	9	6	3	3	3	6

els, since the number of people on duty on any one day is never constant. For example, Table 9.3 shows a typical distribution of patrol workload for the night shift presented in Table 9.2. It can be seen that the level of activity varies considerably from one day to the next, with Friday experiencing almost twice the amount of activity, on a percentage basis, than Sunday.

If we are using a rotating-day-off duty cycle, the number of people on duty at any one time is constantly changing, and is never the same from one week to the next. Patrol activity, on the other hand, is fairly predictable, and we can tell with a reasonable degree of certainty that some days will be busier than others,

As mentioned, the five-day-on, two-day-off work cycle provides a great advantage in that we can keep the number of people working on any day of the week constant. In this way, it is possible to match up staffing levels with variations in activity. For example, Table 9.4 shows a five-day-on, two-day-off work schedule for a 20-person shift, with days off arranged in such a manner that the number of people working on busy days is greater than the number of people working on days when there is less activity.

Table 9.5 compares the staffing data using the schedule shown in Table 9.4 with the activity data shown in Table 9.3. It can be seen that the staffing levels from day to day compare very favorably with the variations in patrol activity. This would not have been possible using a rotating day-off cycle with the number of people each day varying from one week to the other.

As has been discussed in this chapter, there are many reasons why it might not be practical to implement a proportionate shift schedule. The greatest obstacle would, of course, be a collective bargaining agreement which provides for some other scheduling method. On the other hand, where no such impediments exist, proportionate scheduling should be the ultimate goal. It is a goal that can be achieved by using the techniques described in this chapter.

Patrol Beats

The use of designated patrol beats for the deployment of the patrol force is a standard management practice intended to enhance the deterrent capability of the uniformed

Table 9-5 Comparison of Staffing and Workload

Off No.	*Sun*	*Mon*	*Tue*	*Wed*	*Thu*	*Fri*	*Sat*
On duty	10	11	14	17	17	17	14
Off duty	10	9	6	3	3	3	6
Percent on-duty	10.0	11.0	14.0	17.0	17.0	17.0	14.0
Percent workload	9.8	11.4	14.8	15.7	16.6	17.1	14.6

patrol officer. In addition, patrol beats help to distribute the workload of the patrol force so that some officers are not busy all the time while others have little to do. The beat system also helps supervisors monitor the patrol officers in their command, since it makes it easier to know where they are (or should be) and what they are doing (or should be doing). Finally, the patrol-beat system helps to ensure prompt police response to any location.

There are any number of ways to design a patrol-beat system, ranging from the simple to the complex, depending upon the objective. In general, however, the primary goal of the beat system is to provide a more equitable distribution of the available patrol force over the area to be served. Since calls for service do not always occur proportionately throughout the geographic area, it is generally desirable to distribute the patrol force evenly throughout the area so the officers can respond to any given location without unnecessary delay.

Patrol beats should be designed in a rational and systematic manner on the basis of some logical criteria. Several factors should be considered when designing patrol beats: (a) the size of the area to be patrolled; (b) natural or man-made barriers such as rivers, railroad tracks, interstate highways; (c) workload indicators such as calls for service, criminal incidents, and traffic patterns; and (d) significant characteristics which may impact directly on patrol workload such as taverns, parks, recreation areas, or major commercial centers.

Patrol-beat plans should be designed with as much flexibility possible and should be reviewed and modified from time to time. Changes in traffic patterns, crime patterns, residential construction, and a number of other factors will affect the pattern of calls for service in an area and may require new or revised deployment strategies. In addition, different patrol beat plans should be designed to accommodate different staffing levels. In a small police agency, for example, there may be three officers working the day shift, five or six on the afternoon shift, and two or three working the night shift. Each of these staffing patterns should call for a different beat plan, and that beat plan should be based upon the unique characteristics of that shift, including the distribution of calls for service, crime occurrences, traffic patterns, and other factors.

Permanent beat assignments, such as are used in the Santa Ana, California COP program (18 months duration), make patrol officers more accountable, give officers a greater sense of self-confidence, and afford officers more opportunity to become familiar with the people, places, and things in their assigned area. They get to know what is going on, where the action is, and who is involved. Officers who acquire an intimate knowledge of the beat become more proficient in doing their jobs.[15]

The other side of the coin is that the people in the officer's assigned area get to know the officer simply because they see the same one all the time. They learn what to expect and they know whether they can count on the officer in time of need. Patrol

[15]See Jerome H. Skolnick and David H. Bayley, *The New Blue Line: Police Innovation in Six American Cities* (New York: The Free Press, 1985), p. 24.

beats encourage a much more symbiotic officer-citizen relationship than would otherwise be possible.

RESPONSE TIMES

Maintaining a quick response time to emergency calls for service has always been a top priority of the police, and probably will continue to be a top priority in the years ahead. However, research has shown that a quick response time is not always as important as it seems, and that a delayed response by the police to some calls is not necessarily a poor reflection on the police, but may be favorably accepted by the general public.

Only about 25 percent of all crimes are reported while they are in progress. The vast majority of serious crimes are reported only after they have been discovered. In this majority of cases, a rapid response time by the police has no effect on identifying or apprehending the offender.[16] The authors go on to say that 911 systems are not likely to either increase citizen reporting of crimes or significantly reduce reporting times.

> The findings are clear. Most serious crimes reported to the police are discovery crimes for which there is virtually no chance for response-related arrests. For the remaining crimes, those in which there is citizen involvement, the citizen must call the police within one minute, or the likelihood of response-related arrest drops dramatically.[17]

Police administrators typically associate fast response times with "good" police performance. They assume that the public expects quick response to calls for service. Research has shown, though, that "...the time it takes a citizen to report a crime—not the speed with which the police respond—is the major determinant whether an on-scene arrest takes place and whether witnesses can be located. Furthermore, citizen-reporting delays constitute a significant proportion of the total recorded police response time."[18]

Since rapid response to all calls for the police is not as important as we once thought it was, there is good reason to review the way in which we have traditionally responded to citizen calls for service. The dial-a-cop syndrome has been the driving force in the way our police patrols have been managed for too long. Police administrators must begin to think in terms of managing calls for service rather than allowing calls for service to manage the police department.

The idea of differential police response has been studied and experimented within several police departments including Garden Grove, California; Greensboro,

[16]See William G. Spelman and Dale K. Brown, "Response Time," in Klockars, Thinking About Police, p. 161.

[17]*Ibid.,* p. 162.

[18]Joan Petersilia, "The Influence of Research on Policing," in Robert G. Dunham, and Geoffrey P. Alpert, *Critical Issues in Policing: Contemporary Readings* (Prospect Heights, Ill.: Waveland Press, 1989), p. 233.

North Carolina; and Toledo, Ohio.[19] The idea behind differential police response (DPR) is simply that not all calls for service require a rapid response by the police, and that some may not require a response by the police at all. For example, some calls can and should be referred to other community or social service agencies. Some calls for the police may be handled by a civilian community service officer rather than a uniformed police officer. Some calls can be handled over the telephone and require no response at all.

By carefully studying calls for service, it should be possible for any police agency to distinguish those calls that require immediate or delayed response, from those that do not require a uniformed-officer response. Figure 9.1 shows one typical classification scheme that has been used to prioritize police response to calls for service.

FIRST CHARACTER – CALL CATEGORY
1. Crimes Against Persons
2. Disturbances
3. Assistance
4. Crimes Against Property B – Burglary
5. Traffic Accidents T – Traffic Problems
6. Suspicious Circumstances
7. Public Morals
8. Miscellaneous Service
9. Alarms

SECOND CHARACTER – TIME
1. In-Progress
2. Just Occurred
3. Cold

THIRD CHARACTER – INJURY
0. No injury
1. Actual, Probable, or Potential Injury

FOURTH CHARACTER – RESPONSE
0. Immediate Mobile Response
1. Mobile Response Due to Override
2. Expeditor Unit (Telephone Report Unit)

PRIORITIES
99 Immediate – Injury
98 Immediate – Crimes Against Persons
97 Immediate – Crimes Against Property
96 Fifteen (15) Minutes
95 Thirty (30 Minutes)
94 One Hour
94 One Hour
93 Exceeds One Hour or When Available
92 Non-Mobile Response

Example. A "3110" is an Assistance call, in-progress, with injuries, which requires an immediate mobile response. A priority of "99" would automatically be assigned to this call by the CAD system.

Source: Marcia Cohen and J. Thomas McEwen, "Handling Calls for Service: Alternatives to Traditional Policing." NIJ Reports (September 1984), p. 5.k

Figure 9.1 Garden Grove, California Call Classification Codes

[19]J. Thomas McEwen, and others, "Handling Calls for Service: Alternatives to Traditional Policing," *NIJ Reports* (September 1984), pp. 4-8.

There are, in short, a variety of ways in which police services can be provided short of sending a uniformed officer every time. These alternatives need to be explored and, where possible, implemented, in order to make the best possible use of available police resources.

TACTICAL OPERATIONS AND STRATEGIES

Proper utilization and deployment of the patrol force includes the design and implementation of patrol strategies and tactical patrol plans which are designed to meet both day-to-day operational requirements as well as unique conditions that may arise from time to time. Tactical patrol operations have as their primary objective the deterrent of crime and the apprehension of criminal offenders engaged in the commission of crime. This chapter examines a few of the more prevalent tactical operations available to the patrol manager. These can be easily adapted by individual police departments to meet their own operational requirements.

Patrol officers often tend to think of the time between calls for service as "down time" or wasted time, since they feel they have nothing to do but wait for the next call to be dispatched. In fact, however, this time can be very valuable if used properly.[20] The essential task for the police manager or supervisor is to discover ways in which to make this available patrol time as productive and effective as possible.

Despite a growing awareness that traditional patrol methods, emphasizing routine, random patrol activities do not deter crime nor increase citizen fear of crime, no one has seriously suggested that the police abandon patrol efforts altogether. Despite its limitations, the patrol function is the mainstay of the police department and is the core around which all other activities revolve. If traditional patrol efforts do not work as well as we once thought they did, the obvious choice is to find new and better ways of managing our resources. Research, experimentation, and trial and error are what is now needed to improve the efficiency of patrol operations.

As Leonard and More have observed:

> ...while the deterrent value of police patrol activities may be questionable, police presence in the community, in itself, has a positive value as a means of giving citizens at least the minimal feeling of security and safety that is necessary if they are to enjoy freedom of movement. Admittedly, there is an illusory quality in this form of security in that the actual potential of the police to guarantee the safety of residents, their persons and their property, is far less than is suggested by the presence of police personnel, but few doubt the value in generating the illusion.[21]

[20]Robert Sheehan and Gary W. Cordner, *Introduction to Police Administration,* 2nd ed. (Cincinnati, Ohio: Anderson Publishing Co., 1989), p. 365.

[21]V. A. Leonard and Harry W. More, *Police Organization and Management,* 7th ed. (Mineola, New York: The Foundation Press, Inc. 1987), p. 52.

The 1970s and 1980s marked an important era in the development of American policing. A number of innovative programs were designed, implemented, evaluated, modified, improved and, in some cases, discarded. These efforts were somewhat frustrating since they tended to tell us more about what does not work than what does work. If traditional patrol efforts do not do what we hoped they would, what does? The answer to this question has not yet been discovered.

The innovative alternatives of community-oriented policing and problem-oriented policing discussed in Chapter 6 represent a somewhat radical approach to policing that could change forever the way we approach police services and dramatically alter the way in which the patrol force is organized and managed. However widespread these programs become, however, it is unlikely that they will significantly alter the basic function of the police—-to deter crime and to apprehend those responsible for crime. Even though traditional patrol efforts may be ineffective in preventing crime, they remain the best hope we have in dealing with traditional crime problems. Accordingly, it is in our best interest to devise ways in which those efforts can be improved, even while admitting that they will never achieve the ultimate goal of eliminating crime and criminals from our midst.

The remainder of this chapter examines very briefly some of the ways in which the police strive to attain their basic mission of deterring and preventing crime. It is acknowledged that these programs are somewhat specialized and are intended to deal with unique kinds of crime problems. As a result, they are not applicable to all communities or situations. Moreover, in some communities, traditional kinds of patrol activities might work perfectly well. The programs discussed in the remainder of this chapter are simply offered as examples of what can be done by local police administrators as alternatives to more traditional patrol methods.

We know, as a result of the Kansas City Preventive Patrol Experiment that increasing the level of police patrol activity in an area does not necessarily reduce crime, nor does it alter citizen attitudes concerning the police or their fear of crime. Thus, simply adding more people to the patrol force is not the key to solving the crime problem.[22]

The Kansas City Preventive Patrol Experiment left many important questions unanswered. As Goldstein noted, the study did not tell us much about how different kinds of patrol strategies might make a difference in citizen perceptions of crime or in the level of criminal activity.

> While there were no significant differences in the incidence of crime resulting from variations in the *level* of patrol as it was conducted in Kansas City, one wonders if the results would be the same if the form of patrol varied. Are there aspects of preventive patrol—some of which have not been used in Kansas City—that are more effective? Specifically, would officers produce any different results if the time they devote to

[22]George E. Kelling, and others, *The Kansas City Preventive Patrol Experiment: A Summary Report* (Washington, D.C.: Police Foundation, 1974)..

patrol was spent in more aggressive and intensive probing of individuals, places, and circumstances?[23]

High-Visibility and Low-Visibility Patrol[24]

The theory underlying high-visibility patrol is that there are certain kinds of crimes that can be reduced by increasing the "aura of police omnipresence." High-visibility patrol programs have been implemented in a number of major cities in the United States with varying degrees of success. The kinds of crimes they have been directed toward also vary, but are typically those crimes of violence which are usually referred to as street crimes (robbery, purse snatching, sexual assault, and other crimes of opportunity).

High-visibility patrol can be virtually anything designed to increase patrol visibility. It might include saturating a given area with additional kinds of units, walking beats in the downtown area, a park-and-walk program for parks and recreation areas, and similar strategies. The key to high-visibility patrol is the identification of crimes and locations which can be most affected by increased police visibility and the conduct of such efforts on a systematic but irregular basis. High-visibility patrol activities are most successful when changed frequently in terms of their composition, frequency, and location.

While high-visibility patrol is intended to deter street crimes, low-visibility patrol is designed to increase the rate of apprehension of persons engaged in selected types of crimes. A secondary benefit is that other kinds of crimes will be deterred as a result of the greater probability of persons being apprehended in the commission of a crime.

Surprise is a primary element in the design of a low-visibility patrol program. The idea is to provide a police presence in an area where selected crimes are likely to occur without the presence of the police being detected. In some cases, decoy personnel may be used to attract criminal activity. However, it is important that the decoy, if one is used, do nothing to encourage the commission of a crime, but simply provide the opportunity for a crime that would in all likelihood be committed anyway.

Any number of unique and innovative low-visibility patrol strategies may be implemented to deal with particular kinds of crimes. Local resources and the imagination of the patrol commander are the primary limiting factors. Plainclothes patrol officers on bicycles, in taxi cabs, bread trucks, delivery vans, and in other undercover modes can be used to gain a low-visibility, high-impact police presence. Sexual assaults, auto thefts, smash-and-grab thefts from cars, street robberies, and other kinds of crimes are ideal targets for low-visibility patrol efforts.

[23]Herman Goldstein, *Policing a Free Society* (Cambridge, Mass.: Ballinger Publishing Co., 1977), pp. 52-53.

[24]For a discussion of Low-Visibility and High-Visibility Patrol Programs, see Kenneth W. Webb, and others, *Specialized Patrol Projects* (Washington, D.C.: U.S. Government Printing Office, 1977)).

Both high-visibility and low-visibility patrol strategies have much to offer in the deterrence of crime and the apprehension of offenders. Each of these patrol methods has a different objective, but they can both be targeted against similar types of crimes. Neither requires extensive resources or elaborate planning, although the manner in which such methods are used and the specific purposes for which they are used must be carefully considered. Neither high-visibility patrol nor low-visibility patrol should be used to replace routine patrol efforts, for clearly these are supplementary strategies designed to address specific problems for which more traditional patrol methods might not be appropriate.

Directed Deterrent Patrol

Directed deterrent patrol (DDP) differs from traditional patrol methods in that, under DDP, patrol officers perform certain specific, predetermined preventive functions on a planned and systematic basis. These preventive activities are designed on the basis of detailed analysis of crime incidents, offender characteristics, methods of operating, and locations. In essence, DDP attempts to identify certain crime trends and then develop specific patrol methods to interrupt those patterns.[25]

Directed deterrent patrol goes beyond traditional patrol techniques in the way analysis is used to identify crime problems and in the way specific types of patrol strategies are designed to deal with those problems. For example, under more traditional forms of patrol, information about a series of car burglaries in a shopping mall parking lot might be given out at roll call so that the district car will be aware of the problem. However, no specific instructions are provided as to how the problem might be addressed.

Under DDP, though, more detailed information about the time of day, day of week, and type of cars being burglarized might be provided. In addition, specific deterrent strategies might be suggested, such as, "patrol north and east lots for 15 minutes every hour between 4:00 PM and 7:00 PM, alternating routes of travel. After every other drive through, pause behind mall headquarters building for 15 minutes. Reverse patrol cycle every third time."

Directed deterrent patrol is ideally suited for cities which are large enough to experience crime problems in which patterns and characteristics can be identified. However, the basic theory behind DDP can work well in any agency, regardless of size. The primary ingredient in a successful directed-deterrent-patrol effort is the ability to analyze crime information and to devise strategies that will interfere with the offender's ability to successfully commit the crime.

Target-Oriented Patrol

Target-oriented patrol (TOP) strategies are those that, as the name implies, are targeted or directed toward specific persons, places, or events. Target-oriented patrol

[25]See Sheehan and Cordner, *Introduction,* p. 378.

combines the elements of high-visibility patrol, low-visibility patrol, and directed deterrent patrol to identify persons, places, or events which attract or create crime problems. Various means are then used to either deny the opportunity for the crime to occur, or to intercept the criminal in the commission of the offense. Target-oriented patrol programs can be adapted to a great many different situations and crime problems and do not require elaborate design efforts or resource commitments. They do require, however, careful planning for the identification of targets and interdicting strategies.

Target-oriented-patrol programs consist of either location-oriented patrol, offender-oriented patrol, or event-oriented patrol.[26] Location-oriented patrol (LOP) is the process of conducting intensified surveillance over selected areas that have been identified through crime analysis or through intelligence data as being high-risk areas for the commission of selected types of crimes. For example, crime analysis might indicate a crime pattern involving late-model sports cars being stolen from unprotected parking lots near manufacturing plants. This information might be combined with intelligence data gained from a reliable informant or other third party that indicates the cars are being taken just after the mid-day shift change at three major plants in the city. This kind of information could lead to the development of a directed patrol strategy designed to intercept the offenders in the commission of the crime and thereby put an end to the series of offenses.

Offender-oriented patrol (OOP) applies the same basic techniques of location-oriented patrol to an individual rather than to a location. In this case, police intelligence data as well as criminal history information on known offenders will form the basis for the identification of the "targets" against whom the patrol efforts are directed. Essentially, offender-oriented patrol involves the maintenance of intensified surveillance over individuals who are likely to be planning, preparing for, or committing an offense. Surveillance can take the form of either high-visibility or low-visibility patrol activities, depending upon whether the objective is deterrence or apprehension.

Offender-oriented patrol can be highly effective as a deterrent strategy as well as a means of apprehending persons in the act of committing a crime. However, due to the nature of the patrol activity, care must be taken not to violate the fundamental legal rights of the suspected offenders or to engage in any kind of activity that might be termed "harassment." Such allegations can cause irreparable damage to the public image of the police and can also seriously impede the department's crime-control efforts.

Event-oriented patrol (EOP) is simply the identification of events which may require the application of intensified patrol efforts or different kinds of patrol strategies due to the nature of the problems they may create. Event-oriented-patrol strategies, for example, would typically be required for a large carnival in the downtown area; a major parade attended by local as well as visiting dignitaries; a large demon-

[26]For a discussion of location-oriented patrol and offender-oriented patrol, see Tony Pate, and others *Three Approaches to Criminal Apprehension in Kansas City: An Evaluation Project* (Washington, D.C.: Police Foundation, 1976).

stration at a local college, university, or medical clinic; or any other event that poses unusual threats to peace, order, and public safety. Contingency plans are often developed to deal with the outcomes of these events, but specialized patrol strategies also need to be designed to deal with the crime problems that such events may create. The best source of information on planning event-oriented-patrol programs is other police agencies that have experienced similar kinds of events and that might have developed patrol strategies that worked particularly well.

Split-Force Patrol

The split-force-patrol concept was developed in the 1970s as a means of satisfying the needs of both preventive patrol and the demands made by calls for service. Split-force patrol grew out of the recognition that both preventive patrol and responding to calls for service were important, but that one often conflicted with the other. It is very difficult, for example, to conduct preventive patrol in any thorough and systematic manner while at the same time responding to calls for service which occur with varying frequency and which require varying degrees of time on the part of the responding officers.

Split-force patrol involves assigning one part of the patrol force the responsibility of conducting preventive patrol and assigning another part of the patrol force the task of responding to calls for service. The Wilmington, Delaware Police Department was one of the first police agencies to implement split-force patrol on a wide-scale basis and much that is now known about this patrol tactic was learned through its implementation in that department.[27]

Under traditional patrol methods, the entire patrol force is responsible for handling calls for service, conducting routine preventive patrol, enforcing traffic laws, running errands, and performing and endless number of other duties. Under the split-force concept, one part of the patrol force is assigned to respond to calls for service, investigate crimes, and perform other assigned duties. This allows the remaining part of the patrol force to devote itself exclusively to preventive patrol duties. The theory behind the split-force concept is that, by allowing one part of the patrol force to have preventive patrol as its exclusive responsibility, it can perform this duty more effectively than it could if it was simply one of many tasks it was expected to perform.

The initial evaluation of the Wilmington split-force patrol method indicated that several satisfactory objectives were achieved: an increase in patrol productivity, increased accountability of the officers, and a reduction in the rate of certain crimes.[28] One of the chief disadvantages of the split-force patrol concept, on the other hand, is that it relies upon specialization and fragmentation of the patrol force which could have long-term adverse consequences on morale, teamwork, and coordination among members of the patrol force. Periodic rotation of assignments could solve this prob-

[27]James M. Tien, and others, *An Evaluation Report of an Alternative Approach in Police Patrol: The Wilmington Split-Force Experiment* (Cambridge, Mass.: Police Systems Evaluation, Inc.,1977)

.[28]*Ibid.*, pp. vii-viii.

lem. In addition, long periods of protracted patrol efforts, unbroken by the welcome relief of variety and challenge posed by responding to calls for service, could become a difficult assignment if maintained for an extended period of time. Nevertheless, the split-force patrol method, like the other alternative patrol strategies discussed in this chapter, illustrate how creativity and imagination can be used to gain the most effective use of patrol resources.

SUMMARY

The methods of police patrol available to law enforcement agencies are many and diverse and are essentially limited only to the imagination and wisdom of the police executive and patrol commander. Traditional methods of police patrol can no longer be accepted without question. Flexibility and suitability in meeting patrol goals and objectives should be the key considerations when deciding upon which patrol methods to employ.

REVIEW QUESTIONS

1. If most of a patrol officer's time is uncommitted, does this mean that we have too many patrol officers? Why or why not? What are the alternatives?
2. What are some of the community characteristics that influence police patrol-staffing levels? Why is it not possible to have uniform patrol-staffing standards?
3. Describe, in proper order, the steps you would take in conducting a workload analysis for the police patrol force in your community.
4. Explain the difference between systematic and random sampling. What are the advantages and disadvantages of each? Which method is, in your opinion, preferable?
5. Why would it probably not be feasible to have the same number of patrol officers assigned to each shift?
6. What is the availability factor and how is it calculated?
7. What are the goals of proper patrol-force deployment?
8. What are the desired purposes of effective patrol-force scheduling?
9. What are some of the advantages of fixed-day-off work schedules?
10. Describe the advantages and disadvantages of rotating work schedules. What are the circumstances in which a rotating shift schedule might be preferred?
11. What is meant by *proportional scheduling* and why is it important in police patrol operations?

12. What are the reasons for assigning patrol officers to patrol beats? How should beat boundaries be determined? What factors should be considered in establishing beat boundaries?
13. How does directed deterrent patrol differ from more traditional patrol strategies? What are some of the ways in which directed deterrent patrol might be used?
14. What are some of the advantages and disadvantages of split-force patrol?

REVIEW EXERCISES

The following exercises are designed to reinforce or amplify comprehension of the material contained in this chapter.

1. Conduct a survey of five or more police agencies in your area to determine the number of officers assigned to the patrol force. Using municipal population figures, determine the ratio of police officers in the patrol force to the population. Attempt to explain the variations in these figures using information you have obtained about the community.
2. Contact several police agencies in your area to find out (if you can) about their patrol-beat plans (if they use them). Prepare a paper describing the results of your study, including the manner in which the plans are formulated, the basis for forming the beats, and the frequency with which the plans are reviewed and revised.
3. As a class project, survey a large number of police agencies in your state to obtain information concerning their patrol scheduling practices. Find out what kind of schedule they use, how or why it was derived, and whether it provides for regular rotation of shifts or days off. Discuss the results of the survey in class, including any important conclusions reached as a result of the survey.

REFERENCES

A Comparative Analysis of the Lexington-Fayette Urban County Police Division's Home Fleet Program Versus the All Pool Plan. Lexington, Ky.: Lexington-Fayette Urban County Division of Police, 1977.

Cohen, Marcia, and J. Thomas McEwen, "Handling Calls for Service: Alternatives to Traditional Policing," *NIJ Reports* (September 1984), pp. 4-8.

Cordner, Gary W., "The Police on Patrol," in Dennis Jay Kenney, ed., *Police and Policing: Contemporary Issues.* New York: Praeger, 1989, pp. 60-71.

Dunham, Robert G., and Geoffrey P. Alpert, *Critical Issues in Policing: Contemporary Readings*. Prospect Heights, Ill.: Waveland Press, 1989.

Fisk, Donald M., *The Indianapolis Police Fleet Plan*. Washington D.C.: The Urban Institute, 1970.

Goldstein, Herman, *Policing a Free Society*. Cambridge, Mass.: Ballinger Publishing Co., 1977, pp. 52-53.

Goolkasian, Gail A., Ronald W. Geddes, and William DeJong, "Coping With Police Stress," in Robert G. Dunham and Geoffrey P. Alpert, *Critical Issues in Policing: Contemporary Readings*. Prospect Heights, Ill.: Waveland Press, 1989, pp. 498-507.

Kelling, George E., and others, *The Kansas City Preventive Patrol Experiment: A SUMMARY REPORT*. Washington, D.C.: Police Foundation, 1974.

Kroes, William H., *Society's Victim—The Policeman: An Analysis of Job Stress in Policing*. Springfield, Ill.: Charles C. Thomas, 1976.

Leonard, V. A., and Harry W. More, *Police Organization and Management* 7th ed., Mineola, New York: The Foundation Press, Inc., 1987.

McEwen, J. Thomas, and others, "Handling Calls for Service: Alternatives to Traditional Policing," *NIJ Reports* (September 1984), pp. 4-8.

Monk, Timothy, "Advantages and Disadvantages of Rapidly Rotating Shift Schedules: A Circadian Viewpoint," *Human Factors* (October 1986), pp. 553-557.

National Advisory Commission on Criminal Justice Standards and Goals, *Police*. (Washington, D.C.: U.S. Government Printing Office, 1979), p. 201.

Pate, Tony, and others, *Three Approaches to Criminal Apprehension in Kansas City: An Evaluation Project*. Washington, D.C.: Police Foundation, 1976.

Petersilia, Joan, "The Influence of Research on Policing," in Dunham, Robert G., and Geoffrey P. Alpert, *Critical Issues in Policing: Contemporary Readings* (Prospect Heights, Ill.: Waveland Press, 1989), pp. 230-247.

Sheehan, Robert, and Gary W. Cordner, *Introduction to Police Administration* (2nd ed.), Cincinnati, Ohio: Anderson Publishing Co., 1989.

Skolnick, Jerome H., and David H. Bayley, *The New Blue Line: Police Innovation in Six American Cities*. New York: The Free Press, 1985.

Spelman, William, and Dale K. Brown, "Response Time," in Klockars, *Thinking About Police*, pp. 160-166.

Stenzel, William W., and R. Michael Buren, *Police Work Scheduling: Management Issues and Practices*. Washington, D.C.: U. S. Department of Justice, 1983.

The Four-Ten Plan—California Law Enforcement, California Commission on Peace Officer Standards and Training, 1981.

Tien, James M., and others, *An Evaluation Report of an Alternative Approach in Police Patrol: The Wilmington Split-Force Experiment*. Cambridge, Mass.: Police Systems Evaluation, Inc., 1977.

Vanderbosch, Charles G., "24–Hour Patrol Power: The Indianapolis Program," *The Police Chief* (September 1970).

Webb, Kenneth W., and others, *Specialized Patrol Projects*. Washington, D.C.: U. S. Government Printing Office, 1977.

CHAPTER 10
SUPPORTING THE POLICE PATROL EFFORT

CHAPTER OUTLINE

LEARNING OBJECTIVES

After completing this chapter, the student should be able to

1. List the various support activities that aid the police patrol function.
2. Explain the importance of research and planning to the success of the police patrol effort.
3. Describe the relationship between the police patrol function and crime prevention, youth services, and other community relations.
4. Explain the importance of sound procedures for the collection, preservation, and storage of evidence and found property.

While the patrol force is the single most important element of the police organization, it depends heavily for its effectiveness upon the support it receives from other organizational units. This chapter summarizes additional support functions found in most police agencies. The extent to which these functions exist and to what degree of sophistication depends upon the size and requirements of the individual agency. In some departments, for example, planning and research may be conducted on a very informal basis, while in other departments, these functions may be highly formalized within the organizational structure. Likewise, some functions such as evidence and property control may be performed as an additional duty by a staff officer or may require the assignment of someone on a full-time basis.

Whether support functions are conducted informally or formally, and whether they are performed by full-time personnel is not important. What is important is the extent to which they facilitate and enhance patrol operations. It is also important that patrol officers, supervisors, and command officers fully understand the nature of these support functions and the impact they have upon patrol operations. While these functions are important in their own right to ensure the smooth functioning of the police agency, their primary purpose is to support and enhance the patrol effort.

PERSONNEL MANAGEMENT

The efficiency of a police agency, like any other organization, depends directly upon the quality of the personnel. Its personnel are a police agency's most valuable resource. A well-designed personnel system to properly manage recruitment, selection, promotion, performance evaluation, discipline, and dismissal is essential if a police agency is to discharge its responsibilities effectively.

In many instances, the police chief executive's responsibilities and authority in personnel management are shared with other individuals and agencies.[1] For example,

[1]George W. Greisinger, Jeffrey S. Slovak, and Joseph J. Molkup, *Civil Service Systems: Their Impact on Police Administration* (Chicago: Public Administration Service, 1979).

a city's personnel department may exercise considerable control over personnel policies and procedures. Local civil service commissions, which are usually appointed by the legislative body for the city, may also exercise considerable control over personnel matters. Nevertheless, it is important that the police chief, or a designated representative, play an active role in the personnel management function. It is simply not realistic to expect a police chief executive to be accountable for the operations of a police department if that same person does not play an active role in the selection, promotion, and discipline of the members of the organization. The captain of a team cannot be held responsible for the team's won-loss record if he or she does not participate in choosing the members of the team.

Police agencies having 50 or more full-time personnel should have one individual, preferably a ranking officer, assigned to coordinate personnel functions. These functions would typically include accepting and reviewing employment applications, conducting background investigations on persons being considered for employment, and working closely with the personnel department or civil service commission on the development and execution of policies and procedures relative to employment, performance evaluation, and promotion. In addition, the police department's personnel-and-training officer should be responsible for coordinating and administering the department's in-service training program and scheduling personnel to attend specialized training courses. See Chapter 13.

Individual personnel files should be maintained on all persons employed by the agency. These will usually be maintained by the chief of police or the chief's secretary, although in larger departments they will probably be maintained by the department's personnel officer. These files should contain copies of all documents relating to the individual's employment, including the employment application, test scores, training information, disciplinary reports, and performance evaluation reports. In some cases, separate training folders may be maintained by the department training officer.

Even though a city personnel department or civil service commission may retain overall responsibility for regulating and overseeing the personnel management function, it is important that the police administrator maintain an active role in personnel matters. Since the police executive is ultimately responsible for the conduct of police personnel, it is important that he or she become actively involved in developing personnel policies and in the execution of those policies.

RESEARCH AND PLANNING

Police work is becoming increasingly more difficult and complex. Literally volumes of new information concerning police operations and management are published each year. While policing certainly cannot be described as a science, the scientific method is being applied to police problems now more than ever before. Police administrators and public officials are beginning to recognize that traditional methods of solving police problems are not necessarily the best. It is becoming increasingly apparent that new approaches are needed if the police are to respond effectively to emerging problems and citizen expectations.

Research and planning has become a fundamental instrument of police management, although formal research and planning units are usually found only in larger police agencies. Regardless of the size of the department, however, it is important that some form of research and planning be conducted for the purpose of enhancing police executive decision making. As one police executive has noted:

> A police administrator who makes decisions without research and planning will inevitably make an appreciable number of faulty decisions. Management of a police agency without research and planning also results in indecision, and indecision may be a worse alternative than making a faulty decision. In addition, effective use of available resources cannot be made without getting facts, analyzing data and other information, considering alternatives, and developing plans and procedures. Every head of a police agency should have research and planning capabilities.[2]

As with other support activities, research and planning is often conducted on a limited basis, often as an additional duty by a command officer. Departments of 50 or more full-time personnel, however, should have a planning and research assistant. He or she may be a qualified civilian employee, assigned to the office of the chief of police on a full-time basis. In larger departments, a research and planning unit staffed by one or more people may exist under the services division or bureau. Figure 10.1 shows the planning and research function in a small police agency, while Figure 10.2 shows how it may be organized in a larger department.

Whatever the size and level of sophistication of the research and planning effort, the following are typical activities that should be performed:

- Administrative analysis. The examination and evaluation of administrative procedures such as information flow, fiscal control, information management, equipment and facility utilization, and the like.
- Operations analysis. The development of procedures to improve operational activities such as criminal investigation, patrol methods, tactical operations, scheduling, and deployment of personnel.
- Grants management. The preparation and administration of state and federal grant programs.
- Planning and evaluation. The evaluation of routine and special operational programs, such as tactical patrol operations.
- Policy development. The development of written policies governing police operations as well as standard operating procedures concerning, for example, use of deadly force, hot pursuits, mutual aid, and the like.

[2]Vernon L. Hoy, "Research and Planning," in Bernard L. Garmire, ed., *Local Government Police Management* (Washington, D.C.: International City Management Association, 1977), p. 367.

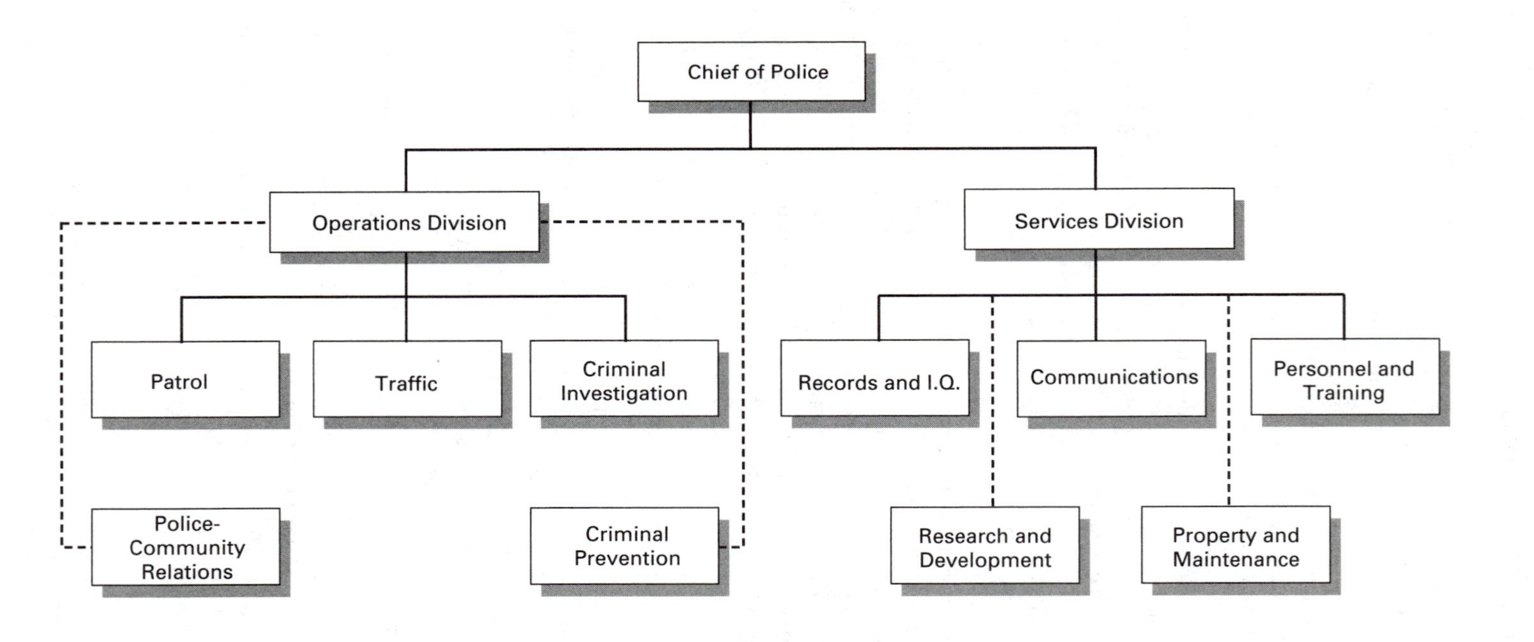

Figure 10.1 Research and development as a staff function in a small police agency

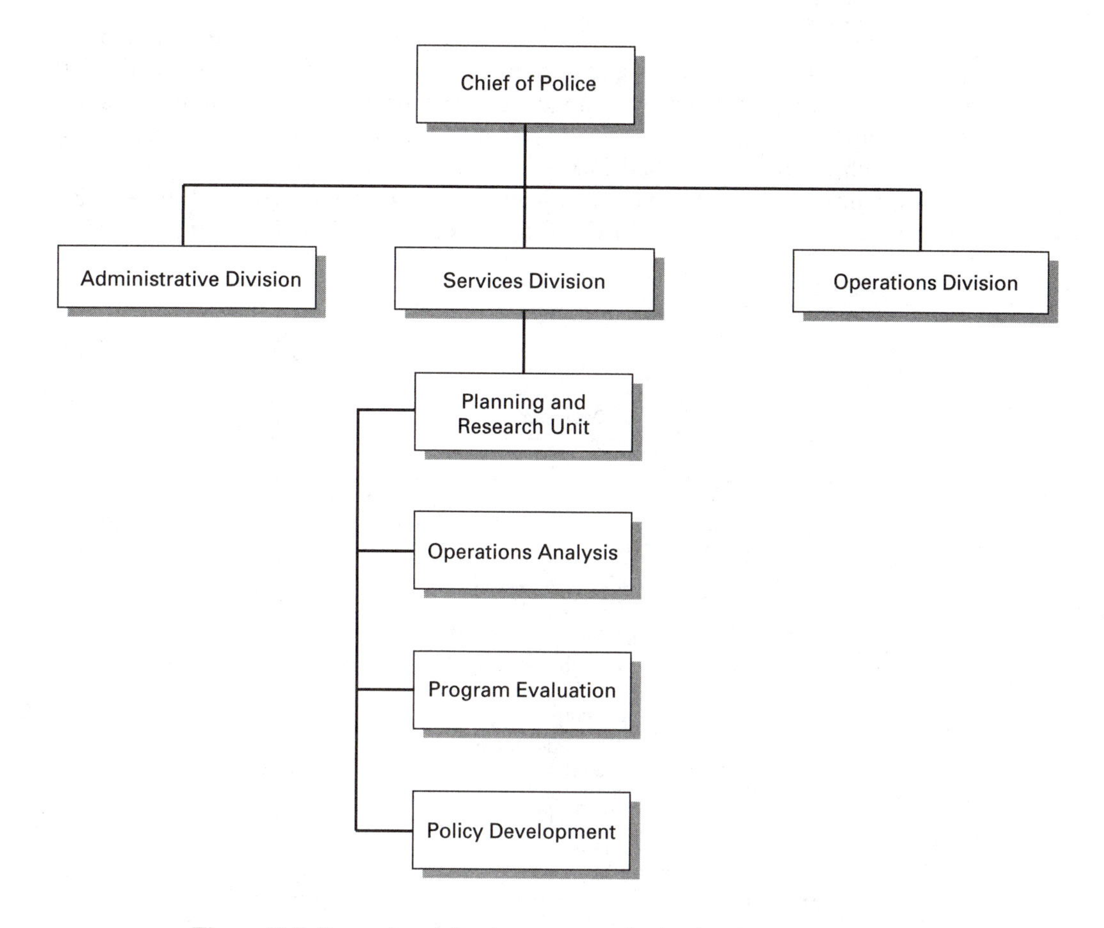

Figure 10.2 Research and development organization in a large police agency

CRIME ANALYSIS

Crime analysis has made an important contribution to the improvement of patrol operations in recent years.[3] While the concept of crime analysis is not new, the extent to which crime analysis has been incorporated into the mainstream of police operations has increased considerably in the last decade. Police departments having 100 or more full-time employees should have at least one individual assigned to perform crime analysis on a full-time basis. In most cases, this will be a qualified civilian rather than a sworn officer. The person assigned to this position should be well-trained in operations research, statistical analysis, computer operations, and related fields.

[3]For an excellent discussion of the crime analysis process, see George A. Buck and others, *Police Crime Analysis Handbook* (Washington, D.C.: Law Enforcement Assistance Administration, 1973).

Crime analysis can best be described as the process by which selected types of information concerning crimes and criminal characteristics are collected and evaluated for the purpose of identifying trends, patterns, or traits. This information is then disseminated to patrol officers for the purpose of developing target-oriented patrol tactics such as those described in Chapter 6. In addition, information developed through crime analysis can be used by investigators to identify criminal suspects and link related cases. The crime-analysis process is illustrated in Figure 10-3.

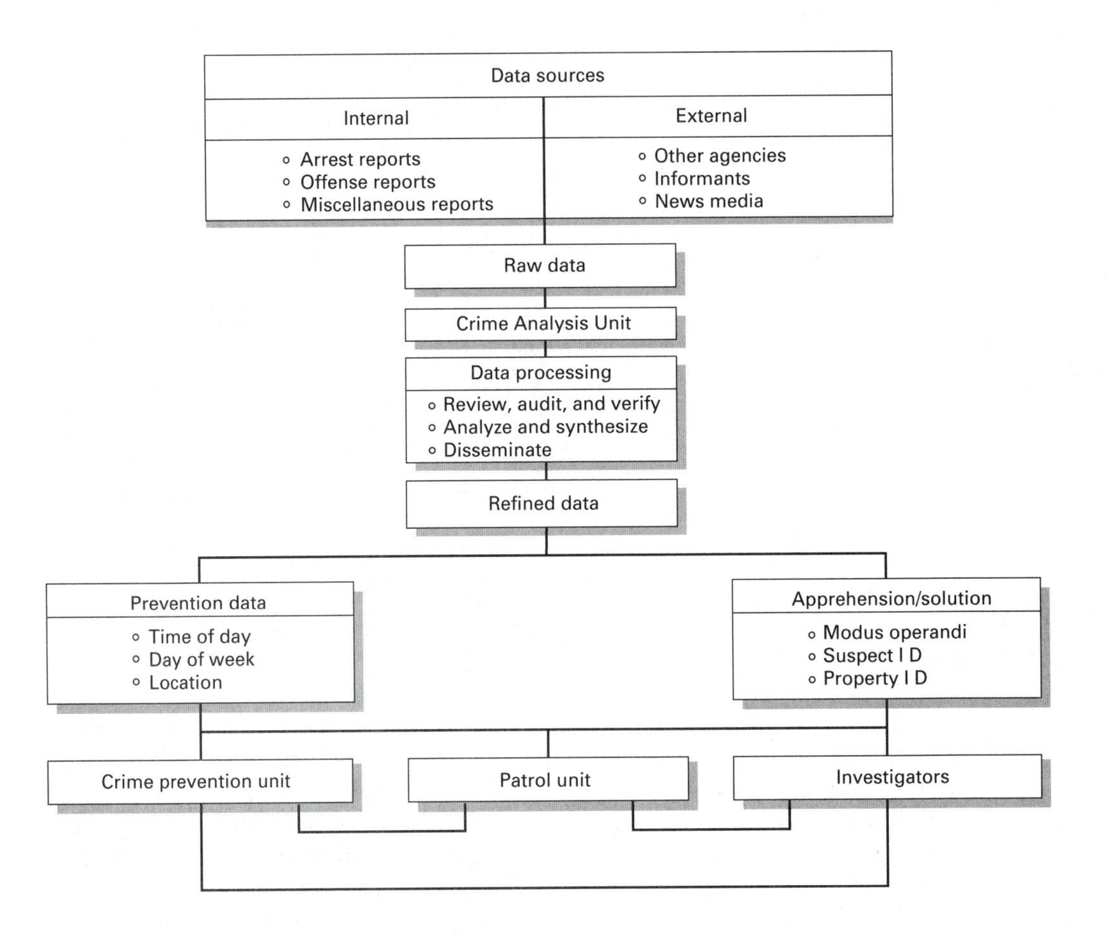

Figure 10.3 The crime analysis process.

An important element in the development of a crime-analysis capability is the determination of the types of information to be collected and analyzed, the sources of such information, the type of analysis to be performed, and the means and regularity with which crime information data are to be distributed to field units. In most instances, existing data sources (reports of offenses, arrests, and field interviews) can provide a wealth of information for crime-analysis purposes. In some cases, however, report formats may need to be modified in order to provide needed information.

Once necessary data sources have been identified and the specific types of information to be collected for analysis have been determined, procedures must be established to ensure the timely flow of information through the crime-analysis unit. This may require a modification in the report-processing system of the department. For example, an additional copy of certain types of reports may need to be furnished to the crime analysis unit.

The manner in which crime-analysis information is disseminated to user units is very important. A variety of methods are available for this purpose. Most police agencies use some form of crime-analysis bulletin such as that shown in Figure 10.4 To develop the appropriate method of dissemination, discussions between the crime analyst, members of the patrol force, and investigators would be useful. In addition, continuing feedback from the users is necessary in order to evaluate the usefulness and reliability of the information being produced, and to recommend improvements in the crime-analysis function and product.

LEGAL SERVICES

Even though most police agencies have available to them some sort of legal advisor such as the city attorney or county prosecutor, members of the patrol force are often left without adequate legal advice concerning their day-to-day activities. As a result, they are often unable to exercise sound judgment in the enforcement of the law and may commit acts that can have a far-reaching impact upon the police agency and upon criminal cases in which they are involved. For this reason, the police legal advisor is an important member of the police management team.

The concept of the police legal advisor has gained widespread acceptance in the police service in recent years.

> Few innovations have gained more rapid acceptance in law enforcement than that of the police legal advisor. With the advent of court decisions that have had a significant impact on the day-to-day operations of even the smallest police agency, the need for competent, thorough, and prompt legal advice has become ruefully evident to the police administrator.[4]

[4]Fred E. Inbau and Wayne W. Schmidt, The Legal Advisor, in Bernard Garmiree, ed., *Local Government Police Management,* (Washington, D.C.: International City Management Association, 1977) p. 382.

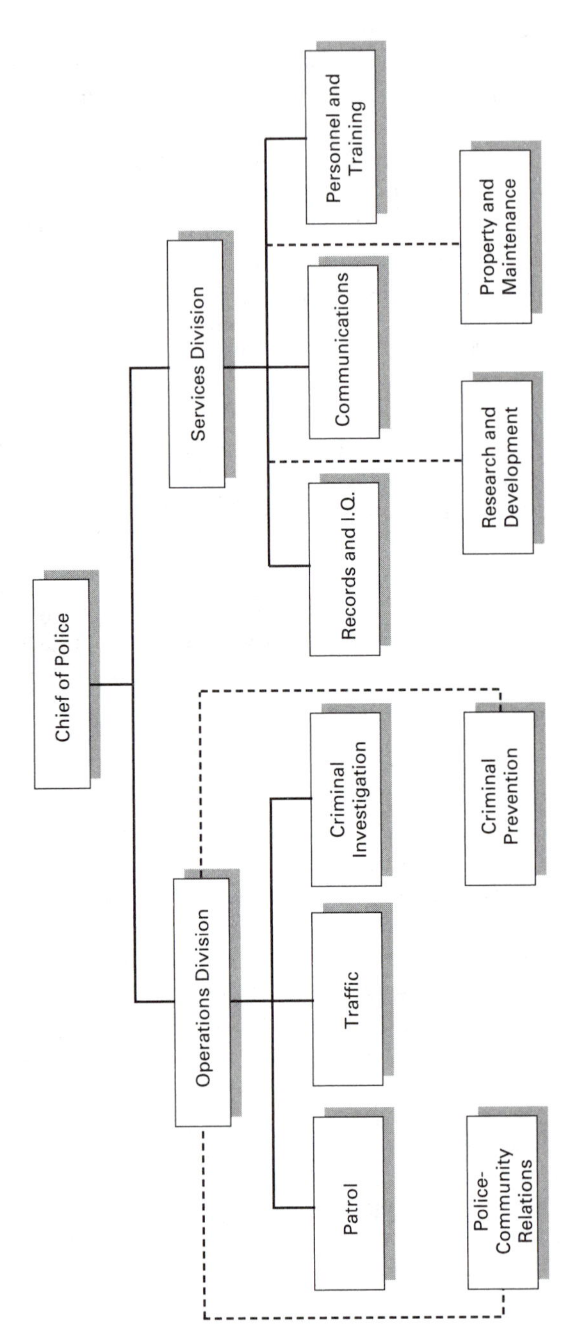

Figure 10-4 Crime analysis {Source:George A. Buck et al., Police Crime Analysis Handbook (Washington, D.C.: National Institute of Law Enforcement and Criminal Justice , Law Enforcement Assistance Administration, U.S. Department of Justice , 1973), p. 70).

While many small police agencies may be unable to afford employing a full-time legal advisor, they should have available to them professional legal counsel whenever it is required. Special arrangements should be made with the local prosecutor or city attorney to fill this need.

In those departments that employ the services of a regular legal advisor, either on a full-time basis or part-time basis, the duties and responsibilities of this position should be carefully defined. Of particular importance is the relationship between the legal advisor and the chief of police, the command staff, and other staff officers. It is usually desirable to designate the legal advisor as a staff assistant to the chief of police.

In some municipalities, the legal advisor may be employed by the office of the city attorney, the law department, or the city manager's office and assigned to the police department as a staff advisor. In any event, it is important that the services to be provided by the legal advisor be articulated throughout the organization in order that all members of the department may have access to those services when necessary.

An important function of the police legal advisor should be to assist in the development and presentation of in-service and pre-service police training programs. Changes in the criminal law occur frequently, and new court decisions often have far-reaching effects on criminal procedure for the police. Few police practitioners are able, by virtue of either training or experience, to interpret these new laws and complex court rulings properly. The police legal advisor can play an important role in the police training program by ensuring that all personnel are thoroughly familiar with current case law and legal requirements.

The police legal advisor may perform a number of other vital functions as well. He or she may advise the chief of police and command staff regarding the development of policy and procedure, provide counsel in the conduct of internal investigations and disciplinary actions, assist in the handling of grievance procedures, and deal with representatives of police employee organizations. In some cases, the legal advisor plays a pivotal role in the investigation and pretrial handling of criminal cases. Such activities could include providing assistance in obtaining search and arrests warrants, participating in the questioning of suspects and witnesses, advising investigators on legal procedures and rules of evidence, and screening cases before they are presented for prosecution.

THE ROLE OF THE POLICE PSYCHOLOGIST

The increased role of the police psychologist may be viewed as an indication of our increased awareness of the personal needs of our police officers and their families, or as a troublesome reminder of the increased strain and stress that has become an integral part of police work. In any event, it is certainly true that police agencies have begun to take advantage of the services that can be offered by police psychologists, counselors, and allied workers in the last few years. There is also no question that these professionals can provide a valuable service to the police organization. Although only the largest police agencies can afford the services of a police psychologist on a full-time basis, police agencies of nearly any size should be able to retain the services of a qualified psychologist on an as-needed basis. The question should not be whether the police agency can afford the services of a psychologist, but whether it can afford not to have these services available to it.

The services of a police psychologist may be generally divided into three categories. The extent to which these categories overlap and take on new meanings will vary from one police agency to another. These three categories are as follows:

1. **Hiring and Dismissal Counseling.** Probably the most common use of psychological services in police work today is in the examination and evaluation of persons being considered for appointment to the police agency, as well as those being considered for retention who may have either suffered emotional trauma or some other job-related psychological disorder. Through various kinds of psychological evaluation techniques, the police agency and the municipality may help to ensure itself against claims of civil liability for hiring or retaining persons who are emotionally incapable of handling the stresses of police work. Smaller and medium-sized police agencies usually contract for these kinds of services with a qualified psychological testing firm, while larger police agencies may use their own staff or municipal personnel specialists to perform these services.
2. **In-house Counseling Services.** As was pointed out in Chapter 4, there are many sources of stress in police work. We are becoming increasingly aware of the kind of strain police officers are under and the effect it has on them, both personally and professionally. We have also come to realize that some of these emotional problems can be identified and dealt with through counseling in such a way as to save an officer's career or perhaps his or her personal life as well. Police agencies in increasing numbers have retained psychological counseling specialists and other professionals to come to the aid of officers, employees, and their families who have suffered severe emotional trauma or any other kind of difficulty in which professional services of this kind may be helpful. Employee assistance programs, offering services counseling and other kinds of assistance, are common in most municipalities today.
3. **Crisis Intervention Services.** Police psychologists and allied professionals can play a vital role as crisis-intervention specialists in dealing with the emotional aftereffects of incidents such as fatal shootings, natural disasters, or other mishaps involving loss of life or serious injury. Although we like to think of police officers as being "cut from a different kind of cloth," the fact of the matter is that they suffer the same emotional distress that the rest of us do. We now know that the emotional trauma associated with many crisis situations does not set in until some time after the event has happened. It is for this reason that teams of crisis-intervention specialists are on call in major metropolitan centers for the express purpose of dealing with the victims of crisis situations. These

teams can provide valuable assistance in treating police officers who have been exposed to traumatic situations, both immediately after the event and on a continuing basis.

COMMUNITY SERVICES

Police agencies must consider the needs of the community as they plan the delivery of police services. The community services unit or representative is responsible for coordinating police services in the areas of community relations, crime prevention, and youth services.

Community Relations

During the late 1960s, in response to increasing tensions between the police and various segments of the population, police agencies around the country developed community relations units. It was the purpose of these agencies to develop a variety of programs designed to foster better relations between the police and community members. Many of these programs were quite successful, while others did little more than try to improve the image of the police. The most successful programs were those that attempted to alter police operations and procedures in order to make them more responsive to citizen sensibilities. The least successful were those that did little more than make cosmetic changes in police operations.

The common failure in many of these programs, at least initially, was that they reflected the notion that community relations was a function to be performed by specialists within the organization, and that it was "business as usual" for other members of the department. Contemporary police administrators, however, have rejected this belief and are now expressing the philosophy that responsibility for good community relations belongs to every member of the department (see Chapter 13). This concept, however, must be fortified by designating someone to oversee and enforce the department's community relations policies.

While it is true that the patrol officer should be the primary community relations representative of the department, it is important to have someone in the department responsible for coordinating a number of important community relations activities. For example, the community relations representative should be responsible for attending community meetings, developing special programs such as citizen ride-along, and for handling particular problems affecting police-community relations.

Every effort should be made to dispel the belief that community relations is a specialized function involving only selected members of the department. Indeed, any community relations program can only be as effective as the support and involvement it receives from the rank and file of the police agency.

Crime Prevention

Crime prevention is another function that was originally developed as a specialty within the police organization but is rapidly becoming a generalized responsibility. Many police agencies today have one or more specially trained crime prevention officers who develop programs designed to increase the effectiveness of the department's crime prevention efforts.

Crime prevention officers may work within the patrol force but are usually members of a special staff unit. Their responsibilities usually include developing special programs to encourage citizen participation in crime-prevention activities. Some successful crime prevention programs have been Operation Identification, Block Watchers, and Citizens on Patrol. Crime prevention officers also conduct residential and commercial security surveys in an effort to help citizens and business people protect themselves from criminal attack.

While crime prevention is still a very specialized field, it can only be effective to the extent that it involves other members of the department. Patrol officers and detectives especially should be familiar with the nature and purpose of the crime-prevention function and should participate in community-based crime-prevention activities.

If a department has a crime-prevention program, members of the patrol force should distribute written information about the program to citizens. It is common practice in many departments, for example, to have patrol officers alert burglary victims to the desirability of having a security inspection conducted by the crime prevention specialist. In addition, patrol officers should be encouraged to advise the crime prevention officer of particular hazards in the community that could be corrected through better security precautions. Even if a department does not have a formal crime-prevention program, patrol officers should be familiar with basic crime-prevention techniques they can apply in their regular patrol duties.

Youth Services

Youth service programs are designed to establish positive relationships between the youth in the community and the police. The purpose of these programs is to create a favorable impression of the police in the minds of young people as well as establish good relations with the rest of the community.

Youth services, as discussed here, is not concerned with investigating offenses committed by or against juveniles. Instead, youth services activities are positive, proactive programs designed to deter and prevent juvenile delinquency. A good example is the school liaison program that many police agencies have adopted. School-liaison-officer programs usually assign a police officer to work in the schools either on a full-time or a part-time basis. These officers typically are involved in counseling, giving lectures on selected subjects, and working closely with school administrators, teachers, and parents in resolving problems concerning the youth of the community.

A more recent successor to the school-liaison-officer programs of the 1970s and early 1980s is the Drug Abuse Reduction Education (DARE) program. Originally developed by the Los Angeles Police Department, DARE has now been adopted by police agencies in every state. The DARE program is a proactive drug abuse education program designed to teach youngsters of all ages the dangers associated with drugs. The program has been so popular that there have been waiting lists for police agencies to send officers to school to become certified as DARE instructors.

A number of police departments around the country have developed highly successful youth programs involving schools, community agencies, religious groups, and civic organizations. Youth services programs may include police athletic leagues, special outings for disadvantaged youth, and a variety of school programs in which members of the department are invited to speak on such things as bicycle safety, drug abuse, and other topics of interest and importance to young people.

As with other community service efforts, it is important to involve other members of the department in youth services programs. Such programs are of limited value if they involve only one or two members of the department.

CRIMINALISTICS SERVICES

Physical evidence is vital to the successful detection of crime, apprehension of offenders, and prosecution of criminals. Patrol officers should always be alert to items of physical evidence when investigating crimes; however, they often lack the training and equipment necessary to do an adequate job of evidence identification and collection.

At one time, a great deal of emphasis was placed upon obtaining self-incriminating statements from persons suspected of committing crimes. Over the years, however, court decisions have greatly limited the extent to which this kind of evidence can be used. As a result, police officers have been required to become much more thorough in their crime scene investigations and in their handling of evidence. This point was made abundantly clear by Supreme Court Justice Goldberg speaking for the majority in the landmark case of *Escobedo v. Illinois.*

> We have learned the lesson of history, ancient and modern, that a system of criminal law enforcement which comes to depend on the "confession" will, in the long run, be less reliable than a system which depends on extrinsic evidence independently secured through skillful investigation.[5]

The reliability of evidence secured by the police depends upon the qualifications, equipment, and training of the person responsible for evidence collection and analysis. Unfortunately, few patrol officers are provided with the necessary equipment and training to adequately identify, collect, and preserve vital evidence. In addi-

[5]*Escobedo v. Illinois* 378 U.S. 478, 488 (1964)..

tion, relatively few police agencies have qualified evidence technicians or laboratory facilities designed to aid in the analysis of evidence.

Even the smallest police department should have at least one officer who has been given special training and equipment for the purpose of evidence collection. Minimum equipment requirements include a camera, a latent fingerprint kit, a plaster-casting kit, and various types of evidence containers. In larger departments, it is advisable to have at least one evidence technician available on a full-time basis. This person may be assigned to the headquarters unit or may be a patrol officer who is designated as evidence technician as an additional responsibility. As the size of the department increases, the need becomes greater for a full-time evidence technician and appropriate laboratory facilities. Departments of over 150 full-time employees normally will require at least one full-time evidence technician, along with sufficient evidence-collection and analysis equipment.

Few police agencies, however, can afford to maintain their own criminalistics laboratories even though most will need criminalistics services from time to time, and it is important that such services be available when the need arises. There are several ways to fulfill such needs. In many states, regional crime labs are maintained by the state police or highway patrol. These may include mobile crime units that are available to local police agencies in important cases. Regional crime labs can assist local agencies in providing expert evidence technicians and by performing various types of scientific analysis.

In some instances, large police agencies may provide criminalistics services to smaller ones either for no charge or on a contract basis. Still other sources of assistance are federal law enforcement agencies such as the Federal Bureau of Investigation, which will provide limited criminalistics support to local law enforcement agencies in major cases.

EVIDENCE AND PROPERTY CONTROL

As indicated in the previous section, the collection and preservation of evidence is extremely important if police agencies are to successfully investigate crimes and prosecute offenders. Regardless of how diligent patrol officers are in their efforts to collect evidence, the results of their efforts will be of little value unless the evidence is stored securely to guard against its being lost or compromised. Thus, it is important that all police agencies have an adequate evidence and property control system.

Since in most police agencies evidence and found or recovered property are covered within the same general guidelines, the following discussion includes the control of both evidence and found or recovered property. Found or recovered property must be distinguished from departmental property such as weapons, equipment, and other items belonging to the police agency. Such items are normally included within the agency's own property inventory and control system and are not subject to the same types of controls usually exercised over found and recovered property and physical evidence.

The primary goal of the evidence-and-property-control system is to guard against the misplacement or theft of evidence or items of property recovered by the department. A second goal is to preclude the possibility that evidence may be tampered with or compromised. This goal is intended to ensure a strong chain of evidence able to withstand judicial scrutiny. A third goal is to permit the prompt and appropriate disposal of evidence and items of found property when their retention is no longer legally required by the agency.

It is often the case that several people handle a particular item of evidence, although procedures should be established to keep the number to a minimum. The patrol officer or the evidence technician will normally be the first to handle an item of evidence. This person will normally be required to mark the evidence properly and enter it into the evidence control system.

Later, others such as laboratory technicians, investigators, and prosecutors may be required to handle the evidence. It is important that a continuous record be maintained of all persons handling the evidence, as well as the eventual disposition of the evidence.

The Evidence-Control System

The first element of the evidence-control system is a procedure for initially recording information concerning the nature of the evidence, its description, the circumstances under which it was obtained, the location obtained, and the name of the officer discovering the evidence. This information may be required later if the evidence is eventually used in court in a criminal case.

All items of evidence should be clearly marked in some way that can later be used for identification purposes by the officer initially discovering the evidence. Evidence tags should be attached to the evidence. Special arrangements need to be made for storing perishable items such as blood or food products. All narcotics should be carefully weighed, stored in glassine bags or other suitable containers, and placed in a designated area where security precautions are especially strict. In some cases, photographs may be taken of some items of evidence. These should show the location from which the evidence was obtained and the officer making the discovery. An evidence record should be prepared for all items marked as evidence. The evidence record should be prepared in multiple copies; one copy should be attached to the offense or arrest report, and one or more copies should remain with the evidence.

All evidence must be placed in a secure place for safekeeping. In addition to the special provisions made for perishable items and narcotics previously described, it is usually wise to segregate narcotics and firearms from other stored items. A system of temporary lockers should be established for securing items of evidence obtained during periods when the evidence officer or evidence technician is not available. It is important that access to the primary-evidence storage area be carefully controlled. One member of the department should be designated as the evidence officer, even if on a part-time basis. In larger departments, of course, there will probably be a full-time evidence-control unit.

Items placed in temporary lockers should be removed each day and transferred to the permanent storage area by the evidence officer. An evidence-control log should be maintained in which to record the description of the evidence, where it was obtained, case number, name of the officer investigating the case, and disposition. The control log should also indicate the bin or shelf number where the evidence is located in the storage area. The storage area should be neatly organized by numbered bins, shelves, drawers, and lockers designed to accommodate and permit easy retrieval of various sizes of evidence.

A system of receipts should be established to preserve the chain of custody when evidence is removed from the storage area for any reason. When evidence is no longer required, it should be removed from storage and disposed of according to local requirements. The date and means of disposition should be noted in the evidence log and on the permanent copy of the evidence record, which should then be placed in an inactive file.

The most important element of the evidence-control system is a set of procedures outlining the proper means of collecting and handling evidence. Police agencies must be able to show that such procedures are in effect if they wish to establish the integrity of the evidence presented in court.

VEHICLE AND PROPERTY MAINTENANCE

Another important support function is the maintenance of property, equipment, and vehicles. In order to perform effectively, patrol officers must have reliable and serviceable equipment, and it must be maintained in good working condition. Periodic inspection of all personal equipment and departmental property should be required, and it is advisable that responsibility for custody and control of all property and equipment be delegated to a senior staff officer. At the same time, however, internal control procedures should be in place that will ensure accountability of individual officers for the care and custody of property and equipment under their control.

The type of vehicle utilized for patrol and other police purposes should also be carefully evaluated. In the past, law enforcement agencies tended to purchase large sedans that were expensive to operate and that offered poor fuel economy. Due to recent gas shortages and the increasing emphasis on energy conservation, however, it is important that police agencies carefully evaluate the types of vehicles selected for patrol use.

An increasing number of law enforcement agencies have begun using smaller automobiles in recent years, and, in most cases, with excellent results. Several automobile manufacturers have introduced a line of small cars designed for police use that feature better fuel economy, high performance, and better-than-average maintenance and repair records.

Every year, the Michigan State Police test more than a dozen vehicles as part of their procurement policy. Representatives from police agencies in the United States and Canada are invited to attend these tests. These extensive tests have produced detailed information about the suitability of several types of automobiles for police work and have provided police agencies with a wealth of information to review when considering the purchase of new police cars.[6] The results of these tests are widely published and provide an excellent source of technical information that can be used by police administrators when evaluating police patrol vehicle purchases.[7]

Vehicle Maintenance

The automobile is one of the most important, and generally the most expensive items of equipment available to the patrol officer. To ensure optimum performance and low operating costs, police vehicles should be periodically serviced and given maintenance as required. A regular preventive maintenance schedule should be established for all police vehicles and the property officer or other ranking officer should be assigned to see that routine maintenance is performed accordingly.

Patrol officers should be required to complete a vehicle inventory record at the beginning and end of each shift, noting all vehicle defects and required maintenance or repairs. Records should also be kept of all fuel and oil consumed as well as all additional expenses for automotive repairs. To ensure proper accountability, these reports should be prepared by all officers for each tour of duty. This information will prove valuable when calculating projected operating expenses for future budget requests.

It is important for a police agency to have an acceptable program of vehicle maintenance. All police vehicles should be periodically inspected and serviced. In smaller departments, the agency may have a contract with a local garage to perform this service; in larger departments, this maintenance will usually be provided by a central maintenance facility operated by the city.

Finally, detailed records should be maintained of all vehicle repairs, maintenance, and fuel consumption. This may be the responsibility of the police agency or the central maintenance facility. This information should be compiled periodically in order to establish normal operating costs for each vehicle as well as for the total police fleet.

Property Care and Maintenance

The policies of local police agencies differ with respect to the ownership and maintenance of items of personal equipment. In some departments, all equipment used by members of the agency may be issued by the department, while in others, individual officers may be required to supply their own uniforms and equipment. Whatever the case, it is important that all items of personal equipment be in good working order and that damaged property be promptly replaced.

[6]National Institute of Justice, *TAP Alert* (October 1986).

[7]Michigan State Police, *1991 Model Year Patrol Vehicle Testing* (Washington, D.C.: National Institute of Justice, 1991).

It is also important that the department issue regulations covering replacement of personal and departmental property that is lost or damaged. Property records should also be maintained on all items of property or equipment issued by the department. Periodic property and equipment inspections should be conducted by supervisory or command personnel to ensure that all issued or personal equipment is in serviceable condition. The departmental property officer should be responsible for ensuring that worn-out or damaged equipment is replaced and that sufficient appropriations are included in the annual budget for new gear.

SUMMARY

The quality and effectiveness of police patrol operations depends upon the availability of a variety of support operations. This chapter has briefly reviewed some of the more common support functions required to facilitate patrol operations in a police agency. No attempt has been made to discuss these functions in depth, inasmuch as their detailed treatment would require several independent volumes. It is important, nevertheless, that patrol supervisors, command officers, and police administrators appreciate the important impact these services have upon police patrol operations.

REVIEW QUESTIONS

1. Examine the personnel management function of your own or another police agency. Identify the key agencies and individuals involved in it.
2. Discuss some of the important functions that should be performed by the research and planning assistant in a medium-sized police agency.
3. Explain why it is important that the police legal advisor be assigned to the office of the chief of police.
4. What are some of the functions the police psychologist may perform to assist the patrol officer?
5. List several purposes of crime analysis; explain how it can be used to enhance patrol operations.
6. Why is it important that community relations not be regarded as just the responsibility of a police specialist?
7. Describe some of the crime-prevention activities that might be performed by the patrol officer.
8. Discuss the three goals of the evidence and property control system.
9. Why is it important that strict accountability be maintained in the handling of evidence?

10. What are some of the advantages of compact police cars? What are some of the disadvantages?

REVIEW EXERCISES

The following exercises are designed to reinforce or amplify comprehension of the material contained in this chapter.

1. Select one of the support functions described in this chapter. Conduct library research to obtain additional information concerning how this function can or should be performed in a variety of situations. Prepare a paper of not more than five pages outlining the results of your work and your findings.
2. Select one of the support functions described in this chapter. Contact several police departments in your area to determine how this function is performed, by whom, and to what extent. Prepare a paper describing the results of your research and any conclusions you have reached.
3. Contact several police departments in your area to request information concerning their evidence-and-property-control systems. If possible, obtain copies of their written procedures governing collection and storage of evidence and property. Evaluate these procedures against the principles described in the chapter.

REFERENCES

Buck, George A., and others, *Police Crime Analysis Handbook.* Washington, D.C.: Law Enforcement Assistance Administration, 1973.

Escobedo v. Illinois, 378 U.S. 478, 488 (1964).

Greisinger, George W., Jeffrey S. Slovak, and Joseph J. Molkup, *Civil Service Systems: Their Impact on Police Administration.* Chicago: Public Administration Service, 1979.

Hoy, Vernon L., "Research and Planning," in Bernard L. Garmire, ed., *Local Government Police Management.* Washington, D.C.: International City Management Association, 1977, p. 367.

Inbau, Fred E., and Wayne W. Schmidt, "The Legal Advisor," in Bernard L. Garmire, ed., *Local Government Police Management.* Washington, D.C.: International City Management Association, 1977, p. 382.

Michigan State Police, *1991 Model Year Patrol Vehicle Testing.* Washington, D.C.: National Institute of Justice, 1991.

National Institute of Justice, *TAP Alert* (October 1986).

CHAPTER 11
PATROL SUPERVISION

CHAPTER OUTLINE

LEARNING OBJECTIVES

After completing this chapter, the student should be able to

1. Explain the importance of effective supervision to the success of police patrol efforts.
2. List the characteristics of effective police supervisors.
3. List the functions typically performed by police patrol supervisors.
4. Describe the manner in which police supervisors are selected and trained.

THE PATROL SUPERVISOR

The patrol supervisor is one of the most important members of the police organization. The first-line supervisor, a sergeant in most forces, is closest to the day-to-day operations of the patrol force and plays a key role in planning, directing, and evaluating the activities and performance of patrol personnel. The first-line supervisor is expected to ensure compliance with established policies and procedures and to initiate corrective action when deficiencies in performance are detected. Moreover, the patrol supervisor plays an important role in developing the skills of subordinates and in motivating them to perform with maximum efficiency and effectiveness.

Unfortunately, the importance of the patrol supervisor is neglected in some police agencies. In some cases, individuals are appointed to supervisory positions on the basis of ill-conceived qualifications or questionable procedures. In some police agencies, promotion to a supervisory position is viewed as a reward for long and faithful service, rather than as a recognition of superior skills and abilities. In still other instances, the duties and responsibilities of supervisors are not clearly defined and they are given insufficient authority to carry out the tasks assigned to them. As a result, patrol supervisors in some police departments are little more than senior patrol officers and perform few, if any, supervisory responsibilities. For these and other reasons, the quality of patrol supervision in some police agencies leaves much to be desired. This ultimately affects the level and quality of patrol operations in general.

There are a number of elements necessary for successful patrol supervision. The patrol supervisor must be capable of making difficult and sometimes unpopular decisions and of exercising sound judgment under trying circumstances. The patrol supervisor must also have exceptional leadership qualities, which will instill in subordinates trust, respect, and confidence. At the same time, the patrol supervisor must be able to evaluate critically the work of subordinates, take corrective action where necessary, issue communications when appropriate, and initiate disciplinary action when the situation so dictates.

This chapter provides a broad overview of the duties and responsibilities of the patrol supervisor and is not intended to be an in-depth discussion of these subjects. Such critical subjects as the selection and training of supervisory personnel and the various aspects of supervisory responsibility deserve more detailed treatment than can be provided here. A number of excellent sources have been cited that should be consulted for more in-depth study of these important topics.

Definition

Supervision is the act of ensuring the proper execution of policies and procedures, and the performance of required duties by subordinates. In most police agencies, the first-line supervisor in the patrol force will be either a corporal or a sergeant. Irrespective of rank or title, the patrol supervisor is responsible for directing the activities of a squad of patrol officers, usually ranging in number from two to eight. As discussed in Chapter 8, the optimum span of control, usually between five and seven subordinates, should not be exceeded if supervisory effectiveness is to be ensured.

In small departments, the patrol supervisor may be the only officer on duty during a particular shift and may be expected to perform the same duties as a patrol officer. In larger departments, the patrol supervisor may be responsible for a number of patrol officers assigned to a particular section of the city or to a specific detail such as traffic enforcement or antiburglary patrol. Regardless of the rank or assignment of the patrol supervisor, the same basic principles of patrol supervision remain in effect. Thus, the material contained in this chapter applies to large departments as well as to small ones.

Selecting and Developing Supervisors

As suggested previously, traditional methods of selecting individuals for supervisory positions in police agencies have often proven unsatisfactory. The problems associated with traditional methods of supervisory selection are many.

> Promotion criteria are often arbitrary and unreliable. Techniques used in the promotion process do not adequately predict future performance in positions of higher responsibility and authority. An element of bias is usually unavoidable. Past performance and tenure tend to overshadow more realistic considerations. At best, traditional promotion policies and procedures are based on guesswork and speculation.[1]

Part of the difficulty associated with traditional methods of selecting police supervisors is that the true dimensions of the supervisory function are not clearly prescribed by the agency or understood by its personnel. Job descriptions, if they exist at

[1]Charles D. Hale, *Fundamentals of Police Administration* (Boston: Holbrook Press, 1977). p. 277.

all, are often vague or do not really define what the supervisor is expected to do. In addition, some police administrators and middle managers fail to bestow confidence in their supervisors and do not regard them as part of the management team.

Other defects have been discovered in the methods used to select supervisory personnel. Many testing and selection practices, for example, have been found to discriminate against minorities and women, who have traditionally been underrepresented in supervisory and command positions.

For this reason, selection and promotion policies must be clearly job related; that is, the criteria by which individuals are selected for promotion must be related to actual job performance, as established through careful and systematic analysis of job requirements. Once again, the lack of adequate performance standards and role definitions for supervisors becomes an issue.

A number of efforts are being made to improve the practices by which individuals are selected for supervisory positions in police agencies. Testing procedures are being revised to remove any discriminatory tendencies they may contain. Attempts are being made to define more clearly the duties and responsibilities of police supervisory personnel. Selection methods that identify individuals with supervisory capabilities are being devised.

One of the most successful methods of selecting persons with supervisory qualities is the assessment-center technique, which is relatively new in American law enforcement but which has proven to be quite reliable in those instances in which it has been utilized. Assessment-center techniques employ a number of situational exercises in which candidates for promotion are expected to demonstrate their analytical, problem-solving, and leadership abilities under circumstances closely approximating those they might expect to confront as supervisors. Candidates are usually observed and rated by professional evaluators or persons who have been trained in assessment-center methods.[2]

Although generally more reliable than traditional testing methods, assessment centers require a great deal of preparation, are usually quite expensive and time consuming, and may require outside assistance in their development and administration. In addition, assessment centers require a value judgment by the evaluators. Nevertheless, the results are usually well worth the cost and effort involved, and they are becoming increasingly popular in the selection and promotion of police personnel.

Civil service reform is a major and necessary key in the effort to upgrade our police organizations by promoting better people to ranking positions in our police organizations. Traditional civil service systems and procedures seem to place more emphasis on protecting the individual than on finding the right people to do the job. As a result, the organization suffers and mediocrity becomes the standard of job performance.

> "But the promotional procedures, which relied heavily on written tests, seniority, oral interviews, and field evaluations, were not reliable indicators of potential leadership,"

[2]See George P. Tielsch and Paul M. Whisenand, *The Assessment Center Approach in the Selection of Police Personnel,* Santa Cruz, Calif.: Davis Publishing Co., Inc., 1977).

> one authority claimed. "Instead of moving up promising commanders, the civil service rewarded run-of-the-mill policemen who were anxious to increase their pay and raise their status and had a knack for taking reports and avoiding trouble."[3]

One of the more innovative methods of solving the civil service dilemma that has been devised in recent years is the exempt-rank concept, which exempts certain designated ranks from normal civil service procedures. Under the exempt-rank concept, persons are assigned to these positions on the basis of merit and qualification and are exempt from usual civil service provisions. The advantage of the exempt-rank concept is that there can be greater flexibility in the assignment of personnel to key positions. In addition, command personnel have greater incentive to perform up to their full potential, since they may be replaced or reassigned if they fail to perform as expected.

Law enforcement is similar to other professions in that experience at lower ranks does not always adequately prepare one for advancement to higher responsibility. The duties of a patrol officer are considerably different from those of a supervisor. While extensive experience as a patrol officer may help a person to understand the hows and whys of supervisory practices, it does not necessarily prepare one to accept the responsibility that goes with the job. Indeed, there are more than a few police officers who decline to take promotional tests because they realize that they either do not have what it takes to be a good supervisor, or they are simply not interested in being a supervisor. In either case, they should be complimented for knowing their own limitations and being realistic about their career goals.

Police training programs tend to concentrate on the legalistic aspects of policing and often do not address the humanistic elements of policing. This is particularly true for supervisory training, since a great part of a supervisor's time is devoted to dealing with human issues. Very few police supervisors are properly prepared to accept the challenges that await them as they make the transition from taking orders to giving orders. This transition has been called one of the most difficult that a person will have to make as he or she moves up the law enforcement career ladder.

It is well known that the demands placed upon the police supervisor are much different from those that the individual faced as a patrol officer. It requires a different mental attitude as well as a different kind of temperament to be an effective supervisor. That is why mediocre police officers sometimes make very good supervisors and excellent patrol officers sometimes make poor supervisors.

> Newly promoted police supervisors often find it difficult to exercise authority over other officers, to give orders, to assume responsibility for the behavior of other officers, and, perhaps most painful, to discipline those who have so recently been their colleagues and peers.[4]

[3]Robert M. Fogelson, "Reform at a Standstill," in Carl B. Klockars, *Thinking About Police: Contemporary Readings* (New York: McGraw-Hill Book Co., 1983), p. 114.

[4]Vincent Henry and Sean Grennan, "Professionalism through Police Supervisory Training," in James J. Fyfe, ed., *Police Practice in the 90's: Key Management Issues* (Washington, D.C.: International City Management Association, 1989), p. 137.

SUPERVISORY ATTRIBUTES

Not all persons make good supervisors. Indeed, not all police officers desire to become supervisors. Some are quite content to be good patrol officers, traffic specialists, investigators, and the like. Moreover, current thinking in police administration holds that police officers should not perceive regular progression through the rank structure as the only means of attaining success in the police profession.

There are a number of qualities generally associated with successful supervisory performance. Among other things, the good supervisor must project a positive attitude toward his or her work, must be prepared to take risks when necessary to defend an unpopular decision, and must be able to communicate effectively. Above all else, a police supervisor must display exemplary leadership qualities. These attributes are discussed in the following pages.

Attitude

One of the most difficult things for new supervisors to accept is that they are not just members of the work force. They are, and should be regarded as, members of the management team.[5] As such, they must project a positive attitude toward their jobs, their superiors, their fellow supervisors, and their subordinates. They must attempt to remain aloof from petty bickering, griping, and squabbling that often characterizes the attitude of the patrol officer. A supervisor who displays a negative attitude toward the job, or who engages in needless recrimination against local policies and procedures, or who takes issue with every directive handed down from above can drastically harm the morale of subordinates and can create an intolerable work environment that will result in poor organizational and individual performance. As Weston has observed:

> Probably the most important attitude of a good supervisor is a favorable attitude toward the people with whom he has significant relationships in the ordinary discharge of his own authority and in the normal acceptance of his responsibilities.[6]

Risk Taking

Supervisors will sometimes be required to take risks in order to gain the respect and confidence of their subordinates. In taking such risks, supervisors must understand

[5]The role of the first-line supervisor as a member of the police "management team" has frequently been ignored by the police labor movement. This is unfortunate in view of the supervisor's unique responsibilities for directing and controlling the bulk of the work of the organization. There is sound justification for excluding supervisors from the same bargaining unit that represents patrol personnel

[6]Paul B. Weston, *Supervision in the Administration of Justice: Police, Corrections, Courts, 3d printing*, (Springfield, Ill.: Charles C. Thomas, 1972), p. 33..

that they stand to lose as much as they gain. Often, they may face significant obstacles, but they must have the courage to stand behind their decisions. For example, a supervisor may receive pressure from above to discipline a subordinate whose work has not been up to standard. If the supervisor believes that disciplinary action would be counterproductive, however, and believes that other means such as counseling or additional training would be more effective, he or she should be prepared to push for one of these alternative courses of action.

Risk taking involves accepting responsibility for one's own actions. Supervisors must also be prepared to accept responsibility for the actions of their subordinates. An error in judgment by a patrol officer will often reflect on the supervisor. When this happens, the supervisor must accept responsibility. At the same time, the supervisor must be management oriented. Supervisors should not attempt to fight unpopular management decisions solely for the purpose of gaining the praise and esteem of subordinates. If a supervisor decides to challenge a management decision, it should be done in a responsible manner and not because the supervisor wishes to curry favor with subordinates. Indeed, the supervisor should avoid taking actions simply for the purpose of maintaining friendships with subordinates.

First-line supervisors are in a very difficult position. Because they are identified closely with their subordinates, members of the upper echelon often distrust them and do not regard them as a part of the management team. Because they are expected to execute the decisions of management, their subordinates often do not accept them and tend to mistrust their motives and actions. In trying to reconcile these opposing views, supervisors must be capable of executing their duties diligently and gaining the trust and confidence of both management and subordinates. Substantial risk may be involved in the process.

Communicating

One of the most important attributes of an effective supervisor is the ability to communicate. **Communication** is essentially the act of transferring information (implying more than mere words or gestures) from one person to another. Communications may be either verbal or written, and both are important in police supervision. Verbal communication is especially important in patrol operations where circumstances often dictate prompt decisions and the immediate transmission of information from supervisors to their subordinates. Whenever possible, however, the written form of communication is preferred simply because it is less subject to distortion and misunderstanding and is generally more reliable.

> The use of verbal orders is indicated when the job is not complex and confusion or misunderstanding is not probable, or when the job has been done before and has become standardized. Verbal orders are used in directing assignments by radio, in the handling of emergencies, and when time is an important element.

> Written orders should be used when orders are standing, when they are nonroutine, when they are complex enough that misunderstanding is reasonably possible, when several persons are involved and all must have the same understanding, and when control and follow-up will be simplified.[7]

Communication is not something that the supervisor can take for granted but is an art that must be developed. A number of factors are important to effective communication. First, the physical and emotional setting in which the act of communicating takes place is a factor. It is much easier to communicate ideas clearly over a casual cup of coffee than during a hot pursuit in which car thieves are being chased. It is also conducive to effective communication if the parties involved in the communication are calm, thoughtful, and not distracted by other events. This, of course, is the ideal and is not always possible in police work. Nevertheless, it is important for supervisory personnel to understand the conditions under which communication is most effective and to attempt to satisfy those conditions whenever possible.

The positional relationship (difference in rank, position, or authority) between the individuals involved in the act of communication is also important. Studies have shown that persons who have something in common (members of the same sex or cultural background) can often communicate more easily than persons who have nothing in common. Two rookie patrol officers will probably be able to communicate more effectively than will two officers, one of whom has 30 years on the job and the other less than 30 days. Similarly, communication is usually easier between persons of like rank than between persons of different ranks.

The personal relationship between persons involved in communicating with each other is also a factor. Friends usually communicate better than enemies for example, since an important element in the art of communicating is trust. Mutual trust and confidence between persons exchanging information is necessary for effective understanding.

> Whether a given communication is viewed with skepticism, doubt, interest, or acceptance may be more a result of who makes the statement than what is said or how it is said. This accounts in part for the fact that some managers whose language skills are minimal achieve a relatively high level of success. Because they are trusted and have a reputation for dependability, they succeed in spite of their inadequacy.[8]

Thus, personal characteristics such as sincerity, honesty, and the ability to instill faith and respect, as well as acquired skills such as clarity of speech and the ability to express one's thoughts concisely and completely are important determinants of effec-

[7]William B. Melnicoe and Jan C. Mennig, *Elements of Police Supervision,* 2nd ed. (Encino, Calif.: Glencoe Publishing Co., Inc., 1978), pp. 165-166.

[8]J. Clifton Williams, *Human Behavior in Organizations* (Cincinnati, Ohio: South-Western Publishing Co., 1978), pp. 249-250.

tive communications. Some individuals have natural communication skills, while in others these skills must be developed. Supervisors must understand the importance of effective communication to the supervisory function and be prepared to develop effective communication skills if they do not already possess them.

Leadership

Leadership is probably the most important characteristic of an effective supervisor, and it is unquestionably the most difficult attribute to develop among supervisors. Leadership in patrol supervisors is extremely important due to the close working relationship between supervisors and their subordinates. Leadership through personal example, demeanor, and self-discipline is a means of directing and controlling the work of subordinates and instilling a sense of purpose in organizational goals.

A supervisor does not automatically become a leader, nor do those with leadership skills always become supervisors. Leadership qualities do not go with the position, but are rather developed in the individual. Thus, when selecting individuals for supervisory assignments, it is important to identify those who either have basic leadership capabilities or are capable of acquiring them.

A large part of supervision is, of course, getting someone else to do what you want them to. This is not always an easy task and may require a different technique in different situations and with different people. Understanding people and what motivates them is an important supervisory attribute. More important is the ability to know when to use coercion and when to use persuasion in getting the job done.

There is a natural tendency among subordinates to resist authority from above, since most workers would prefer to work at their own pace and in their own way rather than in a manner dictated from higher authority. This is especially true in the case of police officers, who themselves are authority figures and who sometimes resent anyone who might try to tell them how their authority should be exercised. In addition, police officers are autonomous workers who have only minimal contact with their supervisors. They therefore tend to resist any authority exercised over them, which further increases the importance of strong leadership skills among supervisors.

The best way for an officer to avoid contact with his or her supervisor is to keep a low profile, do as little as possible, and avoid getting involved in situations which may require the attention of the supervisor. The more active an officer is, the more arrests he or she makes, the more reports he or she writes, and the more incidents he or she becomes involved in, the greater will be the likelihood that the officer will come to the attention of the supervisor. The skilled supervisor will soon learn how to detect which subordinates are trying to avoid contact with them and will take measures to correct this type of behavior.

Leadership is a quality that defies precise definition, although most authorities agree that there are a number of characteristics that are usually associated with successful leadership. Among these are the ability to motivate subordinates, dedication, and skill in accomplishing organizational goals through persuasion rather than through force or command.

> The essence of real leadership is the ability to obtain from each subordinate the highest quality service that he or she has to render. Good leadership does not involve the exercise of authority through commands coupled with threats of punishment for noncompliance, nor is it related to the strength with which commands are barked.[9]

What is a leader? In one sense, the leader is the focal point of organizational development and growth. Success or failure of organizational efforts depends largely upon the capacity and determination of its leaders. Leaders are the motivating force behind organizational activities. They play an important role in inducing compliance with organizational rules, procedures, and standards of performance. They help to make the organization and its members function in an acceptable manner. Leadership, in many cases, distinguishes a highly effective organization from a mediocre one.

> The success of an organization is dependent upon many factors, but none is more important than the impact of its leaders. They make the decisions which determine both organizational purpose and the means by which that purpose is fulfilled. Their actions determine whether the potential of the organization's members will be actualized or lie dormant and whether the emotional tone of the organization will be characterized by warmth and enthusiasm or coldness and apathy.[10]

Leaders influence organizational performance and individual behavior through a combination of their formal authority (that vested in them by virtue of their rank and/or position) as well as through their individual personalities. An important responsibility of the leader is to direct the behavior of subordinates toward desired goals with the least amount of conflict or resistance.

Formal authority does not necessarily accompany leadership. A supervisor, or some other person in a position of authority, may not possess leadership skills, even though they may be able to compel people to follow orders. Leadership implies a special relationship between the leader and his or her followers that may not exist between all supervisors and their subordinates. Leadership denotes a special quality that extends beyond formal authority and involves a personal relationship between leaders and their followers based upon trust, admiration, and respect.

What makes a leader?11 Are there particular and identifiable personality traits that distinguish leaders from followers? What distinguishes a highly effective leader from a mediocre one? Are leadership traits inherent, or can they be developed? These questions have occupied the minds of social scientists for many years, and

[9]Melincoe and Mennig, *Elements*, p. 71.

[10]Williams, *Human Behavior*, p. 214.

[11]For a detailed discussion of leadership styles, qualities and funcitons, see Robert Sheehan and Gary W. Cordner, *Introduction to Police Administration*, 2nd ed. (Cincinnati, Ohio: Anderson Publishing Co., 1989), pp. 335-359.

none have been answered satisfactorily. Indeed, a number of studies have produced conflicting results. One authority did conclude, however, that it was possible to differentiate, on the basis of a number of personality characteristics, between leaders and followers, good leaders and poor ones, and leaders in different levels of the organization hierarchy.[12]

It is also important to recognize that leaders are not all cut from the same cloth but are of widely different types. Some leaders are very colorful and dynamic, while others are quiet and unassuming. Some lead by words while others lead by action and example. A number of studies have been conducted on the subject of leadership styles, and a variety of styles have been identified. Modern leadership theory has identified three basic leadership styles: autocratic, democratic, and free rein.[13]

Autocratic. The autocratic leader is one that relies almost totally on formal organizational power and authority. Autocratic leaders usually place a high value on the accomplishment of assigned tasks but fail to consider the needs of their subordinates. They are generally production oriented rather than people oriented. Autocratic leaders usually succeed best in pressure-type or emergency situations but do not do as well under more leisurely conditions.

Autocratic supervisors are often not well-liked, but they know that a supervisor's job is to make tough decisions and not to win popularity contests. Autocratic leaders often lack much in terms of broad management skills, and usually will not rise to the top of the organization, but they are often the ones to call upon when tough jobs need doing. Thus, one should not assume that, because of their shortcomings, they have no role in the police organization.

Democratic. The democratic leader normally employs a more participative style of management in which subordinates are allowed to offer suggestions and become involved in the making of decisions that affect them. While democratic leaders do not relinquish their authority, they tend to share it with their followers, thereby creating a more open and democratic sense of organizational commitment. Democratic leaders are usually more successful in developing skills and capabilities in their subordinates, and promoting a greater degree of personal commitment in the goals and activities of the organization.

Participative managers often get more out of their subordinates because they share decision-making responsibility with them and make them feel like their ideas are important. This is a characteristic sorely lacking among many police supervisors, who often forget what it was like before they were promoted. Participative leaders are usually respected by their subordinates, since they know that the supervisor has enough self confidence to allow others to share in the decision-making process.

[12]Ralph M. Stodgill, *Handbook of Leadership: A Survey of Theory and Research* (New York: The Free Press, 1974).

[13]N. F. Iannone, *Supervision of Police Personnel,* 2nd ed. (Englewood Cliffs, N.J.: Prentice-Hall, Inc., 1975), pp. 30-31.

These leaders are not worried about someone else usurping their authority or showing them up.

Free rein. The free-rein or laissez faire leader normally exercises a minimum of control over subordinates and in many cases will abdicate supervisory authority and responsibility by allowing subordinates to make decisions by themselves with very little guidance or direction. Free-rein leaders are rarely effective in the organizational setting and often fail to retain control under crisis situations.

This kind of supervisor is often well-liked by subordinates for obvious reasons. Free-rein leaders often create what is referred to as a "country club" atmosphere, which simply means that their subordinates are allowed to do whatever they want to do with a minimum of supervisory interference. Free-rein leaders are popular with their subordinates, but are almost always a liability to the organization and its administration.

Interpersonal Skills

The importance of interpersonal skills among patrol supervisors is due to the direct and intimate contact they have with their subordinates. Supervisors manage people, not inanimate objects. Thus, they must know and understand human behavior, attitudes, feelings, and motivation. To the extent possible, supervisors must attempt to reconcile the human needs of their subordinates with the goals and objectives of the organization.

What has come to be known as the human-relations concept of management is generally recognized to have begun with a series of studies conducted by a group of social scientists at the Western Electric Company's Hawthorne Plant near Chicago in the 1930s.[14] During these studies, a number of methods were used to improve the physical working conditions of workers in order to improve their productivity. While increased productivity was achieved, scientists eventually learned that the increased productivity was associated with factors other than changes in the physical environment of the workers. An important finding was the fact that changes in the social relationships in the work place, as well as in the manner in which workers were treated by management, had a favorable effect on individual productivity. The Hawthorne studies altered earlier conceptions regarding worker attitudes and led to the conclusion that people-oriented improvements in the work setting were just as important, if not more so, as changes in the physical surroundings of the worker.[15] These findings have had an important effect upon the way management views the employee, dramatically changing relationships between supervisors and their subordinates.

In police work, supervisors need to be concerned not only with getting the job done, but also with the effects of the job upon the officer. Police work is often

[14]F. J. Roethlisberger and W J. Dickson, *Management and the Worker* (Cambridge, Mass.: Harvard University Press, 1939).

[15]Williams, *Human Behavior*, p. 3.

demanding, frustrating, and dangerous, and the conditions of the work take a toll on the individual police officer. Good supervisors must consider the stresses placed upon their subordinates when assigning work, supervising, and evaluating their personnel. More is said about this aspect of police supervision later in this chapter.

Example

Subordinates tend to look up to their supervisors and to emulate them. New police officers pattern their behavior and measure their performance by the example set by senior officers, including their supervisors. Thus, first-line supervisors should consider themselves as models of professional conduct and should set an example of high ideals and professional ethics that will be followed by their subordinates.

Supervisors are generally very conspicuous people, and their behavior is usually closely observed by subordinates, superiors, and the general public. Thus, it is important that supervisors set high ideals for themselves and for their subordinates. They should always endeavor to project a positive attitude to their subordinates, even when they may not feel up to it. The attitudes and examples of a supervisor can often mean the difference between a highly energetic, motivated employee and an indifferent, careless one.

THE SUPERVISOR'S ROLE IN ORGANIZATIONAL DEVELOPMENT

One of the supervisor's key responsibilities in the police organization is to promote and stimulate continued and meaningful organizational development. Organizational development in any organization is associated with change, and change in any organization is often threatening and sometimes resisted by employees. The supervisor should be capable of projecting a positive attitude toward change and promoting similar attitudes among subordinates.

In promoting organizational development, supervisors should act as catalysts in the development of new ideas and concepts that will enable the organization to meet new obligations more effectively. Police organizations cannot be allowed to become stagnant or dormant, but instead must continue to evolve and grow in the capacity to meet the challenges of a changing society. New organizational structures, operating procedures, and management styles may be necessary to provide a work setting compatible with emerging knowledge concerning human behavior and employee motivation.

Change in any organization should not be viewed as an end in itself but rather as an instrument through which more important results are achieved. Change should be directed toward overcoming perceived weaknesses in the organization, or toward preparing the organization to meet new conditions or situations. Thus it must be based on careful and deliberate planning and analysis. The police supervisor should endeavor to create a favorable environment for change, in which new ideas are encouraged and carefully considered. Thus, a key element in the process of "organizational development" is "people development." As one writer has stated, organizational development is really:

> ...the process of planned change dealing with people in an organization and the organization as a whole. It focuses on changing the people who make up the organization, how they work together as a unified whole, how they function within their subunits, and what should be changed to make them more effective.[16]

SUPERVISORY FUNCTIONS

The functions performed by first-line supervisors depend largely upon their assignments and upon the customs and traditions observed within the individual police agency. As indicated previously, in some police departments police supervisors play a relatively minor role, while in others the duties and responsibilities are considerably expanded. Patrol supervisors, however, will generally be responsible for ensuring that all calls for service are answered promptly, that patrol officers are courteous and efficient in their contacts with the public, that police procedures are conducted in full accord with existing laws and police regulations, and that patrol activities are conducted in an efficient manner. Even though the exact nature of the police supervisor's duties may vary considerably from one department to the next, there are a number of functions that are typically performed by most supervisory personnel.

Handling Personnel Problems

Human beings are often confronted with a variety of personal problems that sometimes interfere with their performance on the job. These problems may or may not be job related, but they can have the same impact upon job performance. First-line supervisors are in direct, daily contact with their subordinates and are in the best position to recognize personal problems and to initiate appropriate action to resolve them.

Personal problems that are likely to occur may include situations involving the job such as an undesirable shift or work assignment, personality conflicts with other workers, worries stemming from an inability to pass a promotional examination, and similar conditions. Quite often, personal problems not directly related to the job may also interfere with proper performance of duties. Examples of such problems are financial crises, marital difficulties, poor health, and the stress created by a part-time job or attending school.

Supervisory personnel must recognize that they have a responsibility to care for the needs of their subordinates. This responsibility includes detecting the personal

[16]Christine S.Becker, "The Practical Side of OD," *Public Management* (September 1978), p. 4.

problems of subordinates and taking the necessary steps to resolve those problems whenever possible. In addition, police agencies should have available to them the necessary resources, such as marital counseling and psychiatric care, to deal with such problems when they arise. In some cases, the problem may be one that the supervisor is incapable of handling, but he or she should nevertheless see to it that the problem is dealt with by a higher authority in the department.

The police supervisor should also be trained to detect symptoms of stress and mental fatigue that frequently occur in police work, and that, if neglected, may have disastrous organizational and personal consequences (see Chapter 4). Supervisors should, as has been the case in some police agencies, receive instruction in the recognition of mental illness, fatigue, alcoholism, and job stress among their subordinates.[17]

While supervisors should not be expected to fully understand the complex psychological factors that contribute to many personal problems that may be experienced by their subordinates, they should nevertheless be conscious of them and be prepared to take appropriate action when they occur.

> The supervisor must develop a clinical approach in dealing with the emotional problems of his subordinates....He must attempt to understand how these problems arise and deal with them in a practical, common-sensical way. Sometimes a workable solution cannot be found. Perhaps none could be found even by a professional who specializes in the study and treatment of emotional disorders; but ordinarily the degree of success achieved will be directly related to the supervisor's patience and understanding and the effort he expends in helping his subordinates solve their problems.[18]

Motivating Employees

Highly motivated employees help to ensure organizational success. Motivation, however, does not simply occur but must rather be developed through the actions of the individual supervisor. Motivating employees to work diligently toward the accomplishment of organizational goals is an important supervisory responsibility. It is therefore important that supervisors understand the underlying theory of motivation and how that theory applies to the practical world of the police organization.

Early philosophers explained motivation in terms of a person's desire to maximize pleasure and minimize pain, otherwise known as the pursuit of hedonism. During the early 1900s, however, James McDougall and other social scientists advanced the theory that human motivation is not merely the attempt to achieve pleasure and avoid pain, but is, at least in part, instinctive. The underlying assumption of those who advocated the instinctual approach to motivation theory was that each per-

[17]Martin Reiser, Robert J. Sokol, and Susan S. Saxe, "An Early Warning Mental Health Program for Police Sergeants," *The Police Chief* (June 1972), pp. 38-39.

[18]Iannone, *Supervision*, p. 194.

son has a basic, unlearned predisposition to certain types of behavior and that this helps to explain why some people are more motivated than others.[19]

Regardless of the determinants of motivation, it is clear that motivation plays an important role in the performance of people at work. Luthans, for example, describes motivation as the meaning and relationship between needs, drives, and goals.[20] Thus, motivation is neither the end product (result) or the causative factor in a certain type of behavior, but rather is an intermediate condition that is manifested in the behavior it produces. Hunger, for example, is a powerful motive, but it cannot be described except in abstract terms, although the type of behavior it produces (eating) can be described very accurately. Similarly, the motive of fear can be best understood by the various types of behavior it produces such as running away, hysteria, and crying.

It is generally recognized that there are two broad categories of motives: primary and secondary. In addition, there is a more limited third, category called general. The general category includes those motives that do not quite fit into either of the two basic categories. **Primary motives** are those that are related to basic physiological or biological needs such as hunger, thirst, sleep, avoidance of pain, and sex. **Secondary motives** are those that are developed as a result of learning or experience, and are most important in predicting and controlling the behavior of people. Examples of secondary motives are power, achievement, affiliation, security, and status. Examples of general motives, not classified as either primary or secondary, are curiosity, exploration, and love.[21]

Supervisors and managers should obviously be most concerned with the secondary motives, since they are the ones that can most effectively be influenced through supervisory practices and most closely relate to individual performance. Supervisors should understand and appreciate the important contribution that motivation can make toward individual performance and develop management styles that are calculated to achieve a high level of motivation among their subordinates. To accomplish this motivation supervisors should:[22]

Encourage Self-Development. Employees who are involved in their work and who take a personal interest in the results of their efforts will usually be more highly motivated than those who regard their work as just another job. Supervisors should endeavor to stimulate enthusiasm among their subordinates by setting a good example and displaying a positive attitude toward the job.

Delegate Responsibility. The supervisor should be able to delegate sufficient responsibility to encourage subordinates to make decisions for themselves within the

[19]Fred Luthans, *Organizational Behavior: A Modern Behavioral Approach to Management* (New York: McGraw-Hill Book Company, 1973), p. 389.

[20]*Ibid.*, p. 392.

[21]Luthans, *Organizational Behavior*, pp. 396-399.

[22]Melnicoe and Mennig, *Elements*, pp. 63-70.

limits of their own authority and individual capability. Care should be taken, however, not to delegate so much responsibility that employees feel overburdened or incapable of coping with their assigned duties. In addition, supervisors should delegate authority but not avoid the ultimate responsibility for the actions of their subordinates.

Recognize Exemplary Performance. Outstanding performance should always be recognized. Supervisors are in the best position to recognize superior performance and should see to it that subordinates are given credit for exemplary performance. Sincerity is a necessary ingredient. Perfunctory accolades become meaningless and are harmful to organizational morale.

Show Confidence in Subordinates. Supervisors must also make a concerted effort to show confidence in their subordinates. This is not always easy to do, but even the most marginal employee may show improvement if he or she believes that the supervisor has faith in his or her ability. Subordinates should be allowed sufficient latitude in performing their assigned tasks, and the supervisor should avoid the temptation to be unnecessarily critical of employees who fail to measure up to established standards of performance.

Demand Accountability. Subordinates must be made to understand that they will be held accountable for the accomplishment of their assigned tasks. Without the element of accountability, workers often fail to accept responsibility for their actions and may approach their duties in a careless or indifferent manner.

Supervisors should understand the importance of motivation in enabling employees to maximize the potential of their own capabilities.

> In summary, motivation is a question of energy and commitment. We all have untapped sources of energy that are not expressed because we do not see the value of committing ourselves. If we work for an organization that recognizes our worth and gives due weight and dignity to the time we spend working, then we will, naturally, give more of ourselves in return. This helps make the experience of working more fulfilling for each employee.[23]

Developing Subordinates

The relationship between supervisors and their subordinates is a very important one and has been described as the "nucleus of organizational life."[24] This is an apt description and clearly denotes the dramatic role that supervisors play in the develop-

[23]Dave Francis and Mike Woodcock, *People at Work: A Practical Guide to Organizational Development* (La Jolla, Calif.: University Associates, Inc., 1975), p. 64.

[24]John M. Pfiffner and Marshall Fels, *The Supervision of Personnel,* 3rd ed. (Englewood Cliffs, N.J.: Prentice-Hall, Inc. 1964), p. 222.;

ment and performance of their subordinates. Modern organizational theory teaches us that the supervisor today is responsible for more than just issuing orders and seeing to it that they are carried out. Indeed, one of the supervisor's greatest responsibilities is to ensure that subordinates are allowed and encouraged to develop their full potential in the pursuit of organizational goals. They should do this through a sustained and progressive program of performance appraisal and employee training.

Performance Appraisal

Performance appraisal, or employee evaluation, has gained a poor reputation in many police agencies over the years for a number of reasons: (a) employees (including supervisors and subordinates alike) do not fully understand the purpose of performance evaluation, (b) some performance evaluation methods are ill conceived, improperly conducted, and used for the wrong purposes, (c) top management often fails to give full support to the performance-evaluation system, (d) performance-evaluation carries with it neither rewards for exemplary performance nor penalties for unsatisfactory performance, and (e) supervisors are often not adept at evaluating employees in an objective manner.

It is important to understand, first of all, what an employee evaluation form is *not* intended to do. It is *not* (or should not be) designed to be used as a disciplinary measure or to punish individuals whose work is below standard. Rather, an employee evaluation program *is* intended to accomplish several things. First, it should assist employees in learning about their strengths and weaknesses so they may improve their performance. It should also assist management personnel in identifying individuals with promotion potential. Moreover, the system should aid in selecting individuals for specialized assignments such as investigation, traffic, or research and planning. Finally, employee evaluation should provide assistance for evaluating current training programs and identifying future training needs (see Chapter 13).

To be effective in accomplishing these purposes, the employee-evaluation system should be carefully devised and implemented. The system should be based upon a set of performance indicators that accurately reflect acceptable and unacceptable standards of performance. These performance standards should be directly related to the duties and responsibilities of the individual on the job. To the extent possible, they should also be specific and measurable. Vague or abstract qualities such as *loyalty*, should be avoided unless they can be adequately defined and related to actual job performance.

In addition, performance indicators should be both valid and reliable. **Reliability** is demonstrated if the criteria upon which the individual is being evaluated can be measured consistently from one person to the next. **Validity,** on the other hand, refers to the ability of the indicator to measure what it purports to measure.

One of the most common problems associated with performance evaluation in law enforcement is the tendency to measure the **quantity** of the work produced rather than its **quality**. Some police departments, for example, tend to be more concerned with the number of arrests an officer makes than with the quality of those arrests (the number that result in conviction). Unfortunately, many police departments continue to be numbers oriented and often go to the extent of posting charts showing the number of activities (arrests, parking tickets, warnings) produced by each shift, squad, or individual officer. While such practices may promote a favorable atmosphere of competition and esprit de corps, they should not be confused with the real purpose of performance evaluation.

Another factor which often erodes the success of the performance-appraisal system is that of subjectivity, which is the tendency of supervisors to rate an employee on the basis of personal feelings rather than on more objective criteria. For example, if a supervisor has a personal relationship with a subordinate, the supervisor may find it difficult to rate that subordinate's performance objectively. This is a common error in many performance-evaluation systems and can only be overcome through adequate training in proper evaluation techniques, and through the development of a system that minimizes the extent to which personal factors can enter into the evaluation process. The first-line supervisor plays a key role in evaluating individual performance. It is the field supervisor, after all, who is closest to the patrol officer and who is in the best position to judge the performance of patrol personnel. While there are few accepted standards by which individual performance can be evaluated, the following are some questions that should be considered when evaluating the work of a patrol officer.

- Does the officer make a great many arrests in which physical force is necessary to subdue the person being arrested? If so, this might indicate a poor attitude on the part of the officer when dealing with the public.
- Do many of the officer's arrests result in dismissal, acquittal, or reduction of charges? If so, this might indicate a lack of knowledge on the officer's part concerning criminal law and procedure.
- Do many citizens complain that the officer is unnecessarily abrupt or discourteous in handling requests for service? If this is the case, perhaps the officer has a personal problem that is adversely affecting the performance of his or her duties.
- Does the officer issue a great many traffic citations for relatively minor offenses such as registration or equipment violations? This may be an indication that the officer is "ticket happy" and somehow feels that this will enhance his or her service rating. If so, the officer should be counseled regarding the purpose of selective traffic enforcement.[25]
- Do the officer's reports reflect a great deal of thought and care, or are they hastily and sloppily prepared? A carefully prepared report is often an indication of a conscientious and attentive officer.

[25]Unfortunately, such beliefs are often fostered by the actions of supervisors who lead subordinates to believe that a high volume of trafic tickets equates with good job performance.

- Does the officer spend an unusual amount of time in the station, on coffee breaks, at the city garage, and attending to personal affairs? Patrol officers should be evaluated on how well they organize their work and the amount of time they spend in productive activities.
- Does the officer accept criticism and suggestions in a positive manner? If not, he or she may become a serious disciplinary problem in the future.
- Does the officer take proper care of the police unit and other equipment assigned to him or her? Proper use and maintenance of police equipment is an indication of a conscientious and dedicated officer.

These are but a few of the questions that should be considered when evaluating police performance. Performance evaluation is a demanding but necessary task and one that should be approached thoughtfully and objectively. Police officers need and want to know whether they are performing at, below, or above accepted standards. Without this information, they lack the means or motivation to perform to the best of their abilities.

Employee Training

All police agencies, regardless of size, need to have some form of in-service training program if they are to maintain acceptable standards of performance, both individually and organizationally. The subject of in-service training is discussed in more detail in Chapter 13. In addition to the formal in-service training program, however, all supervisors share a responsibility for training and developing their subordinates. While supervisors may not be expected to be teachers in the formal sense, one of their primary duties is to share their own knowledge of law enforcement with their subordinates. As Felkenes has indicated, many of the supervisor's day-to-day duties are related to training.

> When he counsels, he is training; when he gives orders, it is a form of training; when he discusses performance, he is training; and when he corrects or rewards behavior, he is training. In almost everything he does when in contact with his subordinates, the supervisor is a teacher.[26]

As part of their responsibilities as trainers, supervisors must continually observe and evaluate the performance of their subordinates. This evaluation is not the formal process described previously, but is rather an informal activity in which supervisors learn the deficiencies in performance of their subordinates. In some cases, observed performance deficiencies can be corrected through additional or remedial

[26]George T. Felkenes, *Effective Police Supervision: A Behavioral Approach* (San Jose, Calif.: Justice Systems Development, Inc., 1977), p. 225.

training programs. In other cases, observed performance deficiencies can be corrected through informal counseling. Occasionally more concentrated efforts may be required.

Supervisors should also be expected to counsel and guide officers on a daily basis, even when their performance is not below standard. The supervisor should never assume that a subordinate knows the proper way to handle a difficult assignment but should be prepared to explain carefully the requirements of a particular situation so that the officer can have no doubts as to what is expected. If the officer fails to perform the assignment in an acceptable manner, additional counseling may be required.

Since supervisors cannot fulfill all the training requirements of their subordinates, they should be alert to problems in performance that can only be corrected through additional in-service training. For example, officers may not fully understand the techniques involved in making a thorough crime-scene investigation, and the supervisor should be able to recommend additional in-service training in this area.

Unfortunately, many supervisors fail to recognize the importance of their role in the training and development of subordinates and tend to regard training as the responsibility of a special unit or particular individual within the police organization. This should not be the case. Supervisors must recognize that what new officers learn during the first few weeks under the direction of their supervisor may well determine the way they perform in later years. Thoughtful and dedicated supervisors have an excellent opportunity to correct bad habits and to instill in new officers a professional sense of duty and commitment to high ideals. This responsibility should not be taken lightly.

Administering Discipline

Like performance evaluation, the word *discipline* sometimes is viewed with apprehension by police employees. While it is true that discipline is often associated with some form of punishment, this is not always the case. There are also positive forms of discipline that do not involve punishment. In addition, discipline is closely related to performance evaluation and individual development, since it is through proper discipline that high standards of performance are assured and poor work habits are corrected. It should also be kept in mind that the word *discipline* is a Latin word meaning "instruction, teaching, or training."

Depending upon local policy, the administration of discipline may be the responsibility of the chief of police, a division or bureau commander, or the first-line supervisor. In most cases, the level of discipline one can impose depends on one's rank or position in the organization. In nearly all instances, however, it is the first-line supervisor who must either initiate the disciplinary action or recommend the imposition of disciplinary action by higher authority. In any event, the supervisor should play a key role in the formulation and administration of disciplinary policies.

Some supervisors, particularly those who have relatively little time in rank or who have not been properly trained, will avoid imposing discipline on their subordinates because they are afraid of incurring hostility or because they dislike dealing with such sensitive issues. Moreover, some supervisors find it difficult to break the ties that bind them to their fellow workers and find it easier instead to overlook improper conduct and violations of rules. This is not the proper attitude of a good supervisor and creates many problems that can lead to a noticeable decline in morale and performance. Such situations, if allowed to go unchecked, may spread throughout the entire police organization and eventually lead to even more serious problems adversely affecting the reputation and integrity of the police agency.

The nature and purpose of a firm disciplinary policy and the techniques of administering both positive and negative discipline should be taught to all first-line supervisors. In addition, local policies should provide for first-line supervisors to have a direct responsibility for initiating disciplinary actions. It is also important for higher-echelon personnel to support supervisors who initiate disciplinary actions, so long as those actions are based on sound judgment and are not biased in any way. Many unfortunate situations have developed in police agencies where disciplinary actions taken by first-line supervisors have been continually reduced or rescinded by higher authority. This weakens the control of the supervisor over the behavior of subordinates and will eventually lead to a loss of respect for the supervisor by subordinates.

Several things should be done to establish and maintain a strong system of discipline in the police organization. First, individual supervisors must understand the positive aspects of discipline. Second, supervisors should be made to understand the consequences of nonexistent or inadequate discipline on individual and organizational performance. Third, the police supervisor should be encouraged to think of discipline in terms of its organizational consequences rather than in terms of the individual. In police organizations, as in most other organizations, the welfare of the unit must take precedence over the welfare of the individual. Thus, it is sometimes necessary for the individual to suffer in order for the organization to survive.

To have an effective system of discipline in a police agency, disciplinary policies and responsibilities must be clearly stated in writing. All members of the organization should have a clear understanding of what is expected of them, what constitutes improper or unacceptable conduct, and what penalties are prescribed for violations of department rules and regulations.[27] In addition, the disciplinary policies of the police agencies must be consistently, fairly, and impartially applied. Of these characteristics, that of consistency is most difficult to ensure. Although it is quite proper to say that similar breaches of conduct should be dealt with uniformly, in practice this is not always possible or desirable. The experience and past record of the offender, as well as any mitigating circumstances surrounding the incident in question, should be carefully considered when deciding what punishment, if any, to impose. The primary purpose of disciplinary action—preventing further occurrences of the same type—should be a major consideration when deciding the severity of the punishment to be imposed.

> A guiding rule which has been found useful by many supervisors is that just that amount of punishment should be applied which will prevent further derelictions of the

[27] *Managing for Effective Discipline: A Manual of Rules, Procedures, Supportive Law and Effective Management* (Gaithersburg, Md.: International Association of Chiefs of Police, 1976).

same nature. The penalty should fit the offense but it should also fit the individual. This is the ideal state, but it is seldom achieved.[28]

Improving Morale

The condition of morale is an important consideration for any organization in which teamwork and dedication are essential to the accomplishment of organizational purposes. It is especially important in law enforcement, where officers are required to work closely together, often in difficult and sometimes dangerous situations. Iannone defines **morale** as, "a state of mind reflecting the degree to which an individual has confidence in the members of his group and in the organization, believes in its objectives, and desires to accomplish them."[29] With this definition in mind, it should not be difficult to understand the important role morale has in the police service.

If morale is difficult to define, it is even more difficult to measure, although the absence of a high state of morale is often quite evident in a police agency. One often hears such opinions as "The morale around here is at rock bottom," or, "Our morale has never been lower," but precisely what does this mean? How does one assess the existence or absence of morale? Even more importantly, what are the ingredients that contribute to a high state of morale in a police organization?

It would appear that morale is not the result of a single factor but is rather the consequence of a number of elements that should be present in the police organization, including the following.[30]

- Clearly articulated goals and objectives.
- Fair and firm disciplinary policies.
- Reasonable sense of job security.
- Recognition for exemplary performance.
- Job-related and meaningful training.
- Good system of internal communications.
- Strong leadership.

Morale, esprit de corps, and discipline are closely interrelated, and one is usually accompanied by the other. Thus, an effective disciplinary system will usually yield a high state of morale and will lead to great teamwork, dedication, and individual commitment by members of the group to organizational goals. In attempting to resolve problems stemming from or associated with low morale in a police organization, the factors listed previously should be carefully examined.

[28]Iannone, *Supervision,* pp. 288-289.

[29]*Ibid.,* p. 291.

[30]Melnicoe and Menning, *Elements,* p. 101.

SUMMARY

This chapter reviews in very general terms the attributes and functions of effective police supervision. The important part that supervisors play in ensuring compliance with rules and regulations, in promoting a high state of morale, in developing the capacities of subordinates, and in dealing with the personal needs of employees should not be underestimated. In addition, the importance attached to the selection of supervisors based upon their ability to work effectively with people, to display good leadership qualities, and to accept greater responsibilities than their co-workers should also be recognized. Once selected, supervisors should not be immediately thrust into their new jobs but should be provided with proper instruction in supervisory methods and in the nature of their new duties.

Due to the unique nature of patrol operations, patrol supervisors contribute a great deal to the effectiveness and efficiency of the police organization. Given the chance, they should be capable of making decisions that have considerable bearing on the public perceptions of and support for the police. They have a direct influence on the continued training and professional development of their subordinates, many of whom will themselves eventually rise to greater positions of authority and responsibility within the police organization. Just as the highest buildings in the world cannot stand without an adequate foundation, police organizations, both large and small, lean for support on their first-line supervisors.

REVIEW QUESTIONS

1. What are some of the basic problems associated with the manner in which supervisory personnel have been selected in the past?
2. What are some of the basic disadvantages of the assessment center technique?
3. How can negative supervisory attitudes affect group performance?
4. Explain why a supervisor should be prepared to accept risks at times.
5. Give several reasons why the written form of communication is usually preferred.
6. Explain what is meant by the *positional relationship* between persons involved in the act of communication.
7. Explain why leaders are called the motivating force behind organizational activities.
8. Describe several factors that may account for a low state of morale in a police agency.
9. Define the term *organizational development.*

10. What element is most difficult to ensure in the administration of discipline? Why?
11. Why should police supervisors be trained to deal with problems associated with job stress and mental illness?
12. Give several examples of primary and secondary motives and explain the difference between the two types.
13. Describe the primary purposes of performance evaluation.

REVIEW EXERCISES

The following exercises are designed to reinforce or amplify comprehension of the material contained in this chapter.

1. Conduct independent research on police promotional procedures including assessment centers. Compare each in terms of cost, fairness, ability to withstand legal challenge, and overall utility. On the basis of this, develop a model police promotional procedure that you would recommend for a medium-sized police department.
2. Conduct independent research on leadership. On the basis of your research, prepare a short paper on the subject. Include examples of the three types of leadership described in the text.
3. As a group project, conduct additional research on the Hawthorne studies. As a group, discuss the importance of these studies and their relevance to police supervisory practices.

REFERENCES

Becker, Christine S., "The Practical Side of OD," *Public Management* (September 1978).

Felkenes, George T., *Effective Police Supervision: A Behavioral Approach.,* San Jose, Calif.: Justice Systems Development, Inc., 1977.

Fogelson, Robert M., "Reform at a Standstill," in Carl B. Klockars, *Thinking About Police: Contemporary Readings*. New York: McGraw-Hill Book Co., 1983

Francis, Dave, and Mike Woodcock, *People at Work: A Practical Guide to Organizational Development*. La Jolla, Calif.: University Associates, Inc., 1975.

Hale, Charles D.,*Fundamentals of Police Administration*. Boston: Holbrook Press, 1977.

Henry, Vincent, and Sean Grennan, "Professionalism through Police Supervisory Training," in James J. Fyfe, ed., *Police Practice in the 90's: Key Management Issues*. Washington, D.C.: International City Management Association, 1989, p. 137.

Iannone, N. F., *Supervision of Police Personnel* (2nd ed.). Englewood Cliffs, N.J.: Prentice-Hall, Inc., 1975, pp. 30-31.

Luthans, Fred, *Organizational Behavior: A Modern Behavioral Approach to Management.* New York: McGraw-Hill Book Company, 1973.

Managing for Effective Discipline: A Manual of Rules, Procedures, Supportive Law and Effective Management. Gaithersburg, Md.: International Association of Chiefs of Police, 1976.

Melnicoe, William B., and Jan C. Mennig, *Elements of Police Supervision* (2nd ed.). Encino, Calif.: Glencoe Publishing Co., Inc., 1978.

Pfiffner, John M., and Marshall Fels, *The Supervision of Personnel* (3rd ed.). Englewood Cliffs, N.J.: Prentice-Hall, Inc., 1964.

Reiser, Martin, Robert J. Sokol, and Susan S. Saxe, "An Early Warning Mental Health Program for Police Sergeants," *The Police Chief* (June 1972).

Roethlisberger, F.J. and W.J. Dickson, *Management and the Worker.* Cambridge, Mass.: Harvard University Press, 1939.

Sheehan, Robert, and Gary W. Cordner, *Introduction to Police Administration* (2nd ed.). Cincinnati, Ohio: Anderson Publishing Co., 1989.

Stodgill, Ralph M., *Handbook of Leadership: A Survey of Theory and Research.* New York: The Free Press, 1974.

Tielsch, George P., and Paul M. Whisenand, *The Assessment Center Approach in the Selection of Police Personnel.* Santa Cruz, Calif.: Davis Publishing Co., Inc., 1977.

Roethlisberger, F.J. and W.J. Dickson, *Management and the Worker.* Cambridge, Mass.: Harvard University Press, 1939.

Weston, Paul B., *Supervision in the Administration of Justice: Police, Corrections, Courts, 3d printing,* (Springfield, Ill.)

Williams, J. Clifton, *Human Behavior in Organizations* Cincinnati, Ohio: South-Western Publishing Co., 1978.

CHAPTER 12

SPECIAL ISSUES IN POLICE PATROL OPERATIONS

CHAPTER OUTLINE

LEARNING OBJECTIVES

After completing this chapter, the student should be able to

1. Explain the importance of community relations within the context of police patrol operations.
2. Describe the role of the news media in police operations and the relationship that should exist between police representatives and the news media.
3. Explain the importance of providing for a systematic manner of accepting and investigating citizen complaints against the police.
4. Describe the special challenges a police officer faces when serving the elderly, the homeless, the victims of crime, and the deaf.
5. Explain the importance of the proactive role of the police in domestic violence cases.

If there is a common theme that attempts to tie the chapters of this book together, it is that the police exist solely to serve the needs of the community. This, of course is not a new idea, but has been said in one way or another by practically every scholar and eminent police administrator in this century. There can be no question that the police must be a part of the community service system of the local governmental structure. There is also no question—although this point may be debated by some—that service to the community, of whatever kind is required, is the most important product of the police function.

Previous chapters have discussed the importance of understanding the political structures and the forms of government that provide the backdrop for police services, the need for understanding how community influences and values shape demand for police services, and how the "new" concepts of community-oriented policing and problem-oriented policing attempt to transform these noble ideas into practical operating methodologies.

This chapter examines some of the more specific kind of police patrol operations within the context of the community. It discusses the importance of maintaining a sense of openness by establishing cooperative relationships with the news media and by providing fair and thorough procedures by which persons who feel mistreated at the hands of the police may have their allegations thoroughly and fairly investigated. This chapter also addresses some of the unique problems facing police patrol officers: homelessness, the elderly, the deaf, the victims of crime, and the participants in domestic violence cases. These are truly important and sensitive issues that need to be fully discussed within the context of the police patrol operation.

COMMUNITY RELATIONS

The term *community relations* has been misused and abused over the last two or three decades, particularly when used in connection with the police. In all too many cases, community relations has been confused with public relations—a program, or series of activities intended to make the police look good in the eyes of the community. Public relations is a form of advertising. It lets people know how good you are and how reliable and desirable your product is.

Public relations is, of course, very important to any organization—particularly one whose activities involve the community so directly. It is important for the police to have a good public image and professional reputation, but these are not created by public relations programs. A good public image and professional reputation can be created only through sincere effort. This is where community relations comes in, because **community relations** is the sum total of all activities designed to establish and maintain a cooperative and mutually supportive relationship between the police and the community. It is not just making the police look good, but making what the police do something the people in the community can truly appreciate.

An effective and enlightened community relations program must be a long-range, full-scale effort to acquaint the community and the police with each other's problems and to stimulate actions in order to solve the problems.[1] In addition, community relations can only be successful when each member of the police organization understands that community relations is not simply a special assignment, but is rather the assignment of each member of the organization. Good community relations is the result of a team effort. As in other activities involving several persons working together toward a common goal, the effort will be only as successful as the efforts put forth by its weakest member.

Over the years, police agencies have implemented a number of innovative programs designed to open up channels of communication between the police and the public, to educate the public to the policies and procedures of the police department and to acquaint members of the police agency with community sentiments and expectations. The Citizens' Police Academy of Hallandale, Florida is a good example of one such program. The Citizens' Police Academy is a program in which local residents are invited to sign up for a 12-week training program involving three hours of instruction one night a week. Classes are usually taught by off-duty police officers who must prepare their own lesson plans and class materials and who are not compensated for their time. Classes include a variety of topics such as patrol procedures, criminal investigation, tactical operations, drug investigation, and crime prevention. There is a nominal charge for the program to cover the cost of notebooks and other materials.

Students of the Citizens' Police Academy include homemakers, retired persons, college students, married couples, spouses of police officers, city employees and any-

[1]V. A. Leonard and Harry W. More, *Police Organization and Management,* 7th ed. (Mineola, new York: The Foundation Press Inc., 1987), p. 183.

one else who has an interest in police work. Of course, applications are carefully examined to screen out any persons who may have ulterior motives. From its inception, the program has been a great success and there is usually a long waiting list for the next class to start. The program is coordinated by the police department's civilian crime-prevention coordinator.

The News Media

The news media—newspapers, radio, and television—is a force to be reckoned with in just about any community. Due to their involvement in newsworthy local events, police agencies are often the focus of media attention. The local news media can be a vital ally of the police department if a healthy relationship of mutual trust, cooperation, and respect is maintained between the police administrator and media representatives. Police departments have an important message of public service to be carried to the community, and the local news media provides an excellent channel by which this information can be disseminated.

Too often, police administrators view representatives of the news media with skepticism, suspicion, and mistrust. They somehow feel that the news media is out to make the department look bad in order to make headlines. They sometimes feel that the media cannot be trusted with sensitive or confidential information and, as a result, they try to keep as much hidden from the media as possible. This eventually results in a stalemate between the two, and can eventually be damaging to the credibility and reputation of the police agency.

> Police management and its personnel have a tendency on occasion to ignore the good effect of the press and the results that can be achieved through a sound policy of cooperation with the press.[2]

To be sure, there are and have been those few members of the news media who have used unfair and unscrupulous tactics to "get the jump" on their competitors and who have, in so doing, jeopardized their relationship with local authorities. Moreover, there are those few members of the news media who see their mission as being one of keeping the local police "on track" and who perceive themselves to be community watchdogs. They often use underhanded tactics in getting stories and sometimes seem to be more concerned about the sensational value of an incident than about the factual accuracy of the story. These kind of reports are no credit to their profession and can easily undermine the cooperative relationship that should exist between the police and the media.

The fact of the matter is, though, that the interests of both the police and the press can be served through a policy of cooperation and honesty. Nothing is to be

[2]V. A. Leonard and Harry W. More, *Police Organization and Management,* 7th ed. (Mineola, New York: The Foundation Press, Inc., 1987), p. 145.

gained by each party treating the other as "the enemy." This kind of policy is not in the public interest and does little to advance the interests of either party.

> It has been demonstrated over and over again that through a mutual policy of confidence and cooperation, the ends of both news reporting and police administration can be served in a most effective manner. It is seldom that a newspaper reporter has violated the confidence of a police chief in, for example, a sensitive case where premature publicity could prejudice or disrupt the investigation.[3]

All police agencies should have a written policy that outlines the procedures to be followed to ensure that all information that can be released to the public is made available to the local media in a timely manner. Only that information which is so sensitive that its release might compromise an on-going investigation, or which may need to be protected, should be withheld from the press. In addition, police agencies should make it a point to advise the news media of any newsworthy incident of which it might not otherwise be aware. This kind of cooperative attitude will yield worthwhile rewards to all concerned.

Citizen Complaints Against the Police

Patrol officers are the ones who carry out the basic mission of the police and upon whom the reputation of the police department depends, to a very large degree. They are also the first point of contact with the crime victim, the suspect, with witnesses, and other involved parties. In theory, the patrol officer only does what is in accord with departmental policies and procedures, but in fact, the patrol officer exercises a great deal of individual discretion. A patrol officer has the power, through his or her own attitude and temperament, to make the police agency look good, average, poor, or excellent. Too often, patrol officers fail to recognize that they are always in the public eye and that impressions are being formed on the basis of what they do, what they say, how they behave, and the actions they take on the job. Too often, police managers and supervisors fail to remember this fact also. The actions of the individual patrol officer are, however, too important to be taken for granted.

Police work can be, and very often is, adversarial in nature. The police officer must often use his or her authority to get someone to do something they don't want to do. Conflict, frustration, and anger are frequently the byproducts of this kind of interaction. People often feel that they have been wronged by a police officer even when the officer was simply doing what was required under the circumstances. These situations often result in complaints being filed against the officer alleging any number of misdeeds, ranging from bad attitude to unnecessary roughness.

> Every police officer is aware that the complaints are an occupational hazard. As long as the police continue to restrict people in their activity, prevent people from doing illegal

[3]*Ibid.*

> acts, grand or petty and engage in other control or authoritative activity which, at its best, causes inconvenience or resentment, citizen complaints can be expected.[4]

Anyone who has worked as a police officer knows that a majority of the complaints filed against police officers can be easily resolved. Even though the officer may not have been wrong, an apology to the citizen from an understanding and sympathetic superior officer may be all that is required to make the citizen forget the incident.

There are, however, other complaints against police officers that cannot be quite so easily resolved and which will probably need to be handled in a much different way. These are complaints of a much more serious nature which, if found to be true, could mean disciplinary action against the officer. While a significant number of these complaints may eventually be proven untrue, they require, because of their seriousness, more than superficial attention. In many cases, they will require a formal investigation.

The integrity and reputation of the police department depends upon the maintenance of an open system of filing complaints against officers coupled with a written procedure designed to ensure that all such complaints are thoroughly investigated and that appropriate dispositions are made in all cases where a final conclusion can be reached.

> It is imperative to create a positive atmosphere that encourages public cooperation. Public participation is essential, so unless the public is convinced that an agency is truly receptive to complaints, it will not participate in the process.[5]

Many police departments have no written policy concerning the receipt, investigation, or disposition of complaints against the police. This is a serious oversight and may eventually place the department in a very unfavorable light, since it suggests that the department does not think it important to investigate allegations against its officers. A good policy on the receipt, investigation, and disposition of complaints against the police will serve to protect the officer, the department, and the community.

A thorough and well-written policy on citizen complaints against the police should include the following elements:

1. Procedures for the initial receipt and preliminary investigation of all complaints, whether received in writing, by telephone, in person, or anonymously.
2. Responsibility for the supervision of the citizen-complaint process should be firmly affixed within the organization.
3. A complete record of all citizen complaints and their disposition should be maintained.

[4]V. A. Leonard and Harry W. More, *Police Organization and Management,* 7th ed. (Mineola, New York: The Foundation Press, Inc., 1987), p. 181

[5]*Ibid.,* p. 181.

4. A provision for the person filing the complaint to be informed of the final disposition of the complaint.
5. The citizen-complaint procedure should be fully compatible with the departmental disciplinary policy, including any provisions for safeguarding the rights of accused officers.

Citizen Review Boards

One of the methods often touted as a means of bringing about more stringent supervision of police conduct and ensuring police accountability is through the establishment of citizen review boards. These groups charged with reviewing and adjudicating citizen complaints against the police and advising the appropriate authority of recommended disciplinary action in those cases in which the allegations are found to be true. Civilian review boards of one type or another have been established in many cities over the years.

It is difficult to assess their effectiveness, however, since there is no central clearing house or source of authoritative data on the operations or outcomes of citizen review boards, nor have there been any empirical studies conducted to determine their effectiveness compared to other means of reviewing police conduct.

Although the value and efficacy of citizen review boards is not clear, there can be little doubt that police administrators are almost universally opposed to the idea of having a civilian body review citizen complaints against the police. There are a number of reasons why the police resist this kind of review process. Probably the biggest single objection they voice is that the police are supposed to be, or at least want to be "professionals," and as such they do not believe that anyone outside their profession should be capable of judging their actions. To the extent that this is true of other professions such as medicine and the law, this is a valid argument.

The blunt truth is that the police, in general, simply do not like "outsiders poking around in their business." They are naturally attracted to secrecy, internal solidarity, and are suspicious of anything and anyone outside their own environment. In addition, they resist any attempts by politicians, public administrators, or anyone else who is not a member of the fraternity to control them. Their resistance to citizen review boards is a natural corollary to their view of the outside world.

In addition, the police often claim that they think it is unfair for them to be "singled out" for external review.[6] They argue that they are not different from other public employees, but rarely have attempts been made to impose civilian review on these other employee groups. They object to being treated differently, and they have a good point. On the other hand, it is equally true that the police are—and they make this same point themselves—different from other public employees in terms of the job they do and the methods they employ to do it.

[6]James J. Fyfe, "Reviewing Citizen Complaints Against Police," James J. Fyfe, ed., *Police Management Today: Issues and Case Studies* (Washington, D.C.: International City Management Association, 1985), p. 78

With perhaps a few exceptions, citizen review boards have not proven themselves to be a superior method of ensuring proper police conduct or accountability, or of ensuring that citizens who have grievances to file against the police are properly treated. For the most part, citizen review boards can create more problems than they solve because they create controversy and operate outside of the police organization. Because of their natural mistrust of outsiders, police officials will cooperate with members of such panels only when absolutely necessary and then only to the degree required. If citizen complaints against the police are ever to be adequately investigated, it must be with the full concurrence of the police if not by the police themselves. The police are, after all, much better trained and equipped to perform this duty. It remains to be seen whether they are properly motivated to perform it.

In the final analysis, the important thing is that complaints against the police be investigated fully and fairly, with due consideration to all persons concerned. In this case, substance is more important than form. As Fyfe has observed:

> ...it is also clear that the integrity and objectivity of the process of reviewing complaints are far more important than whether the process is staffed by civilians or sworn officers...[7]

DEALING WITH PEOPLE

This book has attempted to provide some very practical guidelines, as well as the basic theoretical perspective to assist patrol supervisors, managers, and police administrators in planning, directing and managing patrol-force operations. This chapter contains a review of some of the special problems associated with maintaining effective patrol operations. We begin with addressing one of the most basic issues confronting the police patrol officer: dealing with the people of the community, and especially those elements of the community which pose special problems for the police or require special understanding and attention.

It is often said that every police officer is a community relations specialist. This means, quite simply, that the task of maintaining favorable relations with the community—all segments of the community—is not the responsibility of a single person or of a specialized unit within the department, but one that rests squarely on the shoulders of the patrol officer. This is as it should be, since it is the patrol officer who comes into contact most often with members of the community. It is the police officer on the beat who will rush to the aid of a crime victim or initiate the search for a crime suspect.

There is really no one better equipped and able to cultivate good public relations with members of the community than the patrol officer on the beat. The patrol officer is the symbol of what the police department represents and is the basic instru-

[7] *Ibid.*, p. 82.

ment by which the police department achieves its goal. The primary burden of crime control, public service, traffic enforcement, and maintaining public order rests with the patrol officer. It is logical that the patrol officer should also assume the burden of representing the department to the public.

The police have to deal with a lot of different people, values, attitudes, perceptions, customs, expectations, likes, and dislikes. In dealing with people, police officers usually try to remain objective and unemotional, which sometimes gets them into trouble because people think they don't care or they are insensitive. On the other hand, to try to identify with the people with whom they come into contact (to put themselves in someone else's shoes) can also get them into trouble. Some people think the police officer is biased or is sympathetic with one point of view in opposition to another. It can sometimes be a no-win situation for the patrol officer.

That is not to say, though, that the police officer needs to be able to understand and be sympathetic to the point of view of all those with whom he or she comes into contact. An officer should not take sides with one person against another. In order to deal effectively with people an officer must, instead, place himself or herself in the shoes of the other person.

The Homeless

One of the negative aspects of police work, if not the primary one, is that the police must deal quite often with the least pleasant aspects of society. Crime victims, drunken persons, drug addicts, suicides, accident victims, runaway children, the mentally ill, and other people who, for one reason or another, do not fit into the mainstream of society, are the people with whom the police officer most often comes into contact.

This is not to say that there are no rewarding moments in being a police officer, for certainly there are, and, on the whole, these probably outnumber the more negative events. Nevertheless, dealing with the coarser side of life is one of the things a police officer must be expected to do. In approaching this responsibility, police officers must strive to maintain composure and not to allow the emotions of the moment to affect their judgement.

We tend to think of "street people" as being somehow different from ourselves. We see them as social outcasts, drunken persons, beggars, trouble makers, rowdies, mentally ill, and generally lacking of the qualities that characterize "normal" people. In short, we see them as inferior to ourselves and a class of people to be avoided, if at all possible. In fact, however, these people are no different, in most respects, from anyone else. They breathe the same air we do; they become ill when deprived of medical care just like most of us; they suffer from the heat or cold when they lack proper shelter; and they experience the same joy, sorrow, bitterness, anger, and greed that all the rest of us do.

The only real difference between these people and the rest of society is the circumstances in which they find themselves. Contrary to popular belief, however, most homeless people did not choose their circumstances, and probably did everything they could to avoid the fate that befell them. Too often, we tend to think that they are where they are either because they want to be or because they deserve to be. In most cases, this is simply not so. More than 2 million persons will find themselves without a place to stay sometime during the year. About one-third of the homeless are families with children and nearly one quarter of them have jobs.[8]

As one researcher has said, these are people who have

> ...experienced economic hard times, or the sudden loss of affordable housing has made it impossible for them to find housing within their means.[9]

The homeless create a problem for the police, not just because of the crimes they commit, but because of the crimes that are committed on them. They are often the targets of assault, rape, robbery, and other crimes because they lack the basic means to defend themselves from predators. Those who choose to live in shelters often find themselves no better off than those who live on the street.

It would appear that the problem of homelessness is a growing one. It is certainly true that we have become more aware of the homeless people in our society in recent years. There are a couple of explanations for this. Neither of these explanations indicates that our society is impoverished, or that we are less concerned than in previous years about homeless people. First, the decriminalization movement of the 1970s resulted in the police being forced to find alternative means to deal with drunken persons and others who would have previously been arrested and lodged in jail. In retrospect, this would have been a better fate than what many of these same people find themselves faced with today. Moreover, the deinstitutionalization of many persons who are mentally ill had the same effect. Finally, shrinking economic resources forced cutbacks in prison construction and forced prison officials to release many people on parole who might have been better served in an institution. Many of these people, unable to care for themselves in a more structured environment, now find themselves "on the street" with no permanent home. Thus, in our desire to correct one social problem we have created another,

Street people are not just an unfortunate fact of modern society, but can be legitimate police concerns. Problems of health care, disease and unsanitary living conditions characterize the homeless way of life. In addition drug use, drunkenness, panhandling, prostitution, fighting, mugging, thievery, and murder are not uncommon among large concentrations of homeless people. Notwithstanding the unfortunate aspects of their situation, they do pose a unique problem for the police which cannot be ignored.

The police do not have a lot of options in dealing with the problem of the homeless. Essentially, they may choose one of three approaches:[10]

[8]Robert Trojanowicz and Bonnie Bucqueroux, *Community Policing: Contemporary Perspective* (Cincinnati, Ohio: Anderson Publishing Co., 1990), p. 241.

[9]Peter Finn, "Street People," in National Institute of Justice, *Crime File Study Guide,* n.d. p. 1.

[10]*Ibid.,* p. 2.

1. *Strict Enforcement.* Under this option, the police can attempt to drive the homeless out of an area by making it as uncomfortable as possible for them, by letting them know they are not welcome, and by invoking whatever legal sanctions they may have available to them. This approach may be acceptable in some situations, but may not be politically or socially acceptable.
2. *Benign Neglect.* In this case, the police may do nothing and hope that the problem will either go away or not grow any worse. In this scenario, the police simply accept the problems associated with the homeless as part of the job and attempt to keep them under reasonable control. Here again, because public officials often insist that the police "do something," this approach may not be acceptable.
3. *Networking.* This approach involves the police interacting with other social-service and government agencies which may be better equipped to deal with the unique problems of the homeless. These other agencies often have better resources and more appropriate ways of dealing with the problems of the homeless, but they need to have the full support and cooperation of the police in order to apply these resources effectively. Networking, where it has been used, seems to offer the most satisfactory means of dealing with this problem.

Although the police may lack the resources to provide a permanent solution for the problems of the homeless, they are in a position to serve as liaison between the homeless and other community and social-service agencies. Police must not only be sensitive to the needs of the homeless, but aware of the programs and services that are available to them.

Police officers are in an excellent position to

> ...enlist and work with community volunteers in improving security in shelters and on the street. The police response must include more than rousting the homeless whenever their unnerving presence inflames taxpayers to demand visual relief or arresting these homeless people who take over abandoned, federally-owned houses.[11]

Networking, where it is possible, seems to hold the most promise for dealing effectively with the problems of the homeless. Good advance planning, involving all participating agencies, is essential to the eventual success of the networking relationship. Each participating agency must have a clear understanding of its own role and responsibilities and how those interact with those of the other agencies.

The goals of networking are as follows:[12]

1. To relieve the police from having to deal with people whose problems are primarily psychiatric, medical, or economic.

[11]*Ibid.*, p. 244

[12]Peter Finn and Monique Sullivan, "Police Respond to Special Populations," National Institute of Justice, *NIJ Reports* (May-June, 1988), p. 3.

2. To ensure that police officers refer only those special populations for which facilities are maintained.
3. To provide the assistance those populations need to prevent them from coming to the attention of the criminal justice agencies again.

Cultural Diversity

For two centuries or so, the United States has been characterized as a "melting pot" of peoples of diverse cultural, racial, religious, and ethnic backgrounds. This diversity has helped to enrich our heritage and enhance our national character. As a people, we are made stronger by our diversity and it is this character that has made us unique among nations.

While the long-term effect of our cultural diversity is positive, the short-term results are not without problems. People of different cultural, racial, ethnic, or religious backgrounds sometimes experience difficulty in gaining entrance into and acceptance by the people of an established area. Neighborhoods in transition are often characterized by tension and sometimes turmoil. People who speak foreign languages experience difficulty in communicating with native individuals and are easily misunderstood. Their cultures and religious customs sometimes clash with the established order and this further contributes to hostility and tension.

Police officers are often called upon to settle disputes between persons of different nationalities and languages. Some investigations require officers to interview members of an ethnic minority as victims, witnesses, or suspects. In such cases, language may be a barrier to effective communication and understanding. Patrol officers are often poorly prepared to handle these situations because they have not been properly exposed to the language or culture of the people with whom they come into contact. As a result, the officer may become frustrated by his or her inability to communicate effectively with the person, and the individual with whom the officer is dealing may also become frustrated for the same reason. The result may very well become a negative encounter between the officer and the resident.

Scenarios such as these can be avoided through proper planning and training by the police department. Whenever it becomes apparent that a particular ethnic minority is growing rapidly in a community, police planners should prepare for potential clashes between established and emerging cultures. Community leaders of ethnic and racial groups should be contacted to establish and maintain liaison with government officials and the police. These leaders might be called upon to provide cultural awareness training or to serve as interpreters in time of need.

Police officers need to be made aware that members of many minority groups have made many contributions to this country in the past and will continue to do so in the future. While their presence in some communities may seem to pose a problem for the police, the overriding concern should be how to assimilate these peoples into our society in a way that will ensure the interests of all concerned. To do so, the police need to be able to communicate with the members of such groups and to deal with them in a manner that will make them feel welcome. Police departments experiencing an influx of immigrants from other countries would be well advised to contact other police departments that have experienced such occurrences to gain the benefit of their experience and apply that information to their own situation.[13]

The Elderly

The elderly pose a very special challenge to the police, but one that is often not recognized. Elderly people have very special needs that are often not fulfilled by society and they are exposed to greater risk than are other segments of the society. They are targeted by con artists because they often have access to greater cash assets than other people. Some older people are easily taken advantage of. They are often forced to rely upon public transportation because of physical infirmities and are therefore susceptible to being attacked on the street.

Due to their age, they often have no one, other than elderly friends to assist them if they become crime victims. Many elderly people have learned they cannot rely too heavily on relatives or family members. Because they are too old to work, the elderly cannot get bank loans if their life savings are stolen or embezzled. The only housing available to some elderly people is in group homes and convalescent centers where the care and treatment is substandard.

The police are poorly equipped to deal with the problems of the elderly for a number of reasons. First, until recently, very little attention was focused on the problems of the elderly and they had very little political clout to bring their special problems to the attention of the American public. As a result, the police are not fully aware of the unique problems faced by the elderly and have given little thought about how to address them.

Second, most police departments are comprised of younger officers. Today's pension plans make it very attractive for an officer to retire long before attaining elderly status. As a result, very few police departments have resident experts who are sensitive to and who understand the problems of the elderly.

Third, in today's economy, police departments, like other city agencies, must deal with priorities. Problems that concern the public the most are the ones that receive the most attention. Problems of the elderly do not rank high among social problems in most American communities. As a result, there is very little pressure for the police to mount much of an effort to deal with the problems of the elderly.

One notable exception is in the Santa Ana, California Police Department where a conscious effort has been made to recognize and deal with the special problems of the elderly. There, special programs have been developed to teach police officers to be more sensitive to the needs of the elderly and to learn the kinds of crime problems

[13]Carole Allen, "The Cultural/Language Barrier," *Law and Order* (April 1987), pp. 44-48. Discusses ways for police to overcome the language barrier. For a discussion of policing in culturally diverse communities, see Herbert E. McLean, "Scaling A Wall of Silence," *Law and Order* (August 1986), pp. 38-42.

that concern them the most. In Santa Ana special arrangements are made to assist elderly crime victims, especially those who have no one else to turn to.[14]

Crime Victims

It has often been said that victims are the forgotten participants in the criminal justice system. Too often, the crime victim is little more than a faceless statistic or a name on a police report—one of a million or more names that are listed as victims of crimes each year in this country. It is well known that victims always seem to be forgotten by the courts, the prosecutors, and probation officers who seem more concerned with dealing with the offender than the persons who have been offended. Unfortunately, this attitude sometimes affects the police officer as well. The crime victim is not regarded as someone in need of care, but as just another name on a police report, another person to interview, another person to deal with under circumstances that are often not pleasant.

It is, unfortunately, all too easy for the police officer to develop a callous and insensitive attitude when dealing with victims of crime. Police officers are taught, after all, not to become emotionally involved, to remain impartial and objective, and to be calm and dispassionate in all situations. To show emotion might indicate a weakness on the part of the officer. In addition, the officer cannot afford to become emotionally involved with every crime victim; the pressures on the officer would simply be too great.

As a result of national legislation and the concern expressed by leaders at the state and national level, we are now beginning to become more sensitive to the plight of crime victims. We are beginning to recognize that they are an important part of the criminal justice system, and that their rights as well as those of the offender, need to be protected.

It has only been in recent years that the victims of crime have begun to receive the attention that many thought they have always deserved.[15] This is due, in part, to the support of national leaders including President Ronald Reagan, who made clear his concern for the victims of crime and who appointed the President's Task Force on Victims of Crime in 1982. The Task Force produced 68 recommendations concerning the rights of crime victims and how those right might be preserved. Two years later, the Victims of Crime Act was passed by Congress which provided financial support for states to enact programs to assist the victims of crime

Today, it is estimated that there are some 4,000 victim-assistance programs in existence. At least 35 states have established comprehensive victim-assistance pro-

[14]Jerome H. Skolnick and David H. Bayley, *The New Blue Line: Police Innovation in Six American Cities* (New York: The Free Press, 1985), p. 44.

[15]For a detailed discussion of the "victim's movement" that became popular during the 1980s see Arthur J. Lurigio, Wesley G. Skogan, and Robert C. Davis, eds., *Victims of Crime: Problems, Policies, and Programs* (Newbury Park, Ca.: Sage Publishing, Inc., 1990), p. 8

grams in the last decade.[16] It is clear that we are becoming more sensitive to the needs of crime victims. We need to recognize them not as merely statistics or names in a police report, but as persons who have been wronged and who deserve to be assisted by society.

> Only recently have people come to realize that victims of crime experience crisis reactions similar to those experienced by victims of war, natural disaster, and catastrophic illness.[17]

The behavior of police officers toward victims of crime is especially important, since the attitude of the police officer often influences how the crime victim will deal with the emotional impact of the crime. Victims of crime—including victims of nonviolent crimes—experience what is called **secondary injury,** which is described as psychological trauma experienced when reliving the crime incident with police officers, judges, prosecutors, and probation officials.[18]

Although they are not trained as professional counselors, police officers can do much to aid crime victims. An officer must maintain an attitude of sensitivity toward victims, and be able to refer them to agencies in a position to assist them. A police officer who is callous or insensitive to a crime victim will often make the victim feel that the crime was unimportant and that no one else cares about it. This is obviously the wrong message to send to the crime victim, and does not reflect favorably on the police agency.

Victims of sexual assault are particularly vulnerable and in need of sensitive care and treatment by police officers. In addition to the shame, fear, and physical pain they suffer as a result of the attack, they often do not know what their legal rights are, or what can or should be done. Callous, indifferent, or apathetic attitudes on the part of responding police officers or investigators can only compound these problems. Police officers have an affirmative role, rather than a passive one, when dealing with victims of sexual assault.

> With very little effort on their parts, practitioners in the criminal justice system are in the position to help rape victims recover. They are also in a position to assess which victims may have more difficulty with recovery and to refer them for treatment.[19]

Many police agencies have made significant efforts to meet the needs of crime victims by establishing programs within their agencies or by identifying community

[16]Peter Finn and Beverly N. M. Lee, "Establishing and Expanding Victim-Witness Assistance Programs," National Institute of Justice, *Research in Action* (August 1988).

[17]Robert C. Davis, "Crime Victims: Learning How to Help Them ," National Institute of Justice, *NIJ Reports* (May/June 1987), p. 2.

[18]Calvin J. Frederick, Karen L. Hawkins, and Wendy E. Abajian, "Victim/Witness Intervention Techniques," *The Police Chief* (October 1990), p. 106.

[19]Patricia A. Rewsick, "Victims of Sexual Assault," in Lurigio, Skogan, and Davis, *Victims of Crime*, p. 79.

resources that can be called upon to assist crime victims. In designing such a program, it is important to understand the kinds of services and assistance that crime victims most often need. These needs usually include some combination of the following:[20]

1. Food, shelter, and clothing
2. Compensation for monetary loss
3. Transportation
4. Medical attention
5. Legal services.
6. Psychological counseling and feedback.

Few victim-assistance programs have the resources necessary to assist all crime victims with all their problems. As a result, it may be necessary to establish priorities by deciding which services are most important in terms of the long-term needs of the clients, and concentrate available resources in those areas.The list of priorities will not be the same for all programs, but will vary according to the needs of the assistance program, the needs of its clients, and the available resources.

As hard as they try and as much as they may want to, police departments rarely have the time or resources the deal effectively with crime victims.[21] That is why it is important to cultivate relationships with other community agencies (the process of networking) so as to broaden the capabilities of the police agency. At the same time, however, it is important that police officers become more aware of and sensitive to the needs of crime victims and not simply see them as "somebody else's problem."

In addition, there are several things the individual police officer can do to aid crime victims in their time of need. These include:[22]

1. Display composure and confidence. The attitude of the officer will impact directly on the emotional state of the victim.
2. Try, as much as possible, to put the crime victim at ease and to make the interview as easy for the victim as possible.
3. Provide honest and accurate information to the victim. This should be done in a discreet and tactful manner.
4. Try to remove the victim from stressful surroundings which may reinforce the trauma of the crime.

[20]Carol A. McClenahan, "Victim Services—A Positive Police-Community Effort," *The Police Chief* (October 1990), p. 10.

[21]See Irvin Waller, "The Police: First in Aid?" in Lurigio, Skogan, and Davis, *Victims of Crime*, pp. 139-156, for an assessment of traditional versus preferred police response to crime victims.

[22]Frederick, Hawkins, and Abajian, "Victim/Witness,", p. 106.

5. Don't rush the interview. Make the victim feel that he or she is the number-one priority of the moment.

In most cases, common sense, sound judgement, and compassion will help the crime victim through the initial period following the crime until social service agencies can be summoned to address the basic needs of the victim.

The Deaf

From time to time we hear of tragic cases in which a person, shot by the police in an apparent attempt to escape or to assault the officer, was found to be deaf. Subsequent investigation revealed that the deaf person had not heard the officer's command to stop, or had not understood the officer's warning, or that the officer had misunderstood the deaf person's intentions or actions. Dealing with a deaf person is not something that most police officers are trained to do, and the experience can create great difficulties for the officer if not handled properly.

When dealing with the deaf, officers should keep in mind that deaf persons can do everything that "normal" people can do, **except** hear. They should also understand that deaf people are not mentally ill or retarded—they simply suffer from a physical defect, as do millions of other people in the United States. Finally, police officers should be aware that deaf people can communicate, in a different way from most other people. Deaf people, for example, can read lips and can understand sign language.

Police officers need to be trained to recognize hearing-impaired people and to use common sense when dealing with them.[23] Police officers should not panic when they realize that they are dealing with a deaf person, but should remain calm; this will help the person they are dealing with to remain calm as well. They should attempt to find a means of communicating lip reading or basic sign language with the person that will accomplish the purpose of the moment. If this is not possible, the officer may be able to obtain the name or telephone number of a friend of the person who can come to the scene to aid in the situation.

If all else fails, there are usually a number of persons trained in sign language who volunteer their time to work with the deaf and who may be able to come to the scene to provide assistance. Each police agency should maintain a list of persons who may be able to aid in dealing with the deaf. In addition, it is not a bad idea to have one or two officers trained in the basics of sign language so that communicating with the deaf can be as effortless as possible.

Domestic Violence

The police role in domestic violence cases is often limited by the legal sanctions provided by the criminal justice system. All too often, the sources of conflict that result

[23]J. Freeman King, "The Law Officer and the Deaf," *The Police Chief* (October 1990), pp 98, 100.

in domestic violence have their roots in problems which are beyond the capacity of the police to solve. They can only deal with the results of the problem, not its causes. There are, however, a multitude of community and social service agencies which are designed to deal with one or more of the problems typically associated with domestic violence; alcoholism, drug abuse, sexual abuse, mental health, and so on. The police need to be able to tap into these agencies when the victims of domestic violence cases indicate that the services they can provide will be useful.[24]

The manner in which police departments respond to domestic violence cases came under scrutiny during the late 1970s and early 1980s. Prior to that time, domestic disturbances were almost universally regarded by the police as one of the most undesirable, unpleasant, and most disliked type of call to which an officer was assigned. There seem to be no winners and only losers in a domestic dispute, and the police officer rarely leaves the scene with any sense of satisfaction that a job has been well done.

Traditional ways of handling domestic disturbances rarely solve problems for very long. Police are reluctant to make an arrest, since the complainant rarely presses charges and might even later blame the police for the arrest. If an arrest is made and charges filed, the complainant frequently refuses to prosecute and the charges are eventually dropped. Too often, the police feel like referees in a fight that never ends, except in tragedy.

It has been suggested that there are several factors that explain the traditional response by the police and the courts to the problem of violence among family members, including

1. A belief in the sanctity of the home and in the family unit.
2. The knowledge that, in those few cases in which an arrest is made, the complainant often decides to drop the charges
3. The belief that counseling and mediation is often more beneficial than an arrest in terms of the long-term impact on the domestic situation.
4. The fact that the assault often did not appear serious when viewed as a single incident rather than as part of a continuing sequence of events.[25]

As researchers have indicated, the traditional method of handling domestic violence cases by the police usually consisted of

> ...doing as little as possible, on the premise that offenders will not be punished by the courts even if they are arrested, and that the problems are basically not solvable.[26]

[24]Wayne B. Hanewica, and others, "Improving the Linkages Between Domestic Violence Referral Agencies and the Police: A Research Note," *Journal of Criminal Justice*, 10 (1982), pp. 493-503.

[25]Lucie N. Friedman and Minna Schulman, "Domestic Violence: The Criminal Justice Response," in Lurigio, Skogan, and Davis, *Victims of Crime*,p. 88.

[26]Lawrence W. Sherman and Richard A. Berk, "The Minneapolis Domestic Violence Experiment," in James J. Fyfe, ed., *Police Management Today: Issues and Case Studies* (Washington, D.C.: International City Management Association, 1985), p. 119.

Domestic violence calls add a significant burden to the police workload, since most involve repeat calls to the same location over some period of time.[27] It is also well-documented that domestic disturbances are situations in which there is a strong likelihood of violence being committed upon the officer. Officers know that each domestic call they respond to increases their chances of being assaulted. Even though an offender may be charged with assaulting an officer, the likelihood of any substantial penalty being imposed is remote. Clearly, for the officer, this is a no-win situation.

In 1981–82, an experiment conducted in the Minneapolis Police Department revealed that, contrary to conventional wisdom, arrest was the most effective way of dealing with domestic abuse.

> The research showed that when the police arrested the suspect, the incidence of repeat violence over the next six moths was only 10%, compared to 19% for advising and 24% when the suspect was sent away.[28]

The Minneapolis Domestic Violence Experiment provided, for the first time, solid, empirical evidence with which to evaluate the methods used to handle domestic violence problems.[29] Additional research is still needed in this area. Although the study results do not indicate that an arrest is **always** appropriate in **all** cases, it does point out the fallacy in our traditional method of "doing as little as possible."

As a result of increased public concern and growing attention in the news media, a number of states have enacted legislation giving the police more power to deal effectively with domestic-violence cases. In addition, many police departments have changed their policies to enable officers to be more aggressive in dealing with domestic violence. Essentially, the effect of these policies is to require the police to take affirmative action against domestic abuse in cases in which they may have taken less stringent action in the past.

The Pittsburgh Police Department's policy effectively eliminated officer discretion in such cases and lifted from the victim the burden of deciding whether to prosecute.[30] During the first six months after the policy went into effect, arrests for domestic violence increased by 300 percent. It is hoped that this more aggressive policy will have the effect of ultimately reducing both the number and severity of domestic violence cases.

[27]Robert Sheehan and Gary W. Cordner, *Introduction to Police Administration,* 2nd ed. (Cincinnati, Ohio: Anderson Publishing Co., 1989), p. 388-389.

[28]Robert Trojanowicz and Bonnie Bucqueroux, *Community Policing: Contemporary Perspective* (Cincinnati, Ohio: Anderson Publishing Co., 1990), p. 113.

[29]See Sherman and Berk, "Minneapolis," pp. 118-131, for a good overview of the study methodology and results.

[30]Anonymous, "Pittsburgh's Aggressive Stand Against Domestic Violence," *The Police Chief* (April 1990), p. 41.

The policy requires police officers to "...investigate thoroughly all instances of domestic abuse and take *positive* and *aggressive* action to prevent its reoccurrence."[31] In addition, officers are required to "...take whatever action is necessary to insure the safety of the victim."[32]

The policy of the Pittsburgh Police Bureau properly reflects the current state of concern about the dangers of domestic violence to the welfare of society. Police can no longer afford to simply separate the parties, allow them to go their own ways, and hope that the next call at the residence comes on another shift. The police have responsibility, now, in light of our increased social consciousness concerning this problem, to take whatever action needs to be taken to solve the problem, not simply delay its next episode. New state legislation provides police agencies with the tools necessary to take the required action.

Ideally, domestic problems should not be referred to the police at all. The police have few tools at their disposal, short of arrest, to remedy these situations. Even when the police do all they can do, it is often not enough and a tragic ending sometimes occurs. Nevertheless, public policy has made this a police problem and the police must respond in an appropriate and responsible manner.

CITIZEN VOLUNTEERS

Few police departments have the resources to do all of the things they would like to do, or to offer each program and service that the community would like. In the last twenty years, most police departments have been under a program of fiscal austerity, and some have suffered significant cutbacks as a result of local financial problems. This is in sharp contrast to the decade of the late 1960s and early 1970s when national attention was clearly focused on the problems of the police and the criminal justice system. The concern in those years resulted in significant amounts of cash being allocated to the police and the other parts of the criminal justice system. While police chiefs in those days did not have blank checks, they did find that local officials were far more sympathetic to their needs than they are today.

The decades of the 1980s and 1990s, however, have forced police administrators to find ways of maintaining services with fewer resources, to add new programs without proportionate increases in personnel, and to respond to increasing demands for police services without increasing the size of their budgets. That they have developed into much better managers goes without saying. In addition, they have had to find alternative resources to assist them in providing the services demanded by the public.

One valuable resource that is often overlooked in many communities is that of citizen volunteers. When we think of citizen volunteers, we often think of uniformed reserve officers who come in on weekends or afternoons to ride along with regular

[31]Pittsburgh (Pa.) Police Bureau Policy No. 124.8, "Domestic Violence."

[32]*Ibid.*

officers, or those who volunteer their time as school crossing guards, or work extra details at parades and high school football games. Although the resources are certainly important, there remains an untapped reserve of other resources in the community that can help make a difference in the level and quality of services provided by the police.

In every community there is a large number of retired persons and others who, for one reason or another, have the time, energy, and ability to volunteer their services to any number of community organizations, projects, and activities. The same time, energy, and talent that is devoted to other community activities could be directed toward selected police programs and services. Some of the best outreach programs offered by police departments are those which are staffed and run entirely by civilian volunteers.

Retired persons can be especially useful in providing alternative staffing for crime prevention, youth services, victim assistance, counseling, and public education programs. They are often anxious to give their time and to make a positive contribution to the community. Retired persons possess a wealth of knowledge, insight, experience, and sensitivity that can be successfully applied to many police activities including counseling assault victims, interviewing complainants, taking police reports, giving crime-prevention lectures, and presenting information on drug abuse to school children. The fact that they are not sworn officers, do not wear badges, or carry guns, does not make them any less effective. The American Association for Retired People (AARP) has been very active in promoting programs involving elderly people in community service activities, particularly in crime-prevention programs.[33]

Police reserve units, consisting of either paid or unpaid volunteers from the community, can provide valuable support to the regular uniformed patrol force. Depending upon local practices, police reserve units may be used to supplement the regular patrol force; security at high school athletic events, parades, demonstrations, and other public events; undercover surveillance operations; traffic control; disaster assistance, and other functions. Police reserve units have also been used to form specialized squads for operations such as diving, marine rescue, and mountain rescue.

> Admittedly, auxiliary police units must be recruited, organized and trained under the closest supervision by regular officers in the department who possess the necessary capability for this important assignment.[34]

With leaner budgets, police administrators need to search out new ways of doing things in order to maximize efficiency and productivity. Citizen volunteers and police reserve units represent an excellent alternative source of personnel to supplement regular police personnel. In addition, this association helps to promote good will between the police and the community. The importance of this added benefit should not be overlooked.

[33]See, for example, Leonard A. Sipes, Jr., "The Power of Senior Citizens in Crime Prevention and Victim Services," *The Police Chief* (January 1989), pp. 45-47.

[34]Leonard and More, *Police Organization,* p. 187.

SUMMARY

Dealing with people is a regular and important part of a police officer's job. Indeed, people are the central focus of the police role in our society. Helping people with their problems is what makes the police officer's job both interesting and frustrating. An officer must have the skills to interact with victims, suspects, witnesses, drunken persons, the homeless, school children, public officials, and community leaders. With each person and in each situation, the police officer must react in a different way. Dealing with people is what the police officer does most—-and what the police officer should be trained to do best.

REVIEW QUESTIONS

1. Why is it so important for the police to develop and maintain a cooperative relationship with the news media?
2. What is the difference between community relations and public relations? How is each important to the police?
3. Why do the police tend to look upon representatives of the news media with mistrust and suspicion?
4. What are the major components that should be included in a written policy on citizen complaints against the police? Why are these elements important?
5. Are civilian review boards a good way of handling citizen complaints against the police? Why or why not?
6. What kind of special problems do the homeless create for the police?
7. Explain the concept of networking and what advantages it offers to the police.
8. What kind of special considerations, if any, do the police need to make in their dealings with the elderly?
9. What kinds of things can the police patrol officer do to ease the burden on the victims of crime?
10. What are some of the reasons for the change in police policies regarding domestic violence cases?
11. What are some of the factors that make it so difficult for the police to deal effectively with domestic violence cases?
12. What are some of the ways that citizen volunteers can be used to help the police do their job more effectively? What limitations, if any, should be placed on the use of citizen volunteers?

REVIEW EXERCISES

The following exercises are designed to reinforce or amplify comprehension of the material contained in this chapter.

1. Select one of the categories of special people with whom the police officer interacts on a regular basis. Conduct library research to obtain more information on this subject. Prepare a paper of not more than five pages summarizing your findings and conclusions.
2. Contact several police departments in your area to find out what programs or policies they have for dealing with crime victims, the elderly, and the homeless. Prepare a brief paper describing the results of your research.
3. Contact several large police agencies regarding their policies on domestic violence. If possible, obtain copies of local department policies as well as existing state laws. In class, discuss the results of your research and any suggestions you may have concerning existing or recommended police practices.

REFERENCES

Allen, Carole, "The Cultural/Language Barrier," *Law and Order* (April 1987), pp. 44-48.

Alpert, Geoffrey P., and Lorie A. Fridell, *Police Vehicles and Firearms: Instruments of Deadly Force*. Prospect Heights, Ill.: Waveland Press, 1992.

Anonymous, "Pittsburgh's Aggressive Stand Against Domestic Violence,": *The Police Chief* (April 1990), p. 41.

Davis, Robert C., "Crime Victims: Learning How to Help Them," National Institute of Justice, *NIJ Reports* (May/June 1987), pp. 2-7.

Finn, Peter, "Street People," in National Institute of Justice, *Crime File Study Guid*e (n.d.).

Finn, Peter, and Monique Sullivan, "Police Respond to Special Populations," National Institute of Justice, *NIJ Reports* (May-June 1988), pp. 2-8.

Finn, Peter, and Beverly N. M. Lee, "Establishing and Expanding Victim-Witness Assistance Programs," National Institute of Justice, *Research in Action* (August 1988).

Fogelson, Robert M., "Reform at a Standstill," in Carl B. Klockars, *Thinking About Police: Contemporary Readings*. New York: McGraw-Hill Book Co., 1983, pp. 113-129.

Frederick, Calvin J., Karen L. Hawkins, and Wendy E. Abajian, "Victim/Witness Intervention Techniques," *The Police Chief* (October 1990), pp. 106, 108-110.

Friedman, Lucie N., and Minna Schulman, "Domestic Violence: The Criminal Justice Response," in Arthur J. Lurigio, Wesley G. Skogan, and Robert C. Davis, eds., *Victims of Crime: Problems, Policies, and Programs*. Newbury Park, Ca.: Sage Publishing, Inc., 1990, pp. 87-103.

Fyfe, James J., ed., *Police Management Today: Issues and Case Studies*. Washington, D.C.: International City Management Association, 1985.

Fyfe, James J., "Reviewing Citizen Complaints Against Police," James J., Fyfe, ed., *Police Management Today: Issues and Case Studies.* Washington, D.C.: International City Management Association, 1985, pp. 76-87.

Hanewica, Wayne B., and others, "Improving the Linkages Between Domestic Violence Referral Agencies and the Police: A Research Note," *Journal of Criminal Justice,* 10 (1982), pp. 493-503.

King, J. Freeman, "The Law Officer and the Deaf," *The Police Chief* (October 1990), pp. 98, 100.

Leonard, V. A., and Harry W. More, *Police Organization and Management* (7th ed.).Mineola, New York: The Foundation Press, Inc., 1987, p. 183.

Lurigio, Arthur J., Wesley G. Skogan, and Robert C. Davis, eds., *Victims of Crime: Problems, Policies, and Programs.* (Newbury Park, Ca.: Sage Publishing, Inc., 1990).

McClenahan, Carol A., "Victim Services—A Positive Police-Community Effort," *The Police Chief* (October 1990), p. 104.

McLean, Herbert E., "Scaling A Wall of Silence," *Law and Order* (August 1986), pp. 38-42.

Pittsburgh (Pa.) Police Bureau Policy No. 124.8, "Domestic Violence."

Rewsick, Patricia A., "Victims of Sexual Assault," in Arthur J. Lurigio, Wesley G. Skogan, and Robert C. Davis, eds., *Victims of Crime: Problems, Policies, and Programs.* (Newbury Park, Ca.: Sage Publishing, Inc., 1990), pp. 69-96.

Sheehan, Robert, and Gary W. Cordner, *Introduction to Police Administration* (2nd ed.). Cincinnati, Ohio: Anderson Publishing Co., 1989, pp. 388-389.

Sherman, Lawrence W., and Richard A. Berk, "The Minneapolis Domestic Violence Experiment," in James J. Fyfe, ed., *Police Management Today: Issues and Case Studies.* Washington, D.C.: International City Management Association, 1985, pp. 118-131.

Sipes, Leonard A., Jr., "The Power of Senior Citizens in Crime Prevention and Victim Services," *The Police Chief* (January 1989), pp. 45-47.

Skolnick, Jerome H., and David H. Bayley, *The New Blue Line: Police Innovation in Six American Cities.* New York: The Free Press, 1985.

Trojanowicz, Robert, and Bonnie Bucqueroux, *Community Policing: Contemporary Perspective.* Cincinnati, Ohio: Anderson Publishing Co., 1990.

Waller, Irvin, "The Police: First in Aid?" in Lurigio, Arthur J., Wesley G. Skogan, and Robert C. Davis, eds., *Victims of Crime: Problems, Policies, and Programs.* Newbury Park, Ca.: Sage Publishing, Inc., 1990, pp. 139-156.

CHAPTER 13

UPGRADING PATROL EFFECTIVENESS

CHAPTER OUTLINE

LEARNING OBJECTIVES

After completing this chapter, the student should be able to

1. Explain the importance of continually searching for and finding ways of improving police patrol effectiveness.
2. Explain the importance of the law-enforcement accreditation process and how it affects the delivery of police services.
3. Define productivity. Explain why productivity is important to the delivery of police services.
4. Explain why training is important to the professional development of all levels of law enforcement personnel.
5. Explain the effect of field training programs on the number of civil lawsuits brought against the police. Describe the manner in which such programs should be organized and conducted.

Despite the tremendous strides that have been made to upgrade, modernize, and professionalize the police in the past thirty years, much remains to be accomplished. There is no question, though, that the battle is being won and that the trend is definitely toward a more progressive and professional police system in this country. This must be done, not because it is the right thing to do (though surely it is), but because there is no other logical or reasonable alternative. Indeed, the risk of not maintaining a modern, well-trained, and responsible police department is simply too great. In recent years, the police have come under increasing legal attack by people who claim that they have been wronged by the police. The police in turn, have had to expend enormous sums of money to defend the officers involved.

Police agencies, their officers, and employees may be held civilly liable for a variety of omissions and commissions under the federal civil Rights Act of 1871 (42 U.S.C. 1983). Among the acts of omission or commission for which a police department can be held liable are the following:[1]

1. Failure to train.
2. Failure to supervise.
3. Failure to have a department policy.
4. Negligent retention.
5. Negligent hiring.
6. Failure to discipline.
7. Failure to screen psychologically.
8. Negligent entrustment.

Section 1983 of the Civil Rights Act can be filed in both state and federal courts. The Civil Rights Act was originally passed to combat the Ku Klux Klan following the Civil War. Section 1983 provides that "any person acting under the color of state or local law who violates the federal constitutional or statutory rights of another shall be liable to the injured party."[2]

For many years, various authorities have engaged themselves in an argument as to whether the police should be regarded as a profession. There are, of course, good arguments to be made on both sides of this issue, and it has never been satisfactorily settled. While the police exhibit many of the characteristics of other professions, they lack the fundamental requirement essential for any occupational group to call itself a profession: a body of knowledge. The laws and principles which are generally accepted as the collective wisdom of its members and which describe to a rather specific degree the manner in which its operations will be carried out is called a **body of knowledge.**

[1]Geoffrey P. Alpert and Lorie A. Fridell, *Police Vehicles and Firearms: Instruments of Deadly Force* (Prospect Heights, Ill.: Waveland Press, 1992), pp. 10-11

.[2]*Ibid.*,p. 11.

Other true professions, such as the medicine, law, theology, engineering, dentistry, and teaching all subscribe to a general body of knowledge which sets forth in some detail the methods and conditions under which its practitioners operate. For the most part, this is not true of the police. There is simply too great a variation in operating practices and methods to say that the police subscribe to a common body of knowledge. Because of the highly localized conditions under which the police operate, this state of affairs may continue long into the future.

There has been, however, in recent years, a genuine and credible attempt to develop what can realistically be termed a central body of knowledge in the form of standards of conduct and procedure. This effort, known as accreditation, is probably the single greatest force in existence which will bring law enforcement close to its goal of being recognized as a profession.

ACCREDITATION

The Commission on Accreditation of Law Enforcement Agencies (CALEA) was created through the joint efforts of four established organizations of law enforcement officials: the International Association of Chiefs of Police (IACP), the Police Executive Research Forum (PERF), the National Sheriffs' Association (NSA), and the National Organization of Black Law Enforcement Executives (NOBLE). The purpose of CALEA was to provide law enforcement with a set of commonly-accepted principles and standards governing organization, operations, administration, and management methods.

Initially, over 900 standards covering virtually all aspects of law enforcement operations were developed. These were divided into 83 different chapters. Each standard is quite specific in nature, leaving no question as to what an agency must do to achieve compliance.

Recognizing that no single set of standards can be applied universally to all police agencies, CALEA established guidelines that depended upon the size of the agency, ranging from less than 10 to more than 1,000 full-time personnel.

Although it is not clear what the long-term impact will be, the initial reaction to the accreditation process has been mostly favorable. Each year, a number of police agencies receive accreditation for the first time, and several have been reaccredited one time or more. There seems to be a growing interest in the process, and at any one time there may be as many as several hundred police agencies in the United States at some point in the accreditation process.

On the other hand, some states, such as California, have opposed the accreditation process for a variety of reasons. Some of the biggest opponents of accreditation seem to feel that it is one more step toward a national police force. This is a rather shallow argument, since for years accreditation has only advanced the professional and ethical standards of schools, hospitals, and other institutions.

Other states reject the philosophy of accreditation because they resent being told that there are standards of professionalism higher than their own. As a result, some states—-such as New York-—have established their own version of accreditation by developing their own set of standards and implementation methods.

There are a number of good reasons why police agencies attempt to become accredited. Perhaps the reason cited most often is the fact that accreditation is seen as a means of reducing a police agency's exposure to liability in the event an officer becomes involved in a situation in which wrongful death, unlawful search and seizure, or unreasonable use of force is alleged. While accreditation will certainly not provide blanket protection to a police agency when such allegations are made, it will make it difficult for a plaintiff to show that such acts grew out of lack of management, poor supervision, or poorly prepared or inadequate departmental policies and procedures.

It is also generally recognized among police administrators, educators, and scholars that an accredited police agency is one that has achieved a high level of professional development. This leads to a higher level of pride and morale among the members of the organization. Police practitioners recognize that accreditation is not an easy process; it requires a great deal of time, commitment, and effort. An accredited police agency has proven that it is capable of carrying through on its commitment to achieve for itself the highest possible standards available to law enforcement agencies.

Reasons for not pursuing accreditation are usually that it is too expensive for the expected benefit, or that it is impossible to impose universal standards on thousands of police agencies, each of which is unique and has its own set of requirements.[3]

Although it is certainly true that the cost of accreditation is significant, the process itself can yield direct, tangible results, the value of which is difficult to assess in dollars and cents. In reality, the few reasons cited for not becoming accredited are, in many cases, little more than excuses.

While it remains to be seen whether an accredited police agency is better than a nonaccredited one, there is little doubt that the goal of accreditation is worthwhile. It has been pointed out that accreditation leads to the development of very specific policy-and-procedure guidelines in such high-risk areas as police pursuits, treatment of prisoners, and the use of deadly force. Without clear guidelines in these areas, a police agency and the community it serves may be exposed to significant liability. As a result, some insurance organizations have volunteered to reimburse police agencies for one-half the cost of accreditation, recognizing that an accredited police agency is a better insurance risk than a nonaccredited one.[4]

As of July 27, 1991, 182 police agencies had been accredited since the process began in 1983. Accreditation is valid for five years, after which agencies seeking to maintain that status must once again show that they meet the approximately 850 criteria promulgated by CALEA. There are at least 23 state and local coalitions regis-

[3]John Kelley, "Negotiating the Accreditation Process," in *The Police Chief* (June 1991), pp. 25-29.

[4]Sheldon Greenberg, "Police Accreditation," in Dennis Jay Kenney, ed., *Police and Policing: Contemporary Issues* (New York: Praeger, 1989), p. 251.

tered with CALEA. "These coalitions provide information about accreditation to other agencies and aid those who decide to pursue the process."[5]

PRODUCTIVITY

Productivity, when used in the context of police patrol operations, is a term that seems to have different meanings to different people. To some police traditionalists, productivity simply means writing more tickets, logging more miles on the daily activity sheet, making more arrests, or finding more doors or windows open during the night shift. To be sure, all of these activities are important and can, in one way or another, contribute to the improvement of productivity. True productivity, however, is not simply doing more of anything, but rather doing it better. In this regard, **productivity** means the accomplishment of goals with a minimum of effort and the optimum utilization of existing resources.

Too often, police administrators tend to put quantity ahead of quality and emphasize numbers instead of results. The number of arrests made by the patrol force means very little if crime increased during the same period, or if a significant number of those persons arrested were either not charged or acquitted. Productivity improvement means working better, not working harder. It means getting the most out of available resources and producing the best results with the least amount of effort possible. It means allowing for the minimum amount of waste and inefficiency in doing any job and producing the best possible results in any endeavor.

The need for productivity improvement in law enforcement should be obvious. Police departments, as part of municipal governments, are tax-supported agencies and as such must rely upon the financial support provided by local taxing bodies. This support does not come readily, particularly in tough economic times. Very few police administrators in recent years have been able to brag about how easy it is to get their budget requests approved. More often than not, police chiefs have to do with less than they would like, but are expected to do the same job anyway.

Public officials, as well as the residents of a community, have a right to know how well (or poorly) their police agency is performing. Citizens have a vested interest in police performance and even if the police are not doing as well as expected, they have a right to know this. By being forthright with the public, police officials can let the people know what problems need to be corrected in order for them to improve their performance. In this way, they may be able to gain the additional support they need from elected officials.[6]

Productivity improvement can be an important management tool in that it can provide a system of evaluating police performance which can lead to improved levels

[5]"CALEA" Adds 13 More to Accreditation Toll," *Law Enforcement News* (September 15, 1991).

[6]Robert Sheehan and Gary W. Cordner, *Introduction to Police Administration,* 2nd ed. (Cincinnati, Ohio: Anderson Publishing Co., 1989), p. 479

and quality of police services. Performance measures can be designed to serve many useful purposes. For example:[7]

1. Current performance levels can indicate the existence of performance problems.
2. Areas in need of attention can be identified.
3. Overtime can be monitored to detect increases or decreases.
4. Specific functional activities, such as traffic, youth services, and investigation can be evaluated and the performance of these functions carefully monitored.
5. Existing performance levels can be used to help create performance goals and objectives.
6. Deficiencies in training for individuals or units in the agency can be identified and remedial programs suggested.
7. Performance incentives can be established as a means of enhancing morale and further improving productivity.
8. Performance measures can be used to ensure an enhanced sense of accountability to the general public as well as to political leaders and local government officials.

Productivity enhancement is made difficult in the police service due to the fact that police goals and objectives are, for the most part, difficult to quantify. While it may be possible to quantify **processes,** such as arrests, citations, and convictions, these are not the same as **results** (for example freedom from crime, peace, safety) which are the real goals of law enforcement. Thus, while it is one thing to measure what the police do, it is quite another to measure what they accomplish. How many crimes, for example, are not committed as a result of the deterrent effect of police patrol? We simply do not know.

Productivity improvement should be an ongoing effort in any police agency, conducted as part of the inspection process. Productivity enhancement need not rely solely upon numbers, although measurement is clearly an integral component of the process.[8] Means need to be developed to evaluate the quality of police services just as easily as we can now measure the level of services provided to the public. Citizen attitude surveys, for example, are an excellent way to identify deficiencies in police performance and to gauge whether improvements have been made in certain areas and activities over a specified period of time. In like manner, citizen complaints against the police provide a good measure of police performance. An excess number of complaints alleging police brutality or rude behavior may very well indicate a

[7]*Ibid.*, pp. 479-480.

[8]For a discussion of police performance measures, see V. A. Leonard and Harry W. More, *Police Organization and Management,* 7th ed. (Mineola, New York: The Foundation Press, Inc., 1987), pp. 276-277.

problem with the training program, a lack of self-discipline in one or more officers, or improper or inadequate field supervision.

An important source of information concerning how to improve performance and productivity is the individual member of the department. We too often overlook this valuable source of ideas, suggestions and recommendations, even though individual employees often are in a good position to make excellent suggestions on how something can be done better, faster, easier, or cheaper. Employee suggestion boxes are a simple way of encouraging employees to comment on performance improvement. An even better way is to form employee committees to advise management on all manner of functions and activities in the department. One suburban fire chief formed over 20 different committees in his 60-person fire department, and each member of the department served on at least three of these committees. Some, of course, were more active than others. The net result, however, was that the employees of the department felt totally involved in all aspects of the organization and were able to make meaningful and positive suggestions concerning organizational performance and productivity.

IMPORTANCE OF A COLLEGE EDUCATION

Improving the educational level of police officers, supervisors, and managers is an integral component of the attempt to upgrade the level and quality of police services. Significant strides have been made since the 1970s to provide educational opportunities and incentives for police officers in this country and the results have been encouraging. There is no question that we have today a better trained and educated corps of police officers than at any other time in our history.

There are several perceived advantages for increasing the educational level of police officers, including the following:[9]

1. We assume that college-educated police officers will bring to the job a greater sensitivity to the social aspects of their role and a greater understanding of the human issues with which they will deal.
2. College-educated police officers will have a greater understanding of and respect for the legal constraints imposed upon them in protecting the civil liberties of all people.
3. College-educated police officers will enter the agency with a better understanding of and appreciation for the complex tasks of managing and administering a police agency, and will eventually rise to assume executive positions within the police agency.

[9]For an expanded discussion of the presumed advantages of advanced education for police officers, see Robert Sheehan and Gary W. Cordner, *Introduction to Police Administration*, 2nd ed. (Cincinnati, Ohio: Anderson Publishing Co., 1989), p. 532.

4. As police officers become more sensitive to community concerns, the public will develop greater respect for the police and there will be less friction between the police and community groups.
5. Advanced education for police officers will enhance the attempt to professionalize the police.

Despite these presumed advantages, though, we have yet been unable to establish a solid empirical link between education and job performance. There is, in other words, no assurance that increased education among police officers increases their level or quality of performance, at least as measured in the more conventional sense. Indeed, there are some, although admittedly not many, who argue that increasing the educational level of police officers is not only expensive and unnecessary, but may actually be counter-productive. One authority suggested three reasons for not requiring police officers to have a college education:[10]

1. Police officers should be representative of the entire community, including its least educated elements.
2. The police department should serve as a ladder into the ranks of the middle class for minorities and other disadvantaged groups in society.
3. The nature of police work is often tedious, routine, and boring and not challenging enough for a person with a college education.

Nevertheless, there is a general consensus that increased education for police officers is desirable and it is likely that this trend will continue well into the future.

PROFESSIONAL DEVELOPMENT

The proper training of employees, at all levels of the organization, is a fundamental, yet often neglected, responsibility of top management. Moreover, it is a responsibility that must be shared with middle management and supervisory personnel, for they too are accountable for the effectiveness and performance of their perspective units and personnel. Proper training is essential to continued organizational change and personal development. Sloppy, unprofessional police work can often be attributed directly to the lack of a modern and energetic program of training and personnel development.

Nowhere in the police organization is adequate training more important than in the patrol force, where the complexity and sensitivity of duties and responsibilities requires the highest level of performance and professional conduct. The remainder of this chapter examines the role of training in the development of personnel and

[10]Robert M. Fogelson, "Reform at a Standstill," in Carl B. Klockars, *Thinking About Police: Contemporary Readings* (New York: McGraw-Hill Book Co., 1983), p. 115.

enhancement of patrol operations. It also provides guidelines that can be used in the planning, development, and administration of police training programs. With minimum modification, these guidelines can be applied to nearly all police agencies, regardless of size or financial resources.

Organizational Commitment and Philosophy

A strong commitment by the police executive and by local public officials is absolutely necessary if an effective police training program is to be maintained. It is a sad commentary on the state of thinking in some jurisdictions that the police training budget is often one of the first elements to suffer when appropriations are to be reduced. One reason for this is that the results of training programs are often difficult to measure. Also the traditional view of training is that it is a luxury rather than a necessity, a view that can no longer be supported in modern police organizations. A modern, comprehensive training program aimed at sharpening the skills of all employees and preparing them to deal with emerging problems is not only desirable but necessary in contemporary society.

> Unfortunately, although training has been given a great deal of lip service by law enforcement executives, it is often relegated to a position of secondary importance because it is expensive, time-consuming, complex, inconvenient, and difficult to coordinate. But it is an essential element of the police system.[11]

It is important that top police executives and public officials make a sound program of continued training and personnel development a priority item when they formulate police budgets. The police administrator must put full support behind an aggressive training program and must ensure that all employees participate in the program. The police administrator should also emphasize the importance of the training program by assigning responsibility for training and personnel development to competent and dedicated individuals within the department. Moreover, all members of the department must be made aware of the need for continued training and its role in their own future in the department.

It is also important that public officials, particularly those responsible for approving the police budget, understand the importance of a strong police training program and provide adequate funds to support comprehensive in-service and specialized training programs. Public officials must also be cognizant of the possibility of civil suits from persons claiming mistreatment or abuse as a result of improperly or inadequately trained police officers. It should be remembered that, as pointed out previously, the failure to train is one of the greatest liabilities facing a police department in the event of a civil suit against an officer or the department.

[11]William J. Bopp, *Police Personnel Administration* (Boston: Holbrook Press, 1974), p. 177.

Role of Police Training in Organizational Development

The role of training in police organizations has changed dramatically in recent years, although many agencies still have a long way to go in this regard. In many progressive police agencies, training is no longer considered a luxury, but is viewed as a fundamental part of organizational life. In addition, the focus of many police training programs is beginning to shift from an emphasis on simplified procedures, techniques, and processes to a more humanistic and theoretical approach to the delivery of police services.

The police operate in a dynamic environment characterized by a high degree of technological, cultural, political, and environmental change. The rapidly evolving nature of society places great demands on the police to keep pace. It is no longer acceptable for the police to follow methods and procedures that have developed slowly over time and that are based primarily on custom and tradition. Change is a fundamental and important characteristic of the operational environment of the police.

The rapid growth of new technologies, the automation of information systems, changing public expectations of the police, and the introduction of entirely new concepts of management have all combined to underscore the need for the orderly management of change in police organizations. The development of a sound training program not only helps to ensure that needed new skills are acquired, but also enables the police manager to control effectively the organization's progress toward the attainment of stated goals. Training is an important management tool and can be utilized to realize the full effectiveness of the organization's most valuable resource: its people.

> A well-conceived training plan is the necessary ingredient for helping police staffs consider the overall concept of change as well as the specific changes necessary in a particular police organization. It helps the police personnel toward a change in attitudes and practices...and in providing more effective police services.[12]

Training and Personnel Development

As noted previously, the most important resource in a police agency is its personnel. People make the organization work or falter; they determine the degree to which organizational goals are achieved. In the police organization, the individual police officer determines the quality and effectiveness of police services. More importantly, the attitudes, behavior, and performance of people can be changed and improved through proper training.

It is the responsibility of the police manager to develop within each member of the agency the ability to perform at his or her maximum potential. Training is one way in which this potential can be realized. Byers indicates the importance of developing individual skills and abilities through a sound training program.

[12]George T. Felkenes, *Effective Police Supervision: A Behavioral Approach* (San Jose, Calif.: Justice Systems Development, Inc., 1977), p. 200.

> If people are the potential greatest asset to an organization, and if it is true (as I am convinced it is) that people in our organization are what really makes the difference, then the development of people and the creation of organizational conditions for full utilization of their developed talents should be of the highest priority and concern to the governing body and the top management of a public organization. This will be their most effective means ultimately to solve or ameliorate their problems and achieve more and better results—even within the limits of scarce fiscal resources.[13]

Establishing Minimum Training Standards

One method of incorporating a progressive training philosophy into the operational mainstream of a police agency is through the development and establishment of minimum training standards for various positions and assignments within the agency. In most states, for example, training commissions, such as the Commission on Peace Officer Standards and Training (POST) in California, have been established to develop minimum entrance-level training requirements for police work. These commissions also assist in the development of police training programs for law enforcement agencies. In most large cities, police departments have developed their own training programs that meet or exceed state minimum standards.

In recent years, several states have enacted legislation providing for mandatory annual in-service training, usually of 40 hours or more. Some states, such as Florida, have adopted incentive programs designed to compensate officers for achieving higher skill levels through advanced training courses in patrol, supervision, and middle management. Minnesota, on the other hand, has developed a very comprehensive and progressive police officer certification program, similar to that implemented in California several years ago, that provides minimum training and education levels for police officers.

It would seem that, if advanced training is to become useful in developing the professional skills of police personnel, minimum standards for advancement and specialized assignment would be both desirable and necessary. With few exceptions, these standards are not set. Some police agencies, however, have incorporated minimum training standards into their promotional policies. In California cities, for example, advancement to supervisory and middle-management positions usually requires the possession of either an intermediate or advanced POST (Police Officer Standards and Training) certificate, which is acquired through a combination of training, education, and time in service. Such policies are a good beginning, but they do not go far enough.

As indicated in Chapter 11, many police officers are appointed to supervisory positions without adequate preparation for their new assignments and responsibilities. A course in police supervision, including a minimum of 80 hours of instruction in such subjects as planning, resource utilization, decision making, and performance

[13]Kenneth T. Byers, ed., *Employee Training and Development in the Public Sector* (Chicago: International Personnel Management Association, 1974), p. 9.

evaluation should be required for all supervisory personnel. Similar programs ought to be developed for personnel assigned to specialized duties. Investigators, for example, should be provided advanced training in such subjects as interviewing techniques, criminal procedure, and case preparation. If possible, these programs should be provided before the individual is assigned to a specialized position within the agency. At the very least, satisfactory completion of such courses of instruction should be required within a specified period after assignment. Moreover, the training should be provided at no cost to the individual.

There can be no universal training standards for advancement and assignment that will apply equally to all police agencies, since the needs of each department are unique. Such standards should be developed around and reflect the actual needs of the individual department. State training commissions, however, can play an important role in the development of supervisory, management, and specialized training programs and should provide such training to local agencies without charge. It should be the responsibility of the local department, however, to determine its own minimum training standards and to ensure that those training standards are observed during the advancement and assignment of personnel.

Responsibility for Training

The ultimate responsibility for personnel training must rest with the police executive. As indicated earlier, however, this responsibility often is, and should be, shared by management and supervisory personnel within the organization. At the very least, management and supervisory personnel should be able to identify training needs in their own personnel, assist in the development of the training program, and serve as resources in the conduct of training. Nevertheless, it is important that the responsibility for developing and managing the department's training program be assigned to an individual or unit within the agency. In smaller departments with less than 50 full-time employees, this responsibility may be assigned as an additional duty, to be performed on a part-time basis. In larger departments, training may be assigned to a staff member who is also responsible for related personnel functions such as recruitment, or background investigations.

In departments having more than 100 full-time employees, the responsibility for training should be assigned to an individual who can devote full time to this effort. He or she may either be a senior staff officer or a well-qualified civilian. Prior police experience, while not mandatory, would be a useful qualification for the person assigned as training director. In many large cities, the police training director is a civilian who is thoroughly familiar with the police function as well as with training techniques.[14]

The number of people assigned to the training function will, of course, increase with the size of the agency. In addition, whether the department is responsible for maintaining its own training academy, participates in a regional police academy or

[14]Edward J. Tully, "The Training Director," *FBI Law Enforcement Bulletin,* 48 (July 1979), pp. 2-4.

has some other arrangement will also affect the size of the training staff. Even a department that relies heavily on other agencies for its training needs, however, should have someone in the agency responsible for coordinating training efforts.

The role of the individual supervisor in the training of personnel (see Chapter 11) is worth repeating. Even though the responsibility of the supervisor in the training and development of subordinates may be only an informal one, it is no less important. As Gammage has observed:

> Let's keep in mind that a recognized component part of supervision is that of guiding and assisting employees to learn how to do their jobs. Supervisors must retain at least a share, and generally a large share, of the responsibility for the instruction of their subordinates.[15]

FUNCTIONS OF THE TRAINING DIRECTOR

Once the responsibility for developing and conducting the department's training program has been established, the training director (or training officer) should consider several tasks that are essential to an effective program. At a minimum, the training director should be responsible for the following.[16]

Development of Training Philosophy, Policy, and Procedures

The training philosophy of the police organization should be clearly stated in the form of written policy to ensure that employees recognize the importance of the training program to their own personal development and professional advancement. The training director, in conjunction with other staff and command officers, should participate in the development of training policies and procedures designed to translate the philosophy and goals of the training program into reality.

Assessment of Training Needs

To be most effective, training must be carefully planned and executed. The development of a sound training program must be based upon a thorough assessment of individual and organizational training requirements. In fulfilling this responsibility, the training director should rely on the expertise of other management and supervisory personnel, as discussed later in this chapter.

[15] Allen Z. Gammage, *Police Training in the United States* (Springfield, Ill.: Charles C. Thomas, 1963), p. 122.

[16] Calvin P. Otto and Rolling O. Glaser, *The Management of Training: A Handbook for Training and Development of Personnel* (Reading, Mass.: Addison-Wesley Publishing Company, 1970), p. 5.

Preparation of Training Budget

The training director should also be responsible for developing a training budget that accurately reflects both the long-term and the short-term training goals of the organization. The budget should be used as a tool for planning the goals of the training program, the methods and procedures to be used, and the probable costs involved.

Administration

The training director should be given full responsibility for administering and controlling all aspects of the agency's training program and for coordinating all training activities. The effectiveness of the training program will be seriously weakened if this responsibility is fragmented throughout the organization. As a staff officer, the training director may have little involvement in command decisions, but he or she should be able to cut through organizational boundaries and overcome personnel problems and organizational conflict in order to assure that training goals are achieved.

Training Goals

The identification of training goals should be the result of a systematic program of training planning and program development. Training goals, both organizational and individual, should be based upon identified training needs and should be stated in terms of desired levels of performance such as improved skills in traffic-accident investigation, human relations, and patrol procedures.[17]

Research

The training director should be responsible for conducting research to ensure that the department's training program incorporates the most advanced training methods, materials, and subject content. The training director should be cognizant of the latest developments in training methods and learning theory, and incorporate this information into the training program.

Supervision

In larger departments, the training director may have one or more part-time or full-time staff to supervise. In addition, outside consultants and instructors may be employed in the training program. The training director should be responsible for supervising the performance of all personnel involved in the training program and for scheduling and coordinating their activities.

[17]Robert F. Mager, *Preparing Objectives for Programmed Instruction* (San Francisco: Fearon Publishers, 1962).

Development of Training Resources

Consistent with budgetary limitations, the training director should be responsible for developing and acquiring all required training resources, including both personnel and equipment. In small police agencies, the training director should be encouraged to develop as many internal resources as possible to assist in the training program.[18] In addition, there are a number of outside resources such as regional police academies, state training commissions, local colleges and universities, as well as representatives from other law enforcement and criminal justice agencies that can be utilized in the training program. Such resources are sometimes invaluable and can often be obtained with a minimum of expense and difficulty.

Scheduling

The training director should also be responsible for the development of department training schedules. In this respect, the training director should work closely with command personnel to ensure that training requirements do not interfere with routine operations of the department. Where possible, training programs should be arranged on a staggered basis (that is, different shifts or days of the week) in order to accommodate as many personnel as possible without interrupting their regular tours of duty.

Evaluation of Training Results

In order to achieve a maximum degree of agency acceptance and to obtain the desired results in terms of improved individual performance, the training program should be carefully evaluated to determine if it is accomplishing its intended purpose. Evaluation results should be carefully studied by the training staff in order to identify areas in which change or improvement is required. The evaluation of training results is discussed in more detail later in this chapter.

PLANNING THE TRAINING PROGRAM

Like other organized activities, the development of an effective training program requires sound and careful planning.[19] Planning is a systematic, continuing process, and the planning of a training program involves a number of tasks, which are shown in Figure 13.1 and described in the following pages.

[18]Gammage, *Police Training,* p. 121.

[19]Otto and Glaser, *Management of Training,* pp. 21-22.

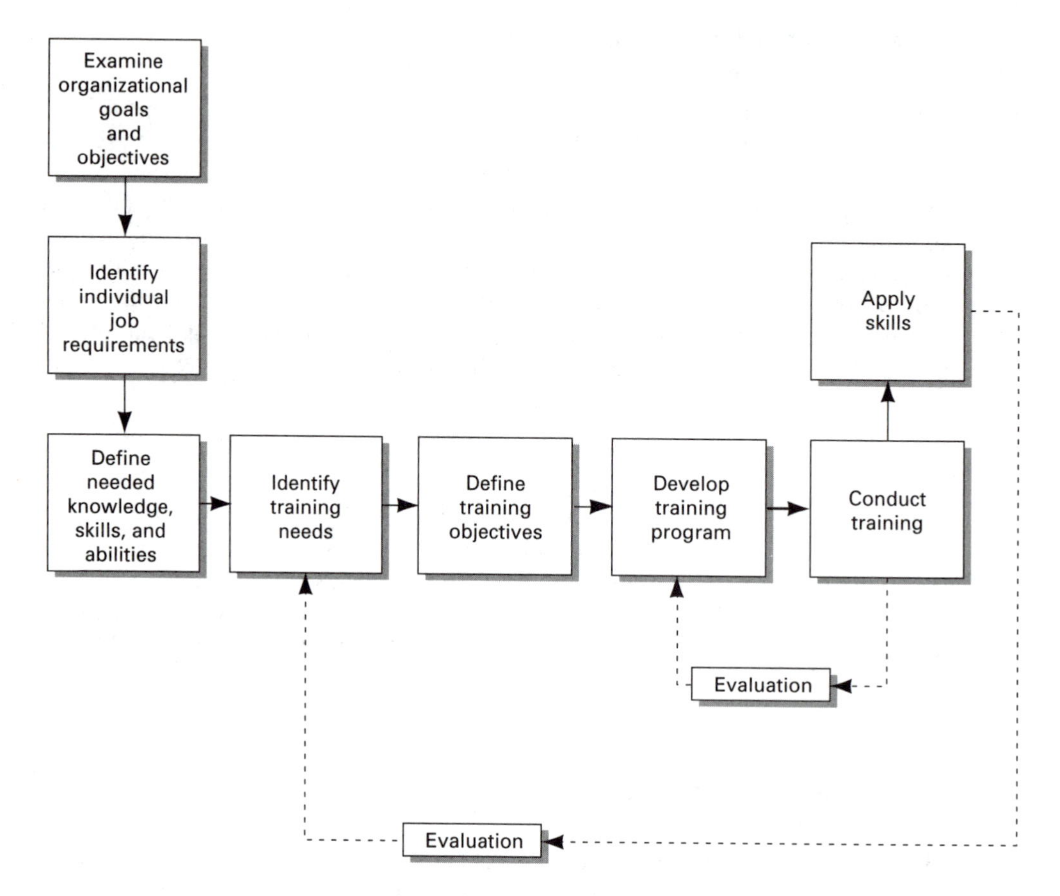

Figure 13.1 Training development and evaluation model

Examining the Organization and its Objectives

Each police department must have a training program that is specifically designed to meet its own unique requirements. In planning the training program, the training staff must clearly understand the primary goals and objectives of the organization and its various units. With this information in hand, training needs can be identified and responsive training programs can be developed.

Evaluating Current Training Programs

As noted previously, current training programs should be continually evaluated and updated. As the needs of the agency change, the scope, methods, and content of the training program should also change. A survey of members of the department, in addition to interviews with key command and supervisory personnel, can yield a great deal of useful information in assessing the value of current training methods and programs.

Identifying Key Organizational Problems

While not all problems concerning organizational effectiveness can be corrected by training, many can be. The training program should be designed to correct as many of these problems as possible, especially those relating to individual performance and job skills.[20] For example, if it is determined that a number of criminal cases are being rejected by the local prosecutor due to poor case preparation, a logical solution might be the development of training programs designed to sharpen the investigative procedures and report-preparation skills of those individuals responsible for preparing cases for prosecution. Similarly, a rise in the number of citizen complaints against the police could indicate the need for improved training in police-community relations or the need for sensitivity training for some officers.

Assessing Training Capabilities

In order to develop a comprehensive training plan, the training director must know what resources are available to the department to support the training program. This assessment should include both personnel and material resources within the department as well as those available from outside sources such as other agencies. Where deficiencies are identified, attempts should be made to correct them. This may require the acquisition of additional training material and equipment, or the addition of staff members with expertise in particular areas.

When selecting personnel to instruct in training courses, it is important that these individuals not only be technically qualified in the subject matter, but also highly motivated to teach others. Members of the department who are simply drafted as instructors because of their technical skills may be very poor instructors and may lack the necessary enthusiasm to motivate others to achieve technical competency.

Preparing the Training Plan

The training plan should be a written document that outlines the goals and objectives of the training program, as well as milestones, methods, resources, and anticipated costs. The training plan should include both long-term and short-term training goals and should be thoroughly reviewed by the command staff of the department prior to its final approval and implementation. Wherever possible, training goals and objectives should be stated in terms of desired levels of performance to give meaning to them.

Establishing Performance Objectives

In order to fully evaluate the training program, planners should establish performance objectives. These objectives should be designed to measure objectively individual or

[20]Robert F. Mager and Peter Pipe, *Analyzing Performance Problems or 'You Really Oughta Wanna'* (Belmont, Calif.: Fearon Publishers/Lear Sigler, Inc., 1973).

organizational performance before and after the training program. Such performance objectives will assist the training staff, supervisory officers, and command personnel in determining what effect, if any, the training program has had on performance within the department. Have police reports improved in terms of their clarity and completeness? Are arrest procedures more thorough? Have the attitudes of officers toward citizens improved? Have personnel complaints decreased? These are all indicators that may be used to measure the effectiveness of the training program.

DETERMINING TRAINING NEEDS

To be most effective, the development of a sound and comprehensive training program must begin with the identification of training needs. Too often, training programs are judged by the number of hours of training provided, instead of whether they actually meet the needs of the department. The process of identifying training needs should proceed on both an organizational and an individual basis. On the organizational level, the goals, activities, and procedures of the department must be examined. Organizational deficiencies should be identified as described previously. On the individual level, specific performance weaknesses such as an inability to investigate traffic accidents properly must be identified.

To a great extent, organizational deficiencies and weaknesses in individual performance may be related. The key to identifying training needs, however, is through the analysis of individual performance. There are a number of informal as well as systematic ways in which the training needs of individual members of the department can be identified.

Performance Appraisal

Most police agencies conduct periodic performance evaluations of one type or another. Some of the problems associated with the system of performance appraisal were discussed in Chapter 11. To reiterate, one of the primary purposes of performance appraisal should be to identify training needs.[21] This purpose should be kept in mind when designing the performance appraisal system, and the results of the appraisal of individual performance should always be examined in terms of improvements in the training program.

Supervisory Evaluation

Another method of identifying training needs is the personal interview, in which individual supervisors are requested to comment on the value of existing or proposed training programs. If personal interviews are not possible, such information may be

[21]Felix Lopez, *Evaluating Employee Performance* (Chicago: Public Personnel Association, 1968)..

obtained through the use of questionnaires by some or all supervisory personnel. Questions should be quite specific. It may be advisable to list a number of possible training subjects (for example, report writing, crime scene investigation) and ask respondents to rate these subjects in terms of their importance to the patrol function as a means of linking them to individual performance.

Since supervisors are in the best position to evaluate the performance of their subordinates, it is important that they play an active role in formulating and prioritizing training goals and objectives. Supervisory input can provide the training staff with invaluable information for the planning and design of the department's overall training program.

Job and Task Analysis

A more formalized method of identifying training needs is through the analysis of the actual jobs and tasks performed by members of the department. This process is usually accomplished through the completion of a form prepared especially for this purpose by the training staff in cooperation with supervisory and command personnel. In the job-and-task-analysis method, supervisors are asked to indicate the **actual** and the **desired** level of all employees in given areas of performance. The difference between **actual** and **desired** performance scores is then used to calculate training needs in various skill areas. An example of a job-and-task analysis form that has been used successfully in both the public and private sector is shown in Figure 13.2.

A separate job-and-task analysis should be used for each position classification or job specialty in the department. For example, a different form would be required for clerk typists, traffic investigators, patrol officers, and the like. Most police agencies have job descriptions that describe the tasks and requirements of a particular position or assignment within the department, and these may be useful in designing the job-and-task-analysis form for various classifications.

Employee Surveys

Another useful method of obtaining information concerning training needs is through periodic surveys of all or a sample of police employees. Some departments survey their employees annually, asking them what kind of training they believe they need in order to improve their performance on the job. It is important, though, that the survey be designed to avoid asking employees merely what kind of training they would "like to have." Rather, the survey questions should concentrate on training that is directly related to actual job performance. Patrol officers, for example, might desire to learn more about the methods and techniques used to investigate homicide cases, but this training might not have a direct impact on their performance if they rarely have the opportunity to investigate such cases. Thus, care should be taken by the training staff when designing such surveys, analyzing their

Job Classification: Patrol Officer	Job Skills Required												
Supervisor: J. C. Devilbiss *JOB SKILL CODES* 1. Limited knowledge of task. 2. Basic knowledge of task; can perform only if assisted. 3. Can perform task with difficulty and requires close supervision. 4. Performs task effectively with minimum supervision. 5. Can perform task well unless special problems are encountered. 6. Performs task completely without assistance. 0. Not a required task.	Report writing	Verbal communication	Human relations	Crime scene invest.	Traffic accident inves.	Pursuit driving	Evidence collection	Public speaking	Court demeanor	Patrol tactics	Crowd control	Arrest procedures	Defensive tactics
Grimmond, Timothy	6/5	6/4	6/6	5/5	5/5	6/4	5/5	5/4	6/5	5/4	6/5	5/5	6/4
Baumrock, Allen	6/4	6/5	6/5	5/4	5/5	6/4	5/5	5/5	6/4	5/5	6/5	5/5	6/5
Johnson, Jimmy	6/6	6/6	6/5	5/4	5/4	6/5	5/4	5/4	6/6	5/5	6/4	5/4	6/6
Lupo, Ronald	6/5	6/4	6/6	5/5	5/4	6/6	5/3	5/5	6/5	5/4	6/5	5/4	6/6
Cosgrove, Michael	6/6	6/4	6/4	5/3	5/4	6/5	5/4	5/5	6/4	5/4	6/4	5/5	6/4
Edward, Howard	6/3	6/6	6/5	5/5	5/5	6/4	5/4	5/4	6/5	5/3	6/3	5/4	6/4
Cumulative Training Discrepancy*	7	7	5	4	3	8	5	3	7	5	10	3	7

*Difference between *desired* (upper left figure) and *actual* (lower right figure) skill level

Figure 13.2 Job and Task Analysis Form [Source: Henry J. Duel, "Determining Training Needs and Writing Relevant Objectives," in Kenneth T. Byers, ed., Employee Training and Development Sector (Chicago: International Personnel Management Association, 1974), pp. 97-101.]

results, and translating the information into actual training programs. An example of a form that can be used for this purpose is shown in the Appendix (Training needs Assessment Form).

EVALUATING THE TRAINING PROGRAM

One of the reasons that training programs do not always receive the financial support they deserve is that the value of training is often difficult to measure. While most authorities will agree that additional and improved training is desirable, it is often not possible to determine with any degree of accuracy what results will be achieved by changing or strengthening the training program. Increasingly, training staffs are being called upon to justify their efforts, and the quality and content of police training programs is being subjected to increasing scrutiny.

The evaluation of a training program should be conducted on two distinct levels: an internal evaluation of the training program itself and an external evaluation focusing on the outcome or results of training efforts.[22] Typically, the internal evaluation of training should concentrate on examining changes in employee knowledge, skills, and attitudes. The external evaluation should focus on changes in behavior and work performance that may be related to the training program.

Few police training programs are evaluated in a rigorous manner. Most training evaluations, if they are conducted at all, use routine trainee evaluation forms. Participants in the training program are asked to describe their attitudes about the adequacy and relevancy of program content and the capabilities of the instructional staff. While such information is of some value, it is not sufficient to satisfy the requirements of a comprehensive program evaluation.[23]

One element that should be included in any evaluation of a training program is an assessment of the degree of learning acquired by the participants in the training program. Evaluation of learning indicates if the trainee's knowledge of the subject has increased or if certain skills have improved. In addition, it may include measures of attitudes toward specific concepts or procedures. An assessment of what has been learned in a training program is very important because changes in knowledge, skills, and attitudes can usually be linked to changes in behavior and performance.

Evaluation of learning will usually be accomplished through performance and written tests. Such tests should be conducted both before and after (and perhaps during) the training program to determine what changes have occurred, if any. Both these tests should be of like difficulty and should cover the same general subject matter, although it is not necessary that exactly the same items or questions appear on

[22]William R. Tracy, *Evaluating Training and Development Systems* (New York: American Management Association, Inc., 1968), pp. 14-23.

[23]Donald L. Kirkpatrick, "Evaluating In-House Training Programs," *Training and Development Journal* (September 1978), pp. 6-9.

both tests. Appropriate statistical tests can be applied to determine whether significant changes have occurred as a result of the training.[24]

Another element that should be included in the evaluation of training is the assessment of changes in on-the-job performance. This, however, is much more difficult to measure than either changes in attitudes or knowledge, since performance itself often cannot be translated into numerical values. Nevertheless, it is essential that some attempt be made to identify changes in behavior or performance that might be attributed to the training program, since improved job performance is the ultimate goal of the training effort.

For the most part, the effects of training on job performance can be assessed on the basis of changes either in the **quality** or the **quantity** of work performed before and after the training program. In police work, however, more emphasis should be placed on quality than on quantity, even though quality is more difficult to measure. To assist in this evaluation, it is important to have some kind of performance evaluation system that provides as much information as possible about the performance of trainees prior to the training program as well as for some time after the completion of training. Such information may be obtained in several ways, including the following:

- Documentation of increased patrol involvement in preliminary investigations as reflected by the number of hours devoted to such activities indicated on officers' daily activity logs or similar records.
- Review of randomly selected field reports to determine the quality of preliminary investigations conducted by patrol officers (that is, how much usable information is produced, whether all possible leads have been checked).
- Personal observation, usually by the immediate supervisor, of trainee work habits and performance. To assist in this effort, some kind of checklist would be desirable to assist the evaluator or observer in identifying particular performance areas that may have been affected by the training program.

There are two basic problems associated with evaluating the results of police training. The first of these, which has already been discussed, is the difficulty in measuring performance. The other problem is caused by intervening factors that may tend to influence behavior and job performance, and thus negate the effects of training. Such intervening factors might include organizational policies (for example, use of deadly force); management styles (for example, participative vs. autocratic); and operational practices (for example, shift scheduling, one-officer vs. two-officer cars). These and other factors might have an unknown effect on the behavior and performance of officers and might alter the impact of the training program.

[24]Treadway C. Parker, "Statistical Measures for Measuring Training Results," in Robert L. Craig, ed., *Training and Development Handbook,* 2nd ed. (New York: McGraw-Hill Book Company, 1976), Chapter 19.

Nevertheless, and despite the possible imperfections encountered, it is important that a concerted effort be made to determine the results of training activities. This information should be used to continually modify and improve the training program to best meet the needs of the organization and the individual.

LEVELS OF POLICE TRAINING

Police training begins with the preservice training provided to recruits in the police academy and continues through several levels in the course of an officer's career. Although the various types of training are quite different in scope and purpose, they should nevertheless be viewed as integral components of the overall police training program.

Contrary to the practice in some police agencies, the need for training does not end when the officer graduates from the police academy. The police academy should be viewed instead as the beginning of the training cycle. If it is to be effective, training should be viewed as it is by many progressive police administrators, as a continual process of professional development and advancement that literally has no end during an officer's police career.

Recruit Training

Recruit training, or preservice training, is the most common and often the most intensive type of training found in most police agencies today, since minimum standards that regulate the content and scope of entry level training for police have been in existence for some time. These standards are gradually being upgraded, and many law enforcement agencies with their own police academies exceed minimum standards established by the state.

One of the problems typically associated with recruit training has been the disparity between what is taught in the classroom and what is practiced in the field. Police recruits, upon receiving their first assignments in the police agency, are often told to forget what they learned in the academy and to follow the example set by their senior partners.[25] One of the successful ways of dealing with this problem is to interrupt the recruit's training program at various intervals with periods of field training in which the officer is taken out of the classroom and put out on the street with a training officer. This seems to have been successful in providing a necessary linkage between the theory taught in the classroom and the practical realities faced by the police officer on the job.[26]

[25]Charles B. Saunders, Jr., *Upgrading the American Police* (Washington, D.C.: The Brookings Institution, 1970), p. 130.

[26]Robert Wasserman and David Couper, "Training and Education," in O. Glenn Stahl and Richard A. Stauffenberger, eds., *Police Personnel Administration* (Washington, D.C.: Police Foundation, 1974), p. 132.

Field Training

Field training, or on-the-job training, is actually an extension of recruit training and equally as important, although its value is underestimated in some police agencies. In some cases, it is viewed merely as a breaking-in process in which the new officer is shown the ropes and introduced into the "real" world of policing. As Wasserman and Couper have observed:

> Such field training is rarely formal, often poorly conceptualized, and usually pitifully meager. So-called field-training officers typically have no special training, nor do they demonstrate any particular skills. Field training experiences are very short; and trainees are given little if any opportunity to reflect on, discuss, and understand their experiences.[27]

Some police agencies, however, have attempted to formalize their field training programs to a considerable degree in order to improve the job performance of new officers. The San Jose, California Police Department is credited with having initiated the first field training program in 1972, and many of those in existence today are said to be modeled after the San Jose program.

A survey commissioned by the National Institute of Justice to determine the state of the art of field training programs in the United States revealed that 64 percent of those responding to the survey reported having some kind of field training program in effect. Of these, about two-thirds are less than 10 years old.[28]

One of the many advantages of field training officer programs is that they seem to have the effect of reducing the likelihood of civil lawsuits against the police. This would seem to indicate that field training officer programs enhance the job preparedness of officers who are then less likely to make critical errors which might result in civil actions against the force.

Proper selection and training of the field training officer (FTO) is critical to the success of the program. Persons assigned as field training officers should be highly motivated and well compensated, since they are being entrusted with the future of the police agency. Their performance will impact directly on the performance of the agency in the years ahead. Serious consideration should be given to providing extra compensation to persons assigned to this important duty.

> Agencies should consider offering extra compensation to ensure that the most qualified persons are attracted to and perform in the position of FTO.[29]

Field training officer programs should not be static, but rather should evolve and adapt as new conditions arise. Any good program should be evaluated at least

[27]*Ibid.*, p. 128.

[28]Michael S. McCampbell, "Field Training for Police Officers: State of the Art," National Institute of Justice, *Research in Brief* (November 1986), p. 2.

[29]*Ibid.*, p. 7.

annually to make sure that it continues to meet its objectives and that it is performing as efficiently as possible.

A key to the success of any field-officer-training program is regular and systematic evaluation of the new officer. This evaluation should take place on a daily basis, with formal evaluations conducted each week. A well-designed evaluation form, such as that shown in Figure 13.3, is essential to this process.

San Diego Police Department
Daily Trainee Evaluation Form

Trainee ______________________ FTO ______________ Date __________

Phase No. __________ No of days in phase ______

Rating instructions: The evaluation should follow the below listed scale. You are encouraged to comment on any behavior you wish. A specific comment is required for all ratings of 1 or 5.

Scale: 1-Not Acceptable 2–4-Acceptable 5-Superior

PERFORMANCE						
1. Report writing	1	2	3	4	5	Not observed
2. Driving skill	1	2	3	4	5	Not observed
3. Self-initiated activity	1	2	3	4	5	Not observed
4. Use of map book	1	2	3	4	5	Not observed
5. Stress control	1	2	3	4	5	Not observed
6. Officer safety	1	2	3	4	5	Not observed
7. Good judgement	1	2	3	4	5	Not observed
ATTITUDE						
1. Attitude toward work	1	2	3	4	5	Not observed
2. Acceptance of criticism	1	2	3	4	5	Not observed
3. Attitude toward public	1	2	3	4	5	Not observed
4. Attitude toward peers	1	2	3	4	5	Not observed
KNOWLEDGE						
1. Department guidelines	1	2	3	4	5	Not observed
2. Penal code	1	2	3	4	5	Not observed
3. Vehicle code	1	2	3	4	5	Not observed
4. Elements of crime	1	2	3	4	5	Not observed
APPEARANCE						
1. General appearance	1	2	3	4	5	Not observed
2. Readiness for duty	1	2	3	4	5	Not observed
PHYSICAL FITNESS						
1. Strength and agility	1	2	3	4	5	Not observed

ADDITIONAL COMMENTS: Use reverse side

Figure 13.3 San Diego Police Department Daily Trainee Evaluation Form

This procedure ensures that the recruit receives immediate feedback and thus learns more quickly. Second, daily evaluation ensures that the FTO remembers how the recruit performed in a specific situation.[30]

In-Service Training

In-service training is perhaps one of the least understood and most confusing of all types of police training. In-service training, for example, is often interpreted to include all training conducted upon completion of the officer's recruit training. In the present context, however, **in-service training** is defined as training conducted by the individual police agency on a regular basis. This would typically include roll-call training, firearms training, special training programs, and the like.

Perhaps the most common form of in-service training, and often the most inefficient in terms of quality, is the form of roll-call training conducted by many police agencies. Roll-call training is typically conducted a few minutes prior to going on duty and is often limited to reading crime bulletins, discussing various shift activities, and general socializing. This is hardly suitable as training but is often the primary form of what some police administrators call their in-service training programs.

A true in-service training program should include books of instruction and be conducted in formal sessions of no less than 30 minutes each. Such instruction may be provided in conjunction with regular field assignments or be held during off-duty hours, in which case overtime pay or compensatory time off should be provided. A minimum of 40 hours of in-service training per officer per year has been recommended, although many police agencies fall far short of this goal.[31]

An excellent training medium that is growing more and more popular is the videotape. Training videotapes are relatively inexpensive and are available from a number of commercial sources. Many progressive police agencies have developed their own videotapes to cover practical topics such as felony traffic stops, field interrogation techniques, crisis intervention, juvenile intervention methods, barricaded-hostage procedures, and others. Training Videotapes can be played before or after roll call or during brief training sessions while officers are on duty.

Another valuable supplement to the in-service training program is the relatively new Law Enforcement Television Network (LETN). This commercial enterprise provides 20-hour a day broadcast of timely and interesting training programs on a variety of important subjects. Agencies must pay an annual subscription fee, usually based on the size of the agency, and purchase a satellite receiver. This service seems to be growing in popularity and is found in hundreds of police agencies throughout the United States.

[30]McCampbell, "Field Training."

[31]National Advisory Commission on Criminal Justice Standards and Goals, *Police* (Washington, D.C.: U.S. Government Printing Office, 1973), p. 404.

Officer availability and the need to pay officers overtime rates for attending training classes on their own time are two problems encountered in trying to mandate regular in-service training in police agencies. One approach that has been taken in some departments is to schedule in-service training classes on a designated day (for example, Monday) of each week and to schedule all personnel to work on this day. Since 100 percent of the work force would be available for duty, approximately 25 percent (those who would normally be off duty on any given day) would be available for an eight-hour training program. In this way, approximately 25 percent of the members of the department can be provided with an eight-hour training course every four weeks. This cycle can be repeated indefinitely throughout the year to provide a minimum of 40 hours of in-service training each year.

It must be recognized, however, that very small departments—those employing less than 25 full-time personnel—would find it extremely difficult to schedule in-service training on this basis. In larger departments where personnel availability is less of a problem, such scheduling can work quite well.

Specialized Training

As the size of a police agency increases, it will develop a greater degree of specialization. Even in small departments, a number of specialized duties (fingerprinting, record keeping, photography) must be performed, even if only on a part-time basis. As individuals are assigned to these functions, they should be given adequate training in their new duties and responsibilities. Whenever possible, this training should consist of both classroom instruction and on-the-job training.

There are a number of training resources that can provide specialized courses of instruction for law enforcement personnel. Many state law enforcement agencies such as the state police or the highway patrol conduct various types of advanced training programs. Some federal agencies such as the Federal Bureau of Investigation and the Drug Enforcement Administration also offer specialized training programs to local police agencies. In addition, many police academies, colleges, and universities provide specialized instruction in a number of subjects. Most of this training can be obtained at little or no cost to the individual department. Such resources should be used to full advantage.

Supervisory Training

As indicated in Chapter 11, it is most desirable to provide newly appointed supervisors with advanced training in a variety of subjects that will enable them to perform their new duties more effectively. Unfortunately, while most police agencies do provide supervisory training of some kind to newly appointed supervisors, the level and quality of this training is often lacking. Rarely does it properly prepare the new supervisor to accept the challenges of supervising, motivating, and evaluating the work of subordinate officers. In all too many cases, this training is little more than advanced field training suitable for senior patrol officers.

Progressive police agencies require some level of supervisory training before an officer is promoted to a supervisory position. In a number of other agencies, such training is required within 12 or 18 months after the effective date of appointment. As a general rule, a minimum of 80 hours of classroom training should be required for newly promoted supervisory personnel. This training should include instruction in duties and responsibilities, communication principles, employee motivation, leadership, supervisory decision making, and a number of other subjects.

Management and Executive-Level Training

Police training, if it is to produce the desired results, must be a continuing process. No one in the police agency, regardless of rank or assignment, should be exempt from training. This is particularly applicable to management or executive-level personnel in whose hands the direction and fate of the agency rest. Indeed, the complex skills required of police managers today make it essential that they receive both preparatory and continuing training if they are to keep abreast of the new developments in management theory and modern administrative practices.

> Popular belief once held that successful managers were born, not made. It was assumed managers just naturally had some unique or special qualities that enabled them to succeed while others failed. However, a careful analysis of managerial functions has dispelled this notion, particularly in view of the increasingly complex nature of the duties and skills required of the manager today. While some managers are more skilled than others, it is no longer assumed that managerial success is somehow related to an inherent trait or quality. Instead, managers must be created.[32]

Police management and executive-level training must be presented on a higher level than other forms of police training. Training of management personnel should highlight the underlying philosophies of the police service as well as the social and humanistic aspects of managing organized activities. Training provided at this level is more theory and concept oriented than that provided to lower-echelon officers; the "why" of policing is stressed more than the "how to."

Except in larger cities, such training can rarely be provided by the local department and is usually obtained from other sources. A number of colleges and universities offer excellent training programs in police management.[33] In addition, the Federal Bureau of Investigation has developed an excellent training program for police executives at its national academy in Quantico, Virginia. Other organizations, such as the International Association of Chiefs of Police, the International City Management Association, the Northwestern University Traffic Institute, and the Southern Police Institute, all provide excellent training programs for police command and executive-level personnel.

[32]Charles D. Hale, *Fundamentals of Police Administration* (Boston: Holbrook Press,1977), p. 96.

[33]Hale, *Fundamentals,* p. 340-345.

MANAGING THE TRAINING PROGRAM

Once the training program has been implemented, it is necessary for the training director or staff to carefully monitor the program, evaluate results, and endeavor to continually upgrade the level and the quality of the training being offered. In order to effectively manage the training program, accurate records must be kept of the training status and future training requirements for all members of the department. Without such records, it would be very difficult to plan and organize a sound training program.

The training-records system should include individual training records on each member of the department. These records are often kept in the individual's personnel file which is not always available to the training staff. If necessary, duplicate records should be maintained in the training unit. These records should indicate the date, type, and length of training each member of the department has received since being appointed to the department. In addition, any pertinent training received prior to appointment, such as in other law enforcement agencies or the military, might also be included. College transcripts and training certificates should be included in the training file as well.

Another important element of the training-record system is the training inventory file, which indicates the specific training needs of each member of the department. This information may be contained on five-inch by eight-inch file cards, organized alphabetically by subject matter, and listed on cross-reference cards by the name of the employee. In this way, all members of the department who may require training in a specific subject can easily be determined. These files can be readily automated by using one of several software systems currently available.

In addition to these records, it is also useful to construct some sort of visible display of the department's training record. Some police agencies have developed impressive status boards that reflect the training accomplishments of all members of the department. These charts can be displayed either in the training office or in the roll-call room and provide graphic evidence of the department's training achievements. They are also useful in planning future training programs.[34]

SUMMARY

As the American police service nears the end of the twentieth century, police educators, researchers, and police executives continue to look for new and better ways of doing things in order that we may yet achieve the goal of professionalism. Continued professional development and training is, of course, the keystone of any program having as its goal the upgrading of the police. Training is clearly the most useful tool

[34]George P. Tielsch, "Planning Training," *The Police Chief* (July 1978), pp. 28-29.

available to the police manager. While law enforcement has lagged behind other professions in terms of its training efforts, progress is gradually being made to correct this deficiency. Training is now acknowledged to be a vital necessity rather than a luxury, and training budgets are being increased to reflect this realization. If police training programs are to realize their intended purposes, however, they must concentrate not merely on quantity but on quality as well. To accomplish this goal, training programs in police agencies must be carefully planned and continually evaluated.

REVIEW QUESTIONS

1. What is one important reason that police training is not often given the financial support it deserves?
2. Explain the role of police training in organizational development.
3. What would be one important advantage of having minimum training standards for advancement and specialized assignment?
4. Why is it not practical to develop universal training standards for all police agencies?
5. List some of the training resources that may be available to a small police department.
6. What is the primary purpose of evaluating the training program?
7. Describe three methods for determining training needs. Can you think of others?
8. Why is it difficult to measure the effects of training on job performance?
9. Why is the emphasis in many police departments focused primarily on preservice training?
10. Describe one useful method of recruit training which coordinates classroom instruction with actual experience.
11. What is a basic weakness in many police in-service training programs?
12. Describe some of the records that should be maintained by the training unit.

REVIEW EXERCISES

The following exercises are designed to reinforce or amplify comprehension of the material contained in this chapter.

1. Conduct library research on the subject of the accreditation. Compare the process used by hospitals and colleges with that used by law enforcement agen-

cies. Prepare a brief paper summarizing the results of your research and any conclusions you have drawn.

2. Conduct a literature review on the subject of police productivity. Prepare a paper outlining several basic ways in which police agencies can attempt to increase productivity.
3. As a class project, contact several police agencies in the area to obtain information about their in-service training programs. Find out what type of training is provided, how much money is devoted to training, and the source of training provided (local college or university, police academy, private training organization, etc.). Discuss the results of the research in class. What common elements can be found? What are the characteristics of the more sophisticated programs?

REFERENCES

Alpert, Geoffrey P. and Lorie A. Fridell, *Police Vehicles and Firearms: Instruments of Deadly Force*. Prospect Heights, Ill.: Waveland Press, 1992, pp. 10-11.

Bopp, William J., *Police Personnel Administration*. Boston: Holbrook Press, 1974.

Byers, Kenneth T., ed., *Employee Training and Development in the Public Sector*. Chicago: International Personnel Management Association, 1974.

"CALEA Adds 13 More to Accreditation Toll," *Law Enforcement News* (September 15, 1991).

Felkenes, George T., Effective *Police Supervision: A Behavioral Approach*. San Jose, Calif.: Justice Systems Development, Inc., 1977.

Fogelson, Robert M., "Reform at a Standstill," in Carl B.Klockars, *Thinking About Police: Contemporary Readings*. New York: McGraw-Hill Book Co., 1983.

Gammage, Allen Z., *Police Training in the United States*. Springfield, Ill.: Charles C. Thomas, 1963.

Greenberg, Sheldon, "Police Accreditation," in Dennis Jay Kenney, ed., *Police and Policing: Contemporary Issues*. New York: Praeger, 1989. pp. 247-256.

Hale, Charles D., *Fundamentals of Police Administration*. Boston: Holbrook Press, 1977.

Hartman, Thadeus L., "Field Training Officer (FTO): The Fairfax County Experience," *FBI Law Enforcement Bulletin* (April 1979), pp. 22-25.

Kelley, John, "Negotiating the Accreditation Process," in *The Police Chief* (June 1991), pp. 25-29.

Kirkpatrick, Donald L., "Evaluating In-House Training Programs," *Training and Development Journal* (September 1978).

Leonard, V. A. and Harry W. More, *Police Organization and Management,* (7th ed.) Mineola, New York: The Foundation Press, Inc., 1987.

Lopez, Felix, *Evaluating Employee Performance* Chicago: Public Personnel Association, 1968.

Mager, Robert F., *Preparing Objectives for Programmed Instruction.* San Francisco: Fearon Publishers, 1962.

Mager, Robert F, and Peter Pipe, *Analyzing Performance Problems or 'You Really Oughta Wanna'*. Belmont, Calif.: Fearon Publishers/Lear Sigler, Inc., 1973.

McCampbell, Michael S., "Field Training for Police Officers: State of the Art," National Institute of Justice, *Research in Brief* (November 1986).

National Advisory Commission on Criminal Justice Standards and Goals, *Police*. Washington, D.C.: U.S. Government Printing Office, 1973.

Otto, Calvin P., and Rollin O. Glaser, *The Management of Training: A Handbook for Training and Development Personnel*. Reading, Mass.: Addison-Wesley Publishing Company, 1970.

Parker, Treadway C., "Statistical Measures for Measuring Training Results," in Robert L. Craig, ed., *Training and Development Handbook,* (2nd ed.). New York: McGraw-Hill Book Company, 1976.

Saunders, Charles B., Jr., *Upgrading the American Police*. Washington D.C.: The Brookings Institution, 1970.

Sheehan, Robert, and Gary W. Cordner, *Introduction to Police Administration* (2nd ed.). Cincinnati, Ohio: Anderson Publishing Co., 1989.

Tielsch, George P., "Planning Training," *The Police Chief* (July 1978), pp. 28-29.

Tully, Edward J., "The Training Director," *FBI Law Enforcement Bulletin,* 48 (July 1979).

Tracy, William R., *Evaluating Training and Development Systems*. New York: American Management Association, Inc., 1968.

Wasserman, Robert, and David Couper, "Training and Education," in O. Glenn Stahl and Richard A. Stauffenberger, eds., Police *Personnel Administration.* Washington D.C.: Police Foundation, 1974.

APPENDIX A

Law Enforcement Code of Ethics

As a law enforcement officer, my fundamental duty is to serve the community; to safeguard lives and property; to protect the innocent against deception, the weak against oppression or intimidation, and the peaceful against violence and disorder; and to respect the constitutional rights of all to liberty, equality, and justice.

I will keep my private life unsullied as an example to all and will behave in a manner which does not bring discredit to me or my agency. I will maintain courageous calm in the face of danger, scorn, or ridicule; develop self-restraining; and be constantly mindful of the welfare of others. Honest in thought and deed, I will be exemplary in obeying the law and the rules and regulations of my department. Whatever I see or hear of a confidential nature or is confided to me in my official capacity will be kept ever secret unless revelation is necessary in the performance of my duty.

I will never act officiously or permit personal feelings, prejudices, political beliefs, aspirations, animosities, or friendships to influence my decisions. With no compromise for crime and with relentless prosecution of criminals, I will enforce the law courteously and appropriately without fear or favor, malice or ill will, never employing unnecessary violence and never accepting gratuities.

I recognize the badge of my office as a symbol of public faith, and I accept it as a public trust to be held so long as I am true to the ethics of the police service. I will never engage in acts of corruption or bribery, nor will I condone such acts by other police officers. I will cooperate with all legally authorized agencies and their representatives in the pursuit of justice.

I know that I alone am responsible for my own standard of professional performance and will take every reasonable opportunity to enhance and improve my level of knowledge and competence.

I will constantly strive to achieve these objectives and ideals, dedicating myself before God to my chosen profession—law enforcement.

APPENDIX B

Police Pursuit Policy

Police Department
St. Charles, Illinois

General Order	Standard Operating Procedure
Number: 87-06	Code: E-1
Revised/Effective:	April 23, 1987
Indexed As:	Emergency Driving
	Pursuit Driving
	Vehicle Use During Emergencies
	Road Blocks

Policy:

It is the policy of the St.Charles Police Department to insure that emergency operation of Police vehicles is done in a manner that provides a reasonable amount of safety for the general public and the Police officers involved.

Purpose:

The purpose of this Order is to establish Departmental policy in the operation of Police vehicles in response to emergency assignments/pursuits.

The emergency operation of Police vehicles is one of the most dangerous tasks Police officers are asked to perform. Death or permanent injury to Police officers and/or citizens can result. Lawsuits against Police officers involved in emergency assignment/pursuit driving are second only to use of deadly force.

Officers will, at all times, consider the external factors which may have a bearing on the emergency operation of the Police vehicle, including the time of day, road and traffic conditions, weather, speeds involved, nature of the incident and their personal ability to control the vehicle.

Definition:

A. *Routine Response Driving*:

1. Officers employing routine responses will utilize the normal traffic flow, obeying all traffic patterns, signs and signals.
2. Examples of routine response incidents are:
   a. Service calls
   b. Discovered burglary w/no perpetrator present.

B. *Pursuit Driving:* An active attempt by a law enforcement officer operating a motor vehicle and utilizing simultaneously all emergency equipment to apprehend one or more occupants of another moving vehicle, when the driver of the fleeing vehicle is aware of that attempt and is resisting apprehension by maintaining or increasing his speed, ignoring the officer, or attempting to elude the officer while driving at speeds in excess of a legal speed limit.

C. *Emergency Driving*: Driving a Police vehicle while responding to the scene of situations involving a forcible felony in progress, officer needs assistance, explosion, hazardous materials incidents and personal injury traffic accidents.

Procedure:

I. *Emergency Driving*

A. The primary concern of personnel assigned to emergency calls is to respond in the most expeditious manner, without endangering the safety of themselves or other persons. An assigned officer is of no assistance to anyone if he does not arrive at the scene of the emergency.

B. Officers employing urgent or emergency responses will utilize emergency audio/visual equipment when tactically feasible, and may do the following with due care:

1. Park or stand irrespective of the provisions of Illinois Statute or City ordinance.
2. Proceed past a red or stop signal or stop sign, but only after slowing down as may be required and necessary for safe operation.
3. Exceed the maximum speed limit by a maximum of 20 miles per hour over the posted limit.
4. Disregard regulations governing direction of movement or turning in specific directions.

C. Personnel may respond to certain emergency calls without using an audible signal, such as robberies or burglaries in progress. Personnel responding on these "silent runs" must clearly understand that they MUST proceed with extreme caution and at reasonable speeds, so as not to endanger the life and property of others.

II. *Pursuit Driving*

A. Pursuit Initiation—Officer Responsibilities

The responsibility for the decision to pursue and the methods to be employed during pursuits rest initially with the individual officer. When making this determination, the officer shall consider the following factors:

1. Nature of the crime.
2. Time of day.
3. Volume of vehicular and/or pedestrian traffic.
4. Location and geographical area that the pursuit will occur in or extend into.
5. Weather conditions.
6. Road conditions.
7. Speeds involved.
8. Vehicle condition.
9. Whether there is a real or apparent emergency.

B. Officers will not pursue subjects that are involved in ordinance violations or traffic misdemeanors unless there is a clear and present danger to the public if the subject is not apprehended.

C. A vehicle pursuit may be initiated when an officer has probable cause to believe the violator has commited a felony or nontraffic misdemeanor.

III. *Pursuit Continuation—Officer Responsibilities*

A. Officers involved in a pursuit will utilize emergency audio/visual equipment when tactically feasible, and may do the following with due care:

1. Park or stand irrespective of the provisions of Illinois Statute or City ordinance.
2. Proceed past a red or stop signal or stop sign, but only after slowing down as may be required and necessary for safe operation.

3. Disregard regulations governing direction of movement or turning in specific directions.

B. Once a pursuit is initiated, the officer shall, as soon as practicable, notify Tri-Com dispatch center and/or the Illinois State Police via I.S.P.E.R.N., relaying information such as the identity of the pursuing unit, location, direction of travel, exact reason for pursuit, description of the fleeing vehicle and other relevant information about the vehicle or occupants).

C. During a pursuit the officer will continually evaluate the internal and external factors involving the pursuit and weigh these factors as to continuing the pursuit or to terminate the pursuit.

D. A safe distance shall be maintained between both vehicles, enabling the pursuing officer to duplicate any sudden turns, stops, or maneuvers by the fleeing vehicle to lessen the possibility of a collision.

E. Pursuing officers shall not pull alongside a fleeing motorist in an attempt to force the subject into a ditch, curb, parked vehicle, or any other obstacle, nor shall any attempt be made to ram the pursued vehicle, unless absolutely necessary for the preservation of life.

F. Police units that have prisoners, witnesses, suspects, complainants, or citizens aboard shall not become engaged in pursuit situations.

G. Unmarked Police units shall not engage in a pursuit, unless the fleeing vehicle represents an immediate and direct threat to life. Whenever a marked unit becomes available to take over the pursuit, the unmarked unit will withdraw immediately from active pursuit.

H. No pursuing unit will continue pursuit if it becomes involved in a collision, unless the collision is with the vehicle being pursued and no other Police units are available to continue the pursuit.

IV. *Pursuit Termination*

A. The pursuing officer shall terminate the pursuit if any of the following events or conditions occur:

1. The emergency equipment on his Police unit ceases to function.
2. It becomes evident that the risks to life and property begin to outweigh the benefit derived from the immediate apprehension or continued pursuit of the offender.
3. Upon the order of a supervisory or command officer.
4. The suspect's identity has been established to the point that later apprehension can be accomplished, and there is no longer any need for immediate apprehension.
5. The environmental conditions indicate the futility of continued pursuit.

B. No officer will be disciplined for discontinuing a pursuit.

V. *Secondary/Assisting Unit's Responsibility*

A. Once the pursuit is engaged, assisting officers shall:

1. Use the radio (transmit) only out of absolute necessity. Assisting offi-

cers and units shall identify themselves and give the communications center a conservative estimate of time of arrival at the scene and provide a status report as soon as possible. Officers will take into consideration distance to travel, traffic and weather conditions.

2. Move into tactically advantageous position to assist with the stop of the felon.
3. Not caravan the pursuit; no more than one (1) Police unit may actively pursue with exception to specific orders from the shift supervisor. A second vehicle may follow the pursuit to act as back-up for the pursuing officer. However, safe operation is mandatory. The second vehicle may not actively pursue the suspect's vehicle unless the primary pursuit vehicle becomes disabled or relinquishes control.
4. The second car shall be spaced appropriately and drive so as to allow for ample reaction time and distance in order to reduce the possibility of officer-involved accidents.

VI. *Supervisor Responsibilities*

A. It shall be the responsibility of the supervisor in charge of the shift to review the facts given by the pursuing officer and to make an independent judgment if the pursuit should be continued.

B. Based on all information available, the ranking shift sergeant will order the termination of the pursuit if, in his opinion, the dangers created by the pursuit outweigh the need for an immediate apprehension of the offender.

C. Continuous monitoring of the pursuit's progress by the shift supervisor will be made until the pursuit ends or the pursuit is terminated, either by the officer or shift supervisor.

1. The shift supervisor will make a determination as to the number of Police units that will be assigned to the pursuit.
2. Unless otherwise directed by the shift supervisor, no more than two (2) Police vehicles will become actively involved in the pursuit. (Other officers should be alert to the pursuit progress and locations).
3. If the pursuit exits the municipal limits, the shift supervisor will make every reasonable effort to ensure that no more than two (2) units continue the pursuit, unless the shift supervisor feels that more units are needed to ensure apprehension of the fleeing vehicle and/or occupants.
4. The shift supervisor will generally remain within the limits of our municipality to ensure proper direction to patrol units not involved in the chase, unless the shift supervisor is the primary pursuing unit or the shift supervisor deems it necessary to leave the municipality, in which case the designated O.I.C. remaining in the municipality will assume command until the supervisor returns.
5. If an officer is involved in a serious situation (i.e., accident, shooting etc.) which requires the shift supervisor being present, the shift supervisor will ensure adequate manpower is left within the municipality.

6. Shift supervisor(s) will notify the respective Division Commander, at the earliest possible time, if an officer is involved in an accident, injured, or deadly force is used, reference General Order 86-06, Code F-1.
7. The shift supervisor will prepare a written report for the respective Division Commander of any emergency/pursuit situations that result in the death or injury to officers, suspects or citizens, or damage to property. The report will be submitted to the respective Division Commander before the shift supervisor ends his tour of duty.

VII. *Telecommunicator's Responsibility*

A. The Tri-Com telecommunicator shall notify the on-duty supervisor of the pursuit.
B. The telecommunicator shall serve as the controller for all messages relevant to the pursuit and shall immediately broadcast all available information relating to the pursuit on all local and IREACH channels.
C. When a pursuit has ended, the dispatcher shall broadcast the termination of the pursuit and notify surrounding jurisdictions when appropriate.

VIII. *Use of Firearms During Pursuits*

A. Officers shall not discharge a firearm at or from a moving vehicle except as the ultimate measure of self-defense or the defense of another when the offender is employing deadly force by means other than the vehicle. Departmental policy regarding deadly force shall be strictly followed.
B. Firing strictly to disable a vehicle is prohibited.
C. In every incident the officer shall take into account the location of vehicular and pedestrian traffic and the potential hazard to innocent persons.

IX. *Use of Roadblocks in Pursuit*

A. The use of a stationary road block *must* be authorized by the shift supervisor
   1. Road blocks will be used ONLY as a last resort and ONLY when all other attempts to terminate the pursuit have failed and the need to terminate the pursuit is imperative.
   2. Roadblocks for the purpose of apprehending wanted suspects or felons shall not be employed when it is *apparent* that innocent persons will be endangered
   3. Once a roadblock has been established and a Departmental vehicle has been stationed as part of a roadblock, no one shall remain in the vehicle.
      a. The Police vehicle(s) will display rotating lights when engaged in a roadblock and flares will be set when appropriate and time permits.
      b. Units establishing a roadblock will leave an escape route for the pursued vehicle.
      c. Units establishing a roadblock will notify the pursuing unit(s) of the location of the roadblock on the primary radio frequency utilized during the pursuit.

d. Privately-owned vehicles will not be used to establish a roadblock, except as a last resort when the need to terminate the pursuit is imperative and all other attempts have failed.

B. The use of a moving road block must be authorized by the shift supervisor. This technique will only be used if the use of deadly force would have been allowed by law to stop the offender or offenders operating the vehicle.

X. *Inter-Jurisdictional Pursuits*

Pursuits initgggtpttiated by an outside agency traveling through the City of St. Charles will be the responsibility of the initiating agency.

XI. *Reporting of Pursuits*

Officers involved in pursuits shall submit a written report through the use of the Department's Offense/Incident Report form. A separate report is not needed if the pursuit is documented by an arrest or a criminal offense When no other report is executed that includes the pursuit, a report will be made for Flee or Attempt to Elude Police Officer or Assist Other Agency.

XII. *Familiarity With State Statute*

A. All officers will be familiar with the following sections of the Illinois Vehicle Code (Chapter 95 1/2) and will adhere to their requirements:

1. 11-205 Public officers and employees to obey Act—exceptions.
2. 11-907 Operation of vehicle and streetcars on approach of authorized emergency vehicles.
3. 12-216 Operation of oscillating, rotating or flashing lights.

This Order supersedes all previous written and unwritten policies of the ST. CHARLES POLICE DEPARTMENT on the above subject.

This Directive is for Department use only and does not apply in any civil or criminal proceedings. This Directive should not be construed as a creation of a higher legal standard of safety or care in an evidentiary sense with respect to third part claims. Violations of this Directive will only form the basis for civil or criminal sanctions in a recognized judicial setting.

APPENDIX C

Training Needs Assessment Form

A questionnaire has been designed to obtain input for the Police Department's training program. The department will identify and prioritize areas of instruction, based upon information obtained through this questionnaire.

Each member of the department is requested to fill out the questionnaire. The results will be used in developing each member of the department to his or her fullest potential. The information can also be used to schedule officers for various outside schools.

Fill in your name and other information in the spaces provided below:

NAME: ______________________________ **RANK:** __________

PRESENT ASSIGNMENT: **PATROL** ☐ **INVESTIGATION** ☐ **OTHER** ☐

SUPERVISORY POSITION ☐ **NONSUPERVISORY** ☐

SECTION I

Below are listed several general areas for which instructional classes can be developed within our training program. Some of the categories are general and may include more specific topics.

Next to each subject are three responses that you can make. Place a check beneath the response that you feel is appropriate. If, for example, you feel that you need training in a particular subject in order to become a better officer, place a check under the column headed "Yes, I feel I need training in this area." Check one of the three columns for each subject.

Make your responses based on what you feel is necessary for your personal development. Another section will ask you for areas in which you are most interested. There will also be a section for supervisors to list areas in which they feel the personnel they supervise need training.

	Yes, I feel that I need training in this area.	No, I do not know if I need training in this area.	No, I do not need training in this area.
Abnormal behavior/mental illness	____	____	____
Accident investigation	____	____	____
Bombs and explosives	____	____	____
Civil rights laws	____	____	____
Community relations	____	____	____
Court testimony	____	____	____
Criminal law	____	____	____
Crime prevention	____	____	____
Defensive driving	____	____	____
Departmental procedures	____	____	____
Domestic/family problems	____	____	____
Drugs/narcotic investigation			
DWI—apprehension	____	____	____
and prosecution	____	____	____
Federal agencies/regulations	____	____	____
Fingerprinting	____	____	____
First aid	____	____	____
Firearms	____	____	____
Homicide investigation	____	____	____
Hostage negotiation	____	____	____
Hit-and-run investigation	____	____	____
Interviewing/interrogation	____	____	____

	Yes, I feel that I need training in this area.	No, I do not know if I need training in this area.	No, I do not need training in this area.
Juveniles	_____	_____	_____
Liquor regulations/enforcement	_____	_____	_____
Laws of arrest, search, seizure	_____	_____	_____
Organized crime	_____	_____	_____
Patrol techniques and tactics	_____	_____	_____
Physical fitness training	_____	_____	_____
Police radio techniques	_____	_____	_____
Photography	_____	_____	_____
Psychology	_____	_____	_____
Public relations	_____	_____	_____
Record keeping	_____	_____	_____
Report writing	_____	_____	_____
Riot training	_____	_____	_____
Self-defense	_____	_____	_____
Sexual crimes/investigation	_____	_____	_____
Stress management	_____	_____	_____
Social problems	_____	_____	_____
Supervision and management	_____	_____	_____
Stolen vehicle investigation	_____	_____	_____
Tactical weapons and maneuvers	_____	_____	_____
Traffic enforcement	_____	_____	_____
Union contract and benefits	_____	_____	_____
Others (specify)	_____	_____	_____

SECTION II

From the above categories, choose the ten areas of training that you feel are most important in your development as a police officer. List them below in order of their importance. Number one, for example, should be the most important and number ten should be the least important.

1 ______________________________ 6 ______________________________

2 ______________________________ 7 ______________________________

3 ______________________________ 8 ______________________________

4 ______________________________ 9 ______________________________

5 ______________________________ 10 ______________________________

SECTION III

List the five areas of training that would be of most personal interest to you. List them in order of interest (number one as the highest interest).

1 ______________________ 3 ______________________

2 ______________________ 4 ______________________

5 ______________________

SECTION IV
(For Supervisors)

From the list of training topics, choose the ten areas that you feel personnel under your supervision need to become better officers. List them in order of importance. Number one should be the area with the greatest needs, and so on.

1 ______________________

2 ______________________

3 ______________________

4 ______________________

5 ______________________

6 ______________________

7 ______________________

8 ______________________

9 ______________________

10 ______________________

SECTION V
(For Supervisors)

A supervisory job requires different talents and training. List, in order of importance, areas of training that you feel you need as a supervisor. These areas may not be listed in the previous list of topics.

1 ______________________

2 ______________________

3 ______________________

4 ______________________

5 ______________________

SECTION VI
(For Supervisors)

List any special supervisory courses that most interest you—in order of personal interest, beginning with number one.

1 ______________________ 3 ______________________

2 ______________________ 4 ______________________

5 ______________________

BIBLIOGRAPHY

Allen, Carole, "The Cultural/Language Barrier," *Law and Order* (April 1987), pp. 44–48.

Alpert, Geoffrey P., "Questioning Police Pursuits in Urban Areas," in Robert G. Dunham and Geoffrey P. Alpert, *Critical Issues in Policing: Contemporary Readings* (Prospect Heights, Ill.: Waveland Press, 1989), pp. 216–229.

Alpert, Geoffrey P., and Lorie A. Fridell, *Police Vehicles and Firearms: Instruments of Deadly Force* (Prospect Heights, Ill.: Waveland Press, 1992).

Anonymous, "Pittsburgh's Aggressive Stand Against Domestic Violence," in *The Police Chief* (April 1990), p. 41.

Arlies, Rebecca M., "Healthy Hearts for New York City Cops," in *The Police Chief* (July 1991), pp. 16–22.

Banton, Michael, *The Policeman in the Community* (New York: Basic Books, Inc.)

Barnett, Camille Cates, and Robert A. Bowers, "Community Policing: The New Model for the Way the Police Do Their Job," in *Public Management* (July 1990), pp. 2–6.

Becker, Christine S., "The Practical Side of OD," in *Public Management* (September 1978).

Berkley, George E., *The Democratic Policeman* (Boston: Beacon Press, 1969).

Bopp, William J., *Police Personnel Administration* (Boston: Holbrook Press, 1974).

Bowman, Barry A., "Commemorating the 30th Anniversary of Airborne Law Enforcement in LA," in *The Police Chief* (February 1989), pp. 17–18.

Boxx, W. Randy, Randall Y. Odom, and Mark G. Dunn, "Organizational Values and Value Congruency and Their Impact on Satisfaction, Commitment, and Cohesion: An Empirical Examination Within the Public Sector," *Public Personnel Management* (Spring 1991), pp. 195–205.

Boydstun, John E. et al., *Patrol Staffing in San Diego: One- or Two-Officer Units* (Washington, D.C.: Police Foundation, 1977).

Bradshaw, Robert V., Ken Peak, and Ronald W. Glensor, "Community Policing Enhances Reno's Image," in *The Police Chief* (October 1990).

Brooks, Pierce R., " . . . Officer Down, Code Three" (Schiller Park, Ill.: Motorola Teleprograms, Inc., 1975), p. 95.

Brown, Lee P. "Community Policing: A Practical Guide for Police Officials," National Institute of Justice, *Perspective on Policing* (September 1989), p. 1.

Buck, George A. et al., *Police Crime Analysis Handbook* (Washington, D.C.: Law Enforcement Assistance Administration, 1973).

Byers, Kenneth T., ed., *Employee Training and Development in the Public Sector* (Chicago: International Personnel Management Association, 1974).

Byrd, Donald A., "Impact of Physical Fitnmess on Police Performance," in *The Police Chief*, 43 (December 1979), pp. 30–32).

"CALEA Adds 13 More to Accreditation Toll," *Law Enforcement News* (September 15, 1991).

Carter, David L., "Methods and Measures," in Robert Trojanowicz and Bonnie Bucqueroux, *Community Policing: A Contemporary Perspective* (Cincinnati: Anderson Publishing Co., 1990), pp. 177–178.

Chapman, Samuel G., *Police Dogs in North America* (Springfield, Ill.: Charles C. Thomas, 1990).

Chapman, Samuel G., *Cops, Killers and Staying Alive* (Springfield, Ill: Charles C. Thomas, 1986).

Chapman, Samuel G. et al., *Perspectives on Police Assaults in the South Central United States, Vol. I* (Norman, Okla.: The University of Oklahoma, 1974).

Chapman, Samuel G., ed., *Police Patrol Readings, 2nd ed.*, (Springfield, Ill.: Charles C. Thomas, 1970).

Cohen, Marcia, and J. Thomas McEwen, "Handling Calls for Service: Alternatives to Traditional Policing," *NIJ Reports* (September 1984), pp. 4–8.

Commission on Accreditation for Law Enforcemnent Agencies, Inc., *Standards for Law Enforcement Agencies* (Fairfax, Va., 1983).

Community Relations Service, U.S. Department of Justice, "Principles of Good Policing: Avoiding Conflict Between Police and Citizens," in Robert G. Dunham and Geoffrey P. Alpert, *Critical Issues in Policing: Contemporary Readings* (Prospect Heights, Ill.: Waveland Press, 1989), pp. 164–204.

Comparative Analysis of the Lexington–Fayette Urban County Police Division's Home Fleet Program Versus the All Pool Plan, A (Lexington, Ky.: Lexington–Fayette Urban County Division of Police, 1977).

Cordner, Gary W., "The Police on Patrol," in Dennis Jay Kenney, ed., *Police and Policing: Contemporary Issues* (New York: Praeger, 1989), pp. 60–71.

Cruse, Daniel, and Jesse Rubin, "Police Behavior, Part I," *The Journal of Psychiatry and Law*, 1 (September 1973).

Davis, Edward M., "Professional Police Principles," *Federal Probation*, 35 (March 1971).

Davis, Robert C., "Crime Victims: Learning How to Help Them," National Institute of Justice, *NIJ Reports* (May/June 1987), pp. 2–7.

Decker, Scott H., "The Impact of Patrol Staffing on Police–Citizen Injuries and Dispositions" *Journal of Criminal Justice*, Vol. 10 (1982), pp. 375–382.

Dunham, Robert G., and Geoffrey P. Alpert, *Critical Issues in Policing: Contemporary Readings* (Prospect Heights, Ill.: Waveland Press, 1989), pp. 498–507.

Eck, John E., and William Spellman, "A Problem–Oriented Approach to Police Service Delivery," in Dennis Jay Kenney, ed., *Police and Policing: Contemporary Issues* (New York: Praeger, 1989), pp. 95–111.

Eck, John E., and William Spellman, "Problem Solving: Problem-Oriented Policing in Newport News," in Robert G. Dunham and Geoffrey P. Alpert, *Critical Issues in Policing: Contemporary Readings* (Prospect Heights, Ill.: Waveland Press, 1989), pp. 425–439.

Elgin (Ill.) Police Department, *Memorandum 91 D 37*, dated 5-22-91.

Elgin (Ill.) Police Department, *Rules and Regulations*, undated.

Escobedo v. Illinois, 378 U.S. 478, 488 (1964).

Farrell, Michael J., "The Development of the Community Patrol Officer Program: Community–Oriented Policing in the New York City Police Department," in Greene and Mastrofski, *Community Policing: Rhetoric or Reality* (New York: Praeger Publishers, 1988), pp. 73–88.

Felkenes, George T., *Effective Police Supervision: A Behavioral Approach* (San Jose, Calif.: Justice Systems Development, Inc., 1977).

Finn, Peter, "Street People," in National Institute of Justice, *Crime File Study Guide*, no date.

Finn, Peter, and Monique Sullivan, "Police Respond to Special Populations," National Institute of Justice, *NIJ Reports* (May–June 1988), pp. 2–8.

Finn, Peter and Beverly N. M. Lee, "Establishing and Expanding Victim–Witness Assistance Programs," National Institute of Justice, *Research in Action* (August 1988).

Fisk, Donald M., *The Indianapolis Police Fleet Plan* (Washington, D.C.: The Urban Institute, 1970).

Fogelson, Robert M., "Reform at a Standstill," in Carl B. Klockars, *Thinking About Police: Contemporary Readings* (New York: McGraw–Hill Book Co., 1983), pp. 113–129.

Forst, Martin, Melinda Moore, and Graham Crowe, "AIDS Education for California Law Enforcement," *The Police Chief* (December 1989), pp. 25–28.

Four–Ten Plan—California Law Enforcement, The, California Commission on Peace Officer Standards and Training, 1981.

Francis, Dave, and Mike Woodcock, *People at Work: A Practical Guide to Organizational Development* (La Jolla, Calif.: University Associates, Inc., 1975).

Frederick, Calvin J., Karen L. Hawkins, and Wendy E. Abajian, "Victim/Witness Intervention Techniques," *The Police Chief* (October 1990). pp. 106, 108–110.

Friedman, Lucie N., and Minna Schulman, "Domestic Violence: The Criminal Justice Response," in Arthur J. Lurigio, Wesley G. Skogan and Robert C. Davis, eds., *Victims of Crime: Problems, Policies, and Programs* (Newbury Park, Calif.: Sage Publishing, Inc., 1990), pp. 87–103.

Friedrich, Robert J., "Police Use of Force: Individuals, Situations, and Organizations," in Klockars, *Thinking About Police*, pp. 302–313.

Fyfe, James J., "Administrative Interventions on Police Shooting Discretion: An Empirical Examination," *Journal of Criminal Justice* (Winter 1979), pp. 309–323.

Fyfe, James J., "Reviewing Citizen Complaints Against Police," in James J. Fyfe, ed., *Police Management Today: Issues and Case Studies* (Washington, D.C.: International City Management Association, 1985), pp. 76–87.

Fyfe, James J., ed., *Police Management Today: Issues and Case Studies* (Washington, D.C.: International City Management Association, 1985).

Gammage, Allen Z., *Police Training in the United States* (Springfield, Ill.: Charles C. Thomas, 1963).

Garmire, Bernard L., ed., *Local Government Police Management* (Washington, D.C.: International City Management Association, 1982), pp. 30–51.

Garner, Joel and Elizabeth Clemmer, "Danger to Police in Domestic Disturbance—A New Look," in Robert G. Dunham and Geoffrey P. Alpert, *Critical Issues in Policing: Contemporary Readings* (Prospect Heights, Ill.: Waveland Press, 1989), pp. 516–530.

Geddes, Ronald W., and William DeJong, "Coping With Police Stress," in Robert G. Dunham and Geoffrey P. Alpert, *Critical Issues in Policing: Contemporary Readings* (Prospect Heights, Ill.: Waveland Press, 1989), pp. 498–507.

Goldstein, Herman, *Problem-Oriented Policing* (Philadelphia: Temple University Press, 1990).

Goldstein, Herman, *Policing a Free Society* (Cambridge, Mass.: Ballinger Publishing Co., 1977).

Goolkasian, Gail A., Ronald W. Geddes, and William DeJong, "Coping With Police Stress," in Robert G. Dunham and Geoffrey P. Alpert, *Critical Issues in Policing: Contemporary Readings* (Prospect Heights, Ill.: Waveland Press, 1989), pp. 498–507.

Green, Jack R., and Stephen D. Mastroski, eds., *Community Policing: Rhetoric or Reality* (New York: Praeger Publishers, 1988), pp. 47–67.

Greenberg, Bernard et al., *Felony Investigation Decision Model—An Analysis of Investigative Elements of Information* (Menlo Park, Calif.: Stanford Research Institute, 1975).

Greenberg, Sheldon, "Police Accreditation," in Dennis Jay Kenney, ed., *Police and Policing: Contemporary Issues* (New York: Praeger, 1989). pp. 247–256.

Greenwood, Peter W., and Joan Petersilia, *The Criminal Investigation Process, Vol. 1—Summary and Policy Implications* (Santa Monica, Calif.: The Rand Corporation, 1975).

Greisinger, George W., Jeffrey S. Slovak, and Joseph J. Molkup, *Civil Service Systems: Their Impact on Police Administration* (Chicago: Public Administration Service, 1979).

Guthrie, C. Robert, and Paul M. Whisenand, "The Use of Helicopters in Routine Police Patrol Operations: A Summary of Research Findings," in S. I. Cohn, ed., *Law Enforcement Science and Technology, Vol. 2* (Chicago: IIT Research Institute, 1968), p. 554.

Hale, Charles D., *Fundamentals of Police Administration* (Boston: Holbrook Press, 1977).

Hale, Charles D., *The Isolated Police Community: An Occupational Analysis* (Master's Thesis, California State University at Long Beach, 1972).

Hammett, Theodore M., "AIDS and the Law Enforcement Officer," National Institute of Justice, *NIJ Reports* (November–December 1987), pp. 2–7.

Hammett, Theodore M., and Walter Bond, "Risk of Infection with the AIDS Virus through Exposure to Blood," National Institute of Justice, *AIDS Bulletin* (October 1987).

Hanewicz, Wayne B. et al., "Improving the Linkages Between Domestic Violence Referral Agencies and the Police: A Research Note," *Journal of Criminal Justice, Vol. 10* (1982), pp. 493– 503.

Hartman, Thadeus L., "Field Training Officer (FTO): The Fairfax County Experience," *FBI Law Enforcement Bulletin* (April 1979), pp. 22–25.

Hatry, Harry P., "Wrestling with Police Crime Control Productivity Measurement," *Readings on Productivity in Policing* (Washington, D.C.: Police Foundation, 1975), pp. 86–128.

Hersey, Paul, and Kenneth H. Blanchard, *Management of Organizational Behavior: Utilizing Human Resources, 5th ed.* (Englewood Cliffs, N.J.: Prentice Hall, 1988).

Hoy, Vernon L., "Research and Planning," in Bernard L. Garmire, ed., *Local Government Police Management* (Washington, D.C.: International City Management Association, 1977), p. 367.

Iannone, N. F., *Supervision of Police Personnel, 2nd ed.*, Englewood Cliffs, N.J.: Prentice–Hall, Inc., 1975), pp. 30–31.

Inbau, Fred E., and Wayne W. Schmidt, "The Legal Advisor," in Garmire, *Local Government Police Management* (Washington, D.C.: International City Management Association, 1977), p. 382.

Kelley, John, "Negotiating the Accreditation Process," in *The Police Chief* (June 1991), pp. 25–29.

Kelling, George L., "What Works—Research and the Police," National Institute of Justice, *Crime File Study Guide*, no date.

Kelling, George L., "Police and Communities: The Quiet Revolution," National Institute of Justice, *Perspectives on Policing* (June 1988).

Kelling, George L., and Mark H. Moore, "The Evolving Strategy of Policing," in National Institute of Justice, *Perspectives on Policing* (November 1988).

Kelling, George E. et al., *The Kansas City Preventive Patrol Experiment: A Summary Report* (Washington, D.C.: Police Foundation, 1974).

Kennedey, Daniel B., Robert J. Homant and George L. Emery, "AIDS and the Crime Scene Investigator" in *The Police Chief* (December 1989), pp. 19, 22.

Kenney, Dennis Jay, ed., *Police and Policing: Contemporary Issues* (New York: Praeger, 1989).

King, J. Freeman, "The Law Officer and the Deaf," *The Police Chief* (October 1990), pp. 98, 100.

Kirkpatrick, Donald L, "Evaluating In–House Training Programs," *Training and Development Journal* (September 1978).

Klockars, Carl, *Thinking About Police: Contemporary Readings* (New York: McGraw–Hill Book Co., 1983), pp. 239–251.

Klockars, Carl B., *The Idea of Police* (Newbury Park, Calif.: Sage Publications), 1985.

Kotulak, Robert, "Is Your Job Driving You Crazy?" *Chicago Tribune*, September 18, 1977, pp. 1, 5, citing study by Michael J. Culligan et al., "Occupational Incidence Rates for Mental Health Disorders, *Journal of Human Stress*, 3 (September 1977), pp. 34–39.

Kroes, William H., *Society's Victim—The Policeman: An Analysis of Job Stress in Policing* (Springfield, Ill.: Charles C. Thomas, 1976).

Leonard, V. A., and Harry W. More, *Police Organization and Management, 7th ed.*, (Mineola, N.Y.: The Foundation Press, Inc., 1987.

Longworthy, Robert H., "Organizational Structure," in Gary W. Cordner and Donna C. Hale, *What Works in Policing? Operations and Administration Examined* (Cincinnati: Anderson Publishing Co., 1992), pp. 87–105.

Lopez, Felix, *Evaluating Employee Performance* (Chicago: Public Personnel Association, 1968).

Lundman, Richard J., "The Traffic Violator: Organizational Norms," in Klockars, *Thinking About Police: Contemporary Readings* (New York: McGraw–Hill Book Co., 1983), pp. 286–292.

Lurigio, Arthur J., Wesley G. Skogan, and Robert C. Davis, eds., *Victims of Crime: Problems, Policies, and Programs* (Newbury Park, Calif.: Sage Publishing, Inc., 1990).

Luthans, Fred, *Organizational Behavior: A Modern Behavioral Approach to Management* (New York: McGraw-Hill Book Company, 1973).

McCampbell, Michael S., "Field Training for Police Officers: State of the Art," National Institute of Justice, *Research in Brief* (November 1986).

McClenahan, Carol A., "Victim Services—A Positive Police-Community Effort," *The Police Chief* (October 1990), p. 104.

McEwen, J. Thomas et al., "Handling Calls for Service: Alternatives to Traditional Policing, *NIJ Reports* (September 1984), pp. 4–8.

McLean, Herbert E., "Scaling a Wall of Silence," *Law and Order* (August 1986), pp. 38–42.

Mager, Robert F., *Preparing Objectives for Programmed Instruction* (San Francisco: Fearon Publishers, 1962).

Mager, Robert F., and Peter Pipe, *Analyzing Performance Problems or "You Really Oughta Wanna"* (Belmont, Calif.: Fearon Publishers/Lear Sigler, Inc., 1973).

Managing for Effective Discipline: *A Manual of Rules, Procedures, Supportive Law and Effective Management* (Gaithersburg, Md.: International Association of Chiefs of Police, 1976).

Mastrofski, Stephen D., "Community Policing as Reform: A Cautionary Tale, in Jack R. Greene and Stephen D. Mastrofski, eds., *Community Policing: Rhetoric or Reality* (New York: Praeger Publishers, 1988), pp. 47–67.

Melnicoe, William B and Jan C. Mennig, *Elements of Police Supervision, 2nd ed.* (Encino, Calif.: Glencoe Publishing Co., Inc., 1978).

Michigan State Police, *1991 Model Year Patrol Vehicle Testing* (Washington, D.C.: National Institute of Justice, 1991).

Misner, Gordon E., "Enforcement: Illusion of Security," *The Nation* (April 21, 1969).

Mitchell, Jeffrey T., "Development and Functions of a Critical Incident Stress Debriefing Team," *JEMS* (December 1988), pp. 43–46).

Monihan, Lynn Hunt, and Richard E. Farmer, *Stress and the Police: A Manual for Prevention* (Pacific Palisades, Calif.: Palisades Publishers, 1980).

Monk, Timothy, "Advantages and Disadvantages of Rapidly Rotating Shift Schedules: A Circadian Viewpoint," *Human Factors* (October 1986), pp. 553–557.

Moore, Mark H., Robert C. Trojanowicz, and George L. Kelling, "Crime and Policing," *Police Practice in the '90s: Key Management Issues* (Washington, D.C.: International City Management Association, 1989), pp. 31–54.

Morris, Norval, and Gordon Hawkins, *The Honest Politician's Guide to Crime Control* (Chicago: University of Chicago Press, 1970).

Muir, William Ker, Jr., "Skidrow at Night," in Klockars, *Thinking About Police: Contemporary Readings* (New York: McGraw-Hill Book Co., 1983), pp. 239–251.

National Advisory Commission on Criminal Justice Standards and Goals, *Police* (Washington, D.C.: U.S. Government Printing Office, 1973).

National Commission on Criminal Justice Standards and Goals, *Police* (Washington, D.C.: U.S. Government Printing Office, 1973).

National Institute of Justice, *TAP Alert* (October 1986).

Niederhoffer, Arthur, *Behind the Shield: The Police in Urban Society* (Garden City, N.Y.: Anchor Books, Doubleday Co., Inc., 1967).

Niederhoffer, Arthur, and Abraham S. Blumberg, *The Ambivalent Force: Perspectives on the Police* (Waltham, Mass.: Ginn & Co., 1970), pp. 293–307.

Oettmeier, Timothy N., and Lere P. Brown, "Developing a Neighborhood-Oriented Policing Style," in Greene and Mastrofski, *Community Policing: Rhetoric or Reality* (New York: Praeger Publishers, 1988), pp. 121–134.

O'Keefe, James, "High-Speed Pursuits in Houston," *The Police Chief* (July 1989), pp. 32–34.

Otto, Calvin P., and Rollin O. Glaser, *The Management of Training: A Handbook for Training and Development Personnel* (Reading, Mass.: Addison-Wesley Publishing Company, 1970).

Parker, Treadway C., "Statistical Measures for Measuring Training Results," in Robert L. Craig, ed., *Training and Development Handbook, 2nd ed.* (New York: McGraw-Hill Book Company, 1976).

Pate, Tony et al., *Three Approaches to Criminal Apprehension in Kansas City: An Evaluation Project* (Washington, D.C.: Police Foundation, 1976).

"Patrol, Patrol, Patrol," Spring 3100 (July–August 1960), in Samuel G. Chapman, *Police Patrol Readings* (Springfield, Ill.: Charles C. Thomas, 1970), pp. 60–61.

Payne, Dennis M., and Robert C. Trojanowicz, *Performance Profiles of Foot Versus Motor Officers* (East Lansing, Mich.: National Neighborhood Foot Patrol Center, School of Criminal Justice, Michigan State University, 1985).

Petersilia, Joan, "The Influence of Research on Policing," in Dunham, Robert G., and Geoffrey P. Alpert, *Critical Issues in Policing: Contemporary Readings* (Prospect Heights, Ill.: Waveland Press, 1989), pp. 230–247.

Pfiffner, John M. and Marshall Fels, *The Supervision of Personnel*, 3rd ed. (Englewood Cliffs, N.J.: Prentice-Hall, Inc., 1964).

Pittsburgh (Pa.) Police Bureau Policy No. 124.8, "Domestic Violence."

Poole, Robert W. Jr., *Cutting Back at City Hall* (New York: Universe Books, 1980).

President's Commission on Law Enforcement and Administration of Justice, *The Challenge of Crime in a Free Society* (Washington, D.C.: U.S. Government Printing Office, 1967).

President's Commission on Law Enforcement and Administration of Justice, *Task Force Report: The Police* (Washington, D.C.: U.S. Government Printing Office, 1967).

Project on Standards for Criminal Justice, *Standards Relating to the Urban Police Function* (Chicago: American Bar Association, 1972).

Reiser, Martin, "Some Organizational Stresses on Policemen," *Journal of Police Science and Administration*, 2 (June 1974).

Reiser, Martin, Robert J. Sokol, and Susan S. Saxe, "An Early Warning Mental Health Program for Police Sergeants," *The Police Chief* (June 1972).

Reuter, Heidi, "Not Everyone Is Sold on Community Policing," *The Milwaukee Sentinel*, November 21, 1991, p. 1A, 8A.

Rewsick, Patricia A., "Victims of Sexual Assault," in Arthur J. Lurigio, Wesley G. Skogan, and Robert C. Davis, eds., *Victims of Crime: Problems, Policies, and Programs* (Newbury Park, Calif.: Sage Publishing, Inc., 1990), pp. 69–96.

Rice, Thomas W., "A Case for MACE," in *The Police Chief* (July 1989), pp. 51–53.

Roethlisberger, F. J., and W. J. Dickson, *Management and the Worker* (Cambridge, Mass.: Harvard University Press, 1939).

Rubenstein, Jonathan, *City Police* (New York: Farrar, Straus and Giroux, 1973).

Saint Charles (Ill.) Police Department General Order 87–06, dated April 23, 1987.

Saunders, Charles B., Jr., *Upgrading the American Police* (Washington D.C.: The Brookings Institution, 1970).

Schack, Stephen, Theodore H. Schell, and William G. Gay, *Improving Patrol Productivity, Volume II: Specialized Patrol* (Washington, D.C.: National Institute of Law Enforcement and Criminal Justice, 1977).

Schell, Theodore H., Don H. Overly, Stephen Schack, and Linda L. Stabile, *Traditional Preventive Patrol* (Washington, D.C.: National Institute of Law Enforcement and Criminal Justice, Law Enforcement Assistance Administration, U.S. Department of Justice, 1976).

Scrivner, Ellen, "Helping Police Facilities Cope With Stress," *Law Enforcement News* (June 15/30, 1991), pp. 6, 8, 10).

Sheehan Robert and Gary W. Cordner, *Introduction to Police Administration, 2nd ed.* (Cincinnati: Anderson Publishing Co., 1989).

Sherman, Lawrence W., "Repeat Calls for Service: Policing the 'Hot Spots,'" in Dennis Jay Kenney, ed., *Police and Policing: Contemporary Issues* (New York: Praeger, 1989), pp. 150–165.

Sherman, Lawrence W., and Richard A. Berk, "The Minneapolis Domestic Violence Experiment," in James J. Fyfe, ed., *Police Management Today: Issues and Case Studies* (Washington, D.C.: International City Management Association, 1985), pp. 118–131.

Siglerm, Jay A., and Robert S. Getz, *Contemporary American Government: Problems and Prospects* (New York: Van Nostrand and Richmond Co., 1972).

Sipes, Leonard A., Jr., "The Power of Senior Citizens in Crime Prevention and Victim Services," *The Police Chief* (January 1989), pp. 45–47.

Skolnick, Jerome H., *Justice Without Trial: Law Enforcement in Democratic Society* (New York: John Wiley & Sons, Inc., 1967).

Skolnick, Jerome H., "Democratic Order and the Rule of Law," in Thomas J. Sweeney and William Ellingsworth, *Issues in Police Patrol: A Book of Readings* (Kansas City, Mo.: Kansas City Police Department, 1973), pp. 3–24.

Skolnick, Jerome H., and David H. Bayley, *The New Blue Line: Police Innovation in Six American Cities* (New York: The Free Press, 1985), pp. 4–5.

Skolnick, Jerome H., and David H. Bayley, *Community Policing: Issues and Practices Around the World* (Washington, D.C.: National Institute of Justice, 1988).

Slovak, Jeffrey, *Police and Their Perceptions* (Chicago: Public Administration Service, 1977)

Smith, Bruce, *Police Systems in the United States, 2nd ed.* (New York: Harper & Row, 1960).

Solomon, Roger, "Police Shooting Trauma," *The Police Chief* (October 1988), pp. 121–123.

Sparrow, Malcolm K., Mark H. Moore, and David M. Kennedy, *Beyond 911: A New Era for Policing* (New York: Bsasic Books, 1990).

Spelman, William, and Dale K. Brown, "Response Time," in Carl Klockars, *Thinking About Police: Contemporary Readings* (New York: McGraw–Hill Book Co., 1983), pp. 160–166.

Stenzel, William W., and R. Michael Buren, *Police Work Scheduling: Management Issues and Practices* (Washington, D.C.: U.S. Department of Justice, 1983).

Stodgill, Ralph M., *Handbook of Leadership: A Survey of Theory and Research* (New York: The Free Press, 1974).

"Their Neighborhoods Are Their Beats," *American City and County* (April 1992), pp. 32 ff.

Thibault, Edward A., Lawrence N. Lynch, and R. Bruce McBride, *Proactive Police Management, 2nd ed.* (Englewood Cliffs, N.J.: Prentice Hall, 1990).

Tielsch, George P., "Planning Training," *The Police Chief*, 45 (July 1978), pp. 28–29.

Tielsch, George P., and Paul M. Whisenand, *The Assessment Center Approach in the Selection of Police Personnel* (Santa Cruz, Calif.: Davis Publishing Co., Inc., 1977).

Tien, James M. et. al., *An Evaluation Report of an Alternative Approach in Police Patrol: The Wilmington Split–Force Experiment* (Cambridge, Mass.: Police Systems Evaluation, Inc., 1977).

Tracy, William R., *Evaluating Training and Development Systems* (New York: American Management Association, Inc., 1968).

Trojanowicz, Robert, and Bonnie Bucqueroux, *Community Policing: A Contemporary Perspective* (Cincinnati: Anderson Publishing Co., 1990).

Trojanowicz, Robert, Bonnie Pollard, Francine Colgan, and Hazel Harden, *Community Policing Programs: A Twenty–Year View* (East Lansing, Mich.: National Neighborhood Foot Patrol Program, Michigan State University, 1985).

Trojanowicz, Robert, and Paul R. Smythe, *A Manual for the Establishment and Operation of a Foot Patrol Program* (East Lansing, Mich.: The National Neighborhood Foot Patrol Center, School of Criminal Justice, Michigan State University, 1984).

Tully, Edward J., "The Training Director," *FBI Law Enforcement Bulletin*, 48 (July 1979).

Uniform Crime Reports: Law Enforcemnent Officers Killed and Assaulted 1991 (Washington, D.C.: U.S. Department of Justice, 1991), Preface.

Vanderbosch, Charles G., "24–Hour Patrol Power: The Indianapolis Program," *The Police Chief*, 37 (September 1970).

Vincent, Henry, and Sean Grennan, "Professionalism through Police Supervisory Training," in James J. Fyle, ed., *Police Practice in the 90's: Key Management Issues* (Washington, D.C.: International City Management Association, 1989), p. 137.

Waller, Irvin, "The Police: First in Aid?", in Arthur J. Lurigio, Wesley G. Skogan and Robert C. Davis, eds., *Victims of Crime: Problems, Policies, and Programs* (Newbury Park, Calif.: Sage Publishing, Inc., 1990), pp. 139–156.

Wasserman, Robert, "The Governmental Setting," in Bernard L. Garmire, ed., *Local Government Police Management* (Washington, D.C.: International City Management Association, 1982), pp. 30–51.

Wasserman, Robert, and David Couper, "Training and Education," in O. Glenn Stahl and Richard A. Stauffenberger, eds., *Police Personnel Administration* (Washington, D.C.: Police Foundation, 1974).

Wasserman, Robert, and Mark M. Moore, "Values in Policing," in National Institute of Justice, *Perspectives on Policing* (November 1988).

Webb, Kenneth W. et al., *Specialized Patrol Projects* (Washington, D.C.: U.S. Government Printing Office, 1977).

Weisburd, David, Jerome McElroy, and Patricia Hardyman, "Maintaining Control in Community–Oriented Policing" in Dennis Jay Kenney, ed., *Police and Policing: Contemporary Issues* (New York: Praeger, 1989), pp. 188–200.

Westley, William A., *Violence and the Police: A Sociological Study of Law, Custom, and Morality* (Boston: Massachusetts Institute of Technology, 1970).

Weston, Paul B., *Supervision in the Administration of Justice: Police, Corrections, Courts, 3d printing* (Springfield, Ill.: Charles C. Thomas, 1972).

Wilson, James Q., "The Police and Their Problems," *Public Policy*, 12 (1963).

Wilson, James Q., *Varieties of Police Behavior: The Management of Law and Order in Eight Communities* (Cambridge, Mass.: Harvard University Press, 1968).

Wilson, James Q., and George L. Kelling, "Broken Windows," in Robert G. Dunham and Geoffrey P. Alpert, *Critical Issues in Policing: Contemporary Readings* (Prospect Heights, Ill.: Waveland Press, 1989).

Wilson, James Q., and George L. Kelling, "Broken Windows," in Robert G. Dunham and Geoffrey P. Alpert, *Critical Issues in Policing: Contemporary Readings* (Prospect Heights, Ill.: Waveland Press, 1989), pp. 369–381.

Witham, Donald C., "Transformational Police Leadership," in James J. Fyfe, ed., *Police Practice in the 90's: Key Management Issues* (Washington, D.C.: International City Management Association, 1989).

Wooldridge, Blue, and Jennifer Wester, "The Turbulent Environment of Public Personnel Administration: Responding to the Challenge of the Changing Workplace of the Twenty–First Century," *Public Personnel Management*, 20 (September 1991), pp. 207–224.

Wycoff, Mary Ann, "The Benefits of Community Policing: Evidence and Conjecture," in Jack R. Green and Stephen D. Mastroski, eds., *Community Policing: Rhetoric or Reality* (New York: Praeger Publishers, 1988), pp. 103–120.

INDEX

A

B

C

D

E

F

G

H

I

J

K

L

M

N

O

P

R

S

T

U

V

W

Y